European Rural Landscapes: Persistence and Change in a Globalising Environment

Edited by

HANNES PALANG

University of Tartu,
Estonia

HELEN SOOVÄLI

University of Tartu,
Estonia

MARC ANTROP

Ghent University,
Belgium

and

GUNHILD SETTEN

Norwegian University of Science and Technology,
Trondheim, Norway

KLUWER ACADEMIC PUBLISHERS

DORDRECHT / BOSTON / LONDON

A C.I.P. Catalogue record for this book is available from the Library of Congress.

ISBN 1-4020-2067-8 (HB)
ISBN 0-306-48512-5 (e-book)

Published by Kluwer Academic Publishers,
P.O. Box 17, 3300 AA Dordrecht, The Netherlands.

Sold and distributed in North, Central and South America
by Kluwer Academic Publishers,
101 Philip Drive, Norwell, MA 02061, U.S.A.

In all other countries, sold and distributed
by Kluwer Academic Publishers,
P.O. Box 322, 3300 AH Dordrecht, The Netherlands.

Cover design and photograph by Vahur Puik

Printed on acid-free paper

Contents

Contributing Authors

Marc Antrop is Professor of geography at the Ghent University, Belgium and vice-president of the Royal Committee for the Protection of Landscapes in Flanders.

Tor Arnesen is Researcher at Eastern Norway Research Institute, Lillehammer, Norway.

Mette Bech Sørensen is PhD student at the Institute of Geography, University of Copenhagen, Denmark.

Elisabeth Gräslund Berg is PhD student at the Department of Human Geography, Stockholm University, Sweden.

Maria Bergström is PhD student at the Department of Physical Geography and Quaternary Geology, Stockholm University, Sweden.

Karl-Martin Born is Assistant Professor at the Department of Geographical Sciences of the Free University Berlin, Germany.

Edmunds V. Bunkše is a cultural geographer who specialises in geosophy at the University of Delaware and is also Adjunct Professor at University of Latvia. He is a member of the Latvian Academy of Sciences and has Dr.h.c. (*honoris causa*) from the University of Latvia.

John Chapman is Lecturer at the Department of Geography, University of Portsmouth, United Kingdom.

Paul Claval is a specialist of the history of geography, is Professor Emeritus after teaching for a long period at the University of Paris-IV (Paris-Sorbonne) in Paris, France.

Chris De Bont is Senior Researcher in historical geography at Alterra Wageningen University and Research Centre, the Nertherlands.

Tomas Germundsson is Senior Lecturer of human geography, at the Department of Social and Economic Geography, Lund University, Sweden.

Per Grau Møller is Historian at the University of Southern Denmark (Odense) working with landscape history and applied research in the field of integrating history into landscape management and planning.

Ann Grubbström is PhD at the Department of Social and Economic Geography, Uppsala University, Sweden. Working as Lecturer in geography and didactics at the same department.

Staffan Helmfrid is Professor Emeritus of human geography at Stockholm University, Sweden.

Maunu Häyrynen is Professor of landscape studies at School of Cultural Production and Landscape Studies, University of Turku, Finland.

Margareta Ihse is Professor of ecological geography at the Department of Physical Geography and Quaternary Geology, Stockholm University, Sweden. She is also vice president of the International Association of Landscape Ecology

Michael Jones is Professor of geography at the Norwegian University of Science and Technology in Trondheim, Norway. He is also a member of the Norwegian Academy of Science and Letters.

Egle Kaur is PhD student at the Institute of Geography, University of Tartu, Estonia.

Hansjörg Küster is Professor of plant ecology, at the Institute for Geobotany at Hannover University, Germany.

Pieter-Jan Lachaert is PhD student at the Department of Medieval History, Ghent University. His research is financed by the Fund for Scientific Research-Flanders in Belgium.

Guri Markhus is Lecturer at the Nord-Trøndelag University College, Steinkjer, Norway.

Tõnu Oja is Professor of geoinformatics and cartography at the Institute of Geography, University of Tartu, Estonia.

Kenneth R. Olwig is Professor at the Department of Landscape Planning of the Swedish University of Agricultural Sciences (SLU), Alnarp, Sweden.

Taavi Pae is PhD student at the Institute of Geography, University of Tartu, Estonia.

Hannes Palang is Senior Researcher at the chair of human geography, Institute of Geography, University of Tartu, Estonia.

Monika Prede is PhD student at the Institute of Geography, University of Tartu, Estonia.

Elle Puurmann is Senior Officer at the Silma Nature Reserve on Vormsi Island, Estonia.

Urve Ratas is Docent at the Chair of Geoecology, and Senior Researcher at the Institute of Ecology, Tallinn Pedagogical University, Estonia

Johannes Renes is Researcher and Teacher in historical geography at Utrecht University and at the Free University Amsterdam, The Netherlands.

Reimo Rivis is PhD student and Researcher at the Institute of Ecology, Tallinn Pedagogical University, Estonia

Gunhild Setten is Associate Professor at the Department of Geography, the Norwegian University of Science and Technology in Trondheim, Norway.

Terje Skjeggedal is Researcher at Nord-Trøndelag Research Institute, Steinkjer, Norway.

Elyze Smeets is PhD student in historical geography at the University of Leeds, United Kingdom.

Katriina Soini is a cultural geographer, working as Researcher at the MTT Agrifood Research, Finland.

Helen Sooväli is PhD student at the Institute of Geography, University of Tartu, Estonia; working parallelly as Lecturer in cultural geography at the same institute.

Marie Stenseke is Associate Professor at the Department of Human and Economic Geography, Göteborg University, Sweden.

Per Gustav Thingstad is Researcher at the Museum of Natural History and Archaeology, Norwegian University of Science and Technology in Trondheim, Norway.

Clas Tollin is human geographer working as Associate Professor at the Department of Economics, History of Agriculture, Swedish University of Agricultural Sciences in Uppsala, Sweden.

Mats Widgren is Professor of geography at the Department of Human Geography, Stockholm University. He is also a member of the Royal Academy of Letters, History and Antiquities in Sweden.

Preface

This book brings together 28 selected papers from two events, the 20th session of the Permanent European Conference for the Study of the Rural Landscape (PECSRL), and a series of workshops, financed by the European Union under the title EURULA. This book focuses on aspects of landscape related to language, representation and power. Together with a special issue of *Landscape and Urban Planning* that discusses planning-related topics, we hope that we can offer a deeper understanding in an interdisciplinary perspective of what is going on with the rural landscapes in Europe.

Compiling the book has been a team effort. We wish to thank all participants of the both meetings – their discussions have lead to this book. The reviewers of the chapters gave helpful comments, and Zac Gagnon did a great job to harmonise the language in the book. Our sincere gratitude to all you!

Tartu, Ghent and Trondheim,
Christmas 2003

Chapter 1

THE PERMANENCE OF PERSISTENCE AND CHANGE
Introduction

Hannes Palang[1], Helen Soov> [1], Marc Antrop[2] & Gunhild Setten[3]
[1]*Institute of Geography, University of Tartu, Estonia*
[2]*Department of Geography, Ghent University, Belgium*
[3]*Department of Geography, Norwegian University of Science and Technology, Trondheim, Norway*

1. PERMANENT PERSISTENCE AND CHANGE

The only thing that is permanent is change. The hidden paradox in the relationship between permanence and change is that "all change implies that something is preserved", and that "in all permanence there is some degree of change" (Jones 1991: 235). Within landscape research, this paradox has in multiple ways been structuring how we have come to understand and interpret both the material and symbolic aspects of landscapes.

For many centuries changes were local and gradual, and existing landscape structures were rarely wiped out completely. Presently we tend to look at such landscapes of the past as being stable and as having a distinct character and identity, forming a basis for the homeland of those who worked it over the centuries (Lowenthal 1985; Antrop 2004). In Europe, with its long and complex history and great cultural diversity, a rich variety of such landscapes was thus created. Present debates on cultural and natural heritage and identity, in fact, rest on the assumption of a stable landscape in the past. Many rapid changes of (post)modern society are consequently seen as a threat to the values attached to these landscapes. Landscapes became considered as an endangered heritage. Hence, care for the landscape was put on the political agenda, and landscapes initiated a boom of many parallel initiatives and activities on national and international levels. The European

H. Palang et al. (eds.), European Rural Landscapes:
Persistence and Change in a Globalising Environment, 1-9.
© 2004 *Kluwer Academic Publishers. Printed in the Netherlands.*

Landscape Convention[1] of the Council of Europe, which opened for
signature in 2000, provides the most recent expression of this growing
concern.

Persistence, change, the meaning and values of cultural landscapes have
for several decades been recurring topics in many international scientific
conferences and workshops. *The Permanent European Conference for the
Study of the Rural Landscape*, with its long tradition, has since 1957 been
one of the cherished forums for discussing these issues.

This book presents a selection of papers from two interrelated events.
First, the 20th session of the Permanent European Conference for the Study
of the Rural Landscape (PECSRL) was held in Tartu, Estonia, from August
25-28, 2002. The conference was followed by a series of workshops in
Otepää, Estonia, from August 29-30, 2002. These events attracted 128
participants representing 20 European countries. The papers presented
covered a broad and very diverse domain of landscape research, with a
special focus upon the cultural and rural aspects and values of European
landscapes.[2]

Within a notion of persistence and change, the 20th session of the
Conference and the workshops hereafter had as their main focus *Rural
Landscapes: Past Processes and Future Strategies*. Four interrelated themes
helped to bring out differing dimensions of how the past and future intersect
with the above-mentioned paradox that is so familiar to landscape research.
The following is a short presentation of these themes and relevant questions
asked and addressed throughout the book.

First: *Lessons of the past: landscape change as a trigger for future
strategies.* Given the continuous material and symbolic changes of
landscapes, the dynamics of landscapes has always been an important issue
in the study of landscape history. An important question asked is: How can
this dynamical view of landscape history be made operational in future
strategies for the protection, management and development of landscapes?
Which conceptualisations and scientific methods can be used to understand
the long-term changes of landscapes, and how can they be applied to
landscape management? In the broadest sense, what lessons can be learned
from the past?

Second: *Language of the rural landscape.* Landscapes "speak" to us in a
variety of ways. As human beings we also "speak" to landscapes. These
"conversations" raise questions such as: Who are the people involved in
landscapes, and how do they express their involvement? How do local

---

[1] http://conventions.coe.int/Treaty/EN/cadreprincipal.htm
[2] In addition to the current volume, a selection of papers from the Conference is being
   published in a special issue on *Rural Landscapes: Past Processes and Future Strategies* in
   the journal *Landscape and Urban Planning*.

people perceive, understand and participate in the planning, care and maintenance of landscapes? Who have the power to decide how landscapes should change, or whether the seemingly unchanged is an option? How are landscapes conceptualised and who have the power over such conceptions?

Third: *Ecology of the rural landscape*. Ecology can be seen as both a scientific discipline as well as a state of mind. This is particularly evident within landscape management. This leads to issues concerned with what is meant by ecology in differing contexts, and how for example political and historical ecology can help us to understand how material landscapes work, and how we then monitor and plan for such workings at present?

Fourth and finally: *Future of the rural landscapes*. Current socio-economic changes put increasing pressure on European rural landscapes. Such processes are becoming more and more global in character. Global processes such as marginalisation, segregation and urbanisation are frequently seen to pose serious threats to historical landscapes that we still have knowledge of today. What are the potential outcomes of such processes? And how might historical knowledge contribute to future management strategies and understandings?

Within these four broadly defined themes, the 28 papers presented in this book reflect some of the multiplicity of landscape studies at present. The wide range of space(s) and time(s) covered by the contributions thus leaves the impression of a vibrant research field. This impressive variety represents very welcome contributions in enabling us to comprehend landscapes, as well as the troubled relationship between nature and culture, a relationship that in many ways has been the very trigger of landscape studies. At the same time, such a variety also represents profound challenges, both within theoretical and more-applied landscape debates. This is very well reflected throughout this book.

The multifarious nature of the contributions has in addition represented a challenge for the editorial team of this volume: what is it that ties the 28 contributions together and keeps them together? We have identified three interrelated perspectives along which we have aimed at creating a narrative of persistence and change in European rural landscapes. These perspectives are: *Language, Representation and Power*.

2.   LANGUAGE, REPRESENTATION, POWER

"The task of scholars is to represent the world to others in speech and print" (Duncan & Ley 1993: 2). Yet, a wider problematising of representation is on the whole a product of the last decade. Within the social sciences, the range of perspectives on how to represent the world has exploded in number since the early 1990s. Debates about representation have, even so, to a large degree centred on the politics of language and power, authority and identity. This is no less true in debates regarding the complexities of representing landscapes (e.g., Barnes & Duncan 1992).

There are three papers in this volume that particularly address issues of language, representation and power. These papers seem well suited to serve as analytical illustrations of the three perspectives respectively, and will thus follow directly after this introductory chapter. Furthermore, they are integrated into the short introductions to follow.

Language is not merely a means of communication. It allows for the construction, imagination and abstraction of the complex world around us, using a powerful linkage between hidden meanings and power and real objects and features that we all can perceive in our environment. Naming things gives us power over them. Language also provides a powerful tool for expressing the representations we make of the perceived, which stimulates a common understanding and the building of a community. Language allows us to tell narratives of places and landscapes, to transmit inherited values and knowledge. In the first paper, *Claval* argues that linguistic models, as one perspective on the reading of landscapes, "are helpful for geographers who try to decipher and interpret specific forms of rural landscapes."

Linguistic models, oral and written language are not the only forms of representing landscape. Visual and (carto)graphical *representations* offer additional and equally powerful forms of communication. New technologies in visualisation and for handling complex geographical data are new challenges to our understanding and communication in regards to landscape values. This becomes particularly important with the increasing demand for inter- and transdisciplinary landscape research and its applications in policy-making, planning, management and conservation. *Olwig*, in the second paper, thus states that "the overview and the map provide useful tools for both research and administration." The crux of the matter is, however, that "they also can overlook salient information and it must be recognized that they represent, literally, a particular perspective on landscape that is not, or should not, be hegemonic" because "we live in a world of property rights that have been carefully plotted on the graticule of the map and proudly viewed, as landscape scenery, from the windows of its proprietors."

Landscape is thus about *power*. In the third paper, *Setten* discusses the role of encyclopaedias in "taming" and standardising the language of geography. *The Dictionary of Human Geography* can be seen as aiming at dictating a hegemonic interpretation of landscape within a human geographical discourse. This raises the crucial question: who has the power to frame discourse? Setten argues that through the ordered knowledge created by writing in general, and through encyclopaedias in particular, one aims at teaching students (of geography) "the necessary precision of terminology based on the intellectual power over objects."

Language, representation and power are interactively linked, resulting in many combinations. Most papers in the book illustrate several of these combinations.

Landscape is about *sense of place*. Landscapes are full of signifiers of beliefs and emotions offering a multitude of mental representations to the individual. "Ordinary" people understand their landscape in another way, *Soini* argues. For them it is everyday practice, and not some academic construct. They understand the landscape subjectively – they have their own favourite places (usually on their own land) and places they tend to avoid (usually somebody else's possession). Soini writes: "The study also revealed that it turned out to be much more difficult than anticipated for the locals, the insiders, to define the pleasant and unpleasant, natural and cultural places. The place experience was a mixture of feelings, senses and memories which people were not necessarily able to locate physically." But, if insiders' perspectives are subjective, how can we make sure that the perspective of what is often seen as "outsiders" – experts, interest groups, decision-makers – is less subjective? Positive image amongst the locals derives from balanced development argue *Oja* and *Prede*, so that both tourism interests and places important for the locals are equally taken care of. At the same time, we face both a loss and a redefining of meanings, and the using of clichés, in handling landscapes. *Pae* and *Kaur* show how the role of churches is changing – a place once important primarily for the locals is more and more becoming a tourism attraction, a remnant of the past that has a certain cultural and historical value, and is no longer functioning as a local religious centre. Similarly, as *Häyrynen* writes, the image of countryside in Finland is losing its previous meaning. Although the image is an intellectual construction, in the current era of detachment and alienation these images become the driving force in nature protection – the fragments of the rural landscape that remind us that clichés are desperately being conserved. Again, landscape is a representation. We do not quite remember, nor do we know clearly what we want to protect – the state of the landscape or the idea behind it, i.e. the rurality (as explained by Gustavsson & Peterson 2003; Peterson 2004). And maybe finding a poet who could mediate the different discourses of power could be the way out, as *Bunkše* shows in this book, but it might not be that easy to find somebody familiar with both poetry and with GIS, mapping men and mapping nature.

Landscapes are *full of borders*. We can interpret these as dividing lines between different territories both in a human and a natural sense, we can handle these as limiting properties, or we can handle these as communication zones, either between different units or between approaches (this has been

described by Palang and Fry (2003)). *Bergström* and *Ihse* show the stability
of borders in Sweden, how three types of functional zones – marginal, access
and conflict zones – have been united into a borderline during time. They
point out: "A fourth variant may also be taken into consideration, i.e. the
wish to weaken a community by dividing it." Borders are about power. Maps
are also about power. *Jones* shows how kings used the mapping of territories
to demonstrate their interest in them – this was how they used their power.
Tollin goes through the huge effort made by the Swedish kings to map their
realm, and wonders why such an effort was needed? *Gräslund Berg* brings
up the gender perspective of historical maps and shows that these maps in
fact display the landscape of the men; what women did in these landscapes
was not regarded as important, at least not for mapping purposes.
Essentially, a map is a representation of the world, but not the world itself.
How can we rely on maps and GIS data in designing our future?

Landscape is a *recording of history* and thus *tells a story*. *Germundsson*
discusses how the social order of feudalism was represented in the landscape
created by the Vittskövle estate owners. The design or layout was to
represent the continuity of the power relations, to show who is who. *Smeets*
presents an account of how the local textile industry created *nouveaux riches*
in Twente, in the Netherlands, who then in turn started to design the
landscape through their estate and example, bringing in new aesthetics, new
ideals. *Lachaert* shows how changing power relations caused the
abolishment of the greens and the commons in Flanders and the subsequent
division of these into privately owned parcels. *Grubbström* shows how the
purchases of farms led to changes in ethnic composition and settlement
pattern. Reprivatisation of land in Germany is also about power – the
experiment of turning a socialist landscape into a capitalist one resulted "in
the general transformation of rural areas into 'homes for the old and the
poor'", *Born* concludes. The enclosure movement in England and Wales was
caused also by changes in power relations and resulted in typical English
countryside so much adored by so many. *Chapman* argues that at the same
time this movement caused the "tragedy of the commons" and the loss of
wide open landscapes, also adored by so many.

Landscape is *heritage* that needs *protection*. So what should we protect,
and how? There are indications that conservation is growing out of the site-
oriented approach. For living landscapes, argues *Renes*, conservation is not
enough – planning should be involved as well. "Protection by planning is
essentially a contextual process, in which coalitions between different
interests are crucial," *Renes* concludes. Can planning be objective? Planning
must be based upon good data. Can these be found in geographical
information systems? Maybe, writes *de Bont*. A GIS system in the
Netherlands called *Histland* enables one, in combination with an inventory

of old relics in the present-day landscape, to understand the rather specific regional landscape histories on quite a detailed scale, and also gives the landscape planner material for making plans with regards to the history of such landscapes. But there are many questions for which the answers do not lie in the GIS, but must still be discovered in archives and historical maps: To what extent was there a kind of physical determination involved here? Is it beyond science to state that in the case of Vriezenveen, the Netherlands, the decline in the soil fertility of an oxidising peat layer on a sandy sub-soil caused (more or less directly) the development of the Russian trade?

Landscape needs *management*. Planning is one thing, but who takes care of the landscape? Landscape is a difficult thing to manage, as it consists of numerous pieces of land owned by many people, who all have particular interests (Antrop 2004). Is it in the hands of the farmers? *Bech Sørensen* shows the nostalgia the Estonian and Lithuanian farmers feel about the past, but now they are the ones who should take care of the rural landscape. The landscape of meaning for them lies in the near future, maybe in the direction as pointed out by *Grau Møller* and *Stenseke*. In Denmark, the future might lie in constructing individual plans for the natural and cultural environments of farmlands, using a dialogue between farmers and authorities, proposes Grau Møller. In Sweden, suggests Stenseke: "future common agricultural and rural policies should acknowledge the semi-natural grasslands as parts of local "lifeworlds" and give good conditions for the development of many locally adapted solutions. Included in this strategy is also a fruitful interaction between landscape planning and rural development. It has to be recognised, however, that knowledge, awareness and social learning are indispensable ingredients in increased local involvement and responsibility. To be able to adjust to local conditions and allow various local solutions, it is, furthermore, an urgent need to develop flexibility in landscape policies."

Landscapes are about *ecology* and *semiosis*, argued Cosgrove (2003). Most of this book deals with the semiosis – are we also aware of the ecology? Can we read about that? *Puurmann et al.* show the way in which humans interact with nature on the narrow coastal area of Estonia. *Skjeggedal et al.* argue that: "the concept of ecological integrity may offer a suitable perspective for describing and analysing landscape changes in rural areas. In contrast to the traditional perspective where nature is considered static, "undisturbed", pure, a biological phenomenon without relation to society, the concept of ecological integrity gives an opportunity to consider nature as dynamic, "disturbed", influenced by society, and also to view it from a transdisciplinary perspective. Hence, management strategies must focus more on maintaining the capacity to adapt to change than predicting and anticipating the consequences of decisions." Finally, *Küster* calls for more co-operation between ecology, historical geography, history, archaeology and palynology. Each of these disciplines has its own tasks in explaining the past and avoiding an undesired future. Separately, they all fail to understand the language of landscape.

To draw the book to an end, we explicitly revisit issues of language, representation and power. *Widgren's* paper takes its departure from maybe the most profound question of them all: Can landscapes be read? His answer is yes, given that we are mindful about the conceptualisations of *form, function, process* and *context*.

The differing contributions in the book have in a variety of ways dealt with aspects of reading landscapes. At the same time, the contributions have shown that landscapes cannot simply be *reduced* to a text for us to read. Reading landscapes by means of language, representation and power is rather a way of broadening our understanding of the relationship between the nature of our mental and material landscapes.

This volume has not only aimed at demonstrating the dynamics of persistence and change in the rural landscapes of Europe and how and why we continue researching these issues. It has also been of vital importance to demonstrate notions of persistence and change within landscape research springing from *The Permanent European Conference for the Study of the Rural Landscape*. The very last contribution is *Staffan Helmfrid's* account of what has been achieved during the 20 sessions of the PECSRL:

> The most relevant way of describing the history of PECSRL is found in the shifts of thematic focus in response to fundamental changes in rural landscapes themselves, and the problems they raise in modern society. ... The main themes have logically moved from the basic questions of origin and evolution to the decision-making processes behind changes, and further towards analysing the recent and ongoing dramatic landscape transformations on the one hand, to issues of landscape management and the application of historical geography in the selection and care of landscapes on the other hand.

REFERENCES

Antrop, M. (2004). Why Landscapes of the Past are Important for the Future. *Landscape and Urban Planning*, in press

Barnes, T. & Duncan, J. (Eds.) (1992). *Writing Worlds: Discourse, Text and Metaphor in the Representation of Landscape*. London: Routledge.

Duncan, J. & Ley, D. (1993). Introduction. Representing the Place of Culture. In J. Duncan & D. Ley (Eds.), *Place/Culture/Representation* (pp. 1-21). London, New York: Routledge.

Cosgrove D. (2003). Landscape: Ecology and Semiosis. In H. Palang & G. Fry (Eds.), *Landscape Interfaces. Cultural Heritage in Changing Landscapes* (pp. 15-21). Boston, Dordrecht, London: Kluwer Academic Publishers.

Gustavsson, R. & Peterson, A. (2003). Authenticity in Landscape Conservation and Management – the Importance of the Local Context. In H. Palang & G. Fry (Eds.), *Landscape Interfaces: Cultural Heritage in Changing Landscapes* (pp. 319-357). Boston, Dordrecht, London: Kluwer Academic Publishers.

Jones, M. (1991). The Elusive Reality of Landscape. Concepts and Approaches in Landscape Research. *Norsk Geografisk Tidsskrift-Norwegian Journal of Geography*, 45, 229-244.

Lowenthal, D. (1985). *The Past is a Foreign Country*. Cambridge: Cambridge University Press.

Palang, H. & Fry, G. (Eds.) (2003). *Landscape Interfaces: Cultural Heritage in Changing Landscapes*. Boston, Dordrecht, London: Kluwer Academic Publishers.

Peterson, A. (2004). Has the Generalization Regarding Conservation of Trees and Shrubs in Swedish Agricultural Landscapes Gone too Far? *Landscape and Urban Planning*, in press.

Chapter 2

THE LANGUAGES OF RURAL LANDSCAPES

Paul Claval
Department of Geography, University of Paris-IV (Paris-Sorbonne), France

1. PROLOGUE: RURAL LANDSCAPES AND LINGUISTICS

Linguistics, in the forms it took during the 20th century, is a fascinating science. It deals with languages seized as wholes and their components, sounds, words and sentences. It delves on the way the components are conditioned by the wholes. It explores the significance of the signs it is made up of. It insists on the two faces of signs, which appear at the same time as signifiers and signified. The explanations linguistics provides are not based entirely on external forms of causation, but also on the inner logic of the relations between the whole and the parts, and between signifiers and signified.

Some analogy exists between languages and landscapes. Both can be considered as wholes and decomposed into parts. It is the reason for which geographers often try to rely on linguistic models for interpreting landscapes.

1.1 Structural Linguistics and the Rural Landscapes of Traditional Societies

In the studies of rural landscapes that geographers began to develop from the very end of the 19th century, they used either a structuralist approach (each type of rural landscape had been shaped by a particular ethnic group from the beginning of history), a functional one (each rural landscape was organized in order to allow for combining cultivation and cattle rearing and a sound crop rotation), or an archaeological one (the observed features were born in the past and reflected the functional conditions which prevailed at

11

H. Palang et al. (eds.), European Rural Landscapes:
Persistence and Change in a Globalising Environment, 11-39.
© 2004 *Kluwer Academic Publishers. Printed in the Netherlands.*

that time). What these approaches shared was the idea that a few types of rural landscapes could be identified: openfields, infield-outfield systems, consolidated farmsteads with enclosed fields etc., in Western and Northern Europe and elsewhere. It meant that rural landscapes were organised according to rules. They obeyed a certain grammar. Even if geographers did not refer to linguistic studies, their approaches were in a way close to the findings of the new structural linguistics of the time.

During the first decades of the 20th century, when these types of approaches were prevalent, most geographers did not explore the way local people perceived the environment in which they worked and lived. When relying on the functional approach, geographers tried to find out the logic behind land-use division, and the reason for which people preferred to lay their fields open or chose to enclose them with hedges, walls or whatever other material barrier. They started from the evidence provided by landscape analysis and the study of cadastral maps. In order to interpret it, they made enquiries among land surveyors, landowners or farmers. They relied on these informants when formulating their interpretations. They did not, however, take advantage of all of the evidence they had collected in their enquiries. They rubbed out the subjective aspect of the testimonies they had gathered and selected only the rational elements of interpretation – they believed in the coherence of the landscape systems their analyses had discovered – their only aim was to understand them. The functional dimension was certainly essential, but it did not preclude other readings of the landscapes by local people – symbolic ones, for instance.

1.2   Semiotics and the Aesthetic Readings of Rural Landscapes by Urban Dwellers

A second way of analysing landscapes developed in the 1960s and 1970s. In Italy, for instance, Emilio Sereni (1951) was struck by the differences between the beautiful landscapes of *bel paesi,* in the vicinity of cities, and the more banal forms which prevailed elsewhere. This contrast traced its origins back to the time when Lorenzetti painted the *Allegoria del buono e del cattivo governo* in the City Hall of Sienna, in the mid-14th century. Local elites had discovered that it was good to spend summertime in a second home somewhere in the countryside – it was safer, since it offered a means of escaping the epidemics that were frequent in congested cities when temperatures were high. Since the heads of families often had to be present in their offices or shops, second homes needed to be located relatively close to the walls of the cities. As a result, the Italian suburban landscapes of *bel paesi* were drawn up according to a logic which differed much from that which was dominant in remoter parts of rural areas. Landowners knew perfectly well that crop rotation was a necessity and did not try to renounce the advantages it gave to those who know how to organise their fields; but they generally escaped the collective discipline which prevailed in regards to ordinary farmers. As a result, they could introduce – in the way they organised their land – a new care for its esthetic quality. This second tradition in the study of rural landscapes was best exemplified in Britain by Denis Cosgrove (1984).

When studying landscapes according to this perspective, functional features cease to appear essential. Geographers try to discover the dreams and artistic models used by the building architects, landscape architects and gardeners responsible for drawing up the villas, halls or castles of the well-to-do landowners. The language landscapes spoke was for them one of beauty and harmony.

This type of approach was close to semiotics – the main problem was to interpret the significance of the signs observed in the landscapes. In this field, geographers often rely on semiotics, especially the forms of topological and spatial semiotics developed by Grémias (Grémias 1979). For them, the landscape itself could be compared to a text. The problem was to understand its content and the way meanings were associated with its specific features. A good example of this type of approach was provided by James Duncan's study on the landscape of Kandy as a text (Duncan 1990).

1.3 What Readings for the Rural Landscapes of Postmodern Societies?

The productivity of cultivation jumped during the 20th century. In the most-developed countries around 1900, each farmer could feed his family and three or four others. Today, he is able to produce enough food for 50, 100, or more persons – hence the dramatic decline in the densities of farm employment, with two possible consequences: either a dramatic decline of the density of rural areas occurs, or a complete reversal in the composition of their populations takes place, with suburbanites or "rurbanites" becoming increasingly numerous.

This demographic change has dramatic consequences for the composition of rural landscapes. The functional logic of crop rotation and the necessary association between tilling and cattle-raising has ceased to rule over most of the territory. The new rural population has ceased to share the dreams of classical or romantic landowners. People who have chosen to leave the central parts or the suburbs of late 20th-century cities for rural areas are not motivated by the visions of Arcadia or of virgin lands, which were so pregnant from the 16th to the 20th centuries. They have ceased to consider shepherds as kinsmen of the gods. For them, traditional farming was a dirty activity. There is nobody now who believes in the purity of attitudes and habits of traditional villagers.

Today people are fond of nature, i.e. "true" nature, with its complexity, the struggle for life which characterises it, and its ensuing strength – the first reason for settling in a rural area is to live close to true nature. It is also to enjoy the possibility of practicing sports, games and any other type of activity which requires much space.

The geographers who analyse these contemporary transformations are looking for models in order to interpret what appears as a patchwork of unrelated uses. Is an order emerging of the multiplicity of private and collective initiatives responsible for contemporary evolution? Is there anything like a consensus among the new suburbanites or rurbanites about what is a "good" landscape?

1.4   The Chosen Perspectives

If there were some analogies between the functional approach to the landscape and the first form of structural linguistics, and other analogies between the aesthetic approach (developed from the 1960s) to the landscape and semiotics, the relevance of linguistics for explaining contemporary evolution is less evident. As a result, the first set of questions we shall cover will consist of the following: Up to what point did first structural linguistics, and later semiotics, help to enlighten the genesis and significance of traditional rural landscapes? And is there a form of linguistics relevant for the contemporary evolution?

We shall also try to cover a second set of questions. Landscape specialists have generally focused on the way landscapes were conceived by land surveyors, landowners, landscape architects and gardeners. They did not explore the way landscapes were read and used by ordinary people, women and men who did not try to shape them, but were eager to find in the forms which surrounded them some relief for their existential anxiety, a shelter from the dangers of life and a meaning for their existence. Are linguistic models also useful for explaining these readings of rural landscapes?

2.   TRADITIONAL RURAL LANDSCAPES

2.1   Traditional Rural Landscapes as Farming Artifacts

At the end of the 19th century and the beginning of the 20th, the majority of the population in rural areas – sometimes more than 90 percent – was made up of farmers and farmhands. In each community, there were also craftsmen, shopkeepers, a priest, and eventually some lawyers. The main opposition was not between these minority groups and the overwhelming mass of farmers, but between sharecroppers or farmers and landowners.

At that time, rural landscape studies had several reasons for ignoring the non-farm part of the population. As a result, as of the first phases of the industrial revolution, many craftsmen have been ruined; for example, spinning and weaving have disappeared from most rural areas. Even in the regions where industrial activities were dispersed – either because of the prevalence of a sweating system, or because production came from small plants using water power – most of the land was used for farming. As a result, the main social categories present in rural areas and interested in the organisation of farmland were big landowners, small independent farmers, sharecroppers or farmhands. Geographers analysed the results of the strategies developed by these different actors when distinguishing great families of landscapes: openfield, infield-outfield and other systems.

Traditional rural landscapes may be studied according to three perspectives: one, as sets of fields, meadows, pastures, and woods organised for agricultural production; two, as expressions of the societies which inhabit them; and three, as a surface where one can read nature and discover the existence of beyonds.

When exploring the agrarian perspective, the processes which were responsible for the genesis of landscapes were not always the same. In some cases, landscapes resulted from planning actions; in other ones, they were generated through the prevalence of a farming system with an inbuilt logic.

2.1.1 Planned Traditional Rural Landscapes and Jakobson's Structural Linguistics

The areas where rural landscapes resulted from the work of planning authorities are certainly more widespread than those which evolved out of the independent choices of landowners and farmers.

A) In many cases, land planning is a state privilege – we have testimonies regarding this type of situation in ancient China or Japan. In the Mediterranean region, the Romans applied their system of geometric land division to extensive areas in the Italian peninsula, Western Europe (Western Germany, Belgium, France, Switzerland, Austria, Spain and Portugal), North Africa (mainly Tunisia and Algeria, but also the northern part of Morocco), and in the Middle East.

To encounter as ambitious a scheme of land division as planned by a state, we have to wait for the 18th century, with the decision taken in the young United States to apply a geometrical grid to all the open land of their western margins – in less than a century, the regular squares of the land survey covered about 80 percent of the whole coterminous United States (Johnson 1976). Similar systems were applied in other colonized countries: Canada, Australia, New Zealand, South Africa, and also Argentina, Algeria, Morocco, Tunisia etc.

In all these cases, the concern of states was to draw plots big enough for a farm to be operated, but not too big, since it seemed important to avoid the concentration of landownership in the hands of a small aristocracy. Farmers were free to use the land they got according to their own needs and preferences; this meant that land planning concerned only the geometry of rural exploitations, and not the crop systems they used.

The operation of land planning schemes offers few possibilities of applying linguistics skills to the genesis of rural landscapes. It is not necessary to invoke structural linguistics to explain the geometries which preside over these operations. There are, however, interesting exceptions. In Japan, Minoru Senda studied the *jori* settlement patterns of central Hondo, in the vicinity of Nara for instance, using models provided by modern linguistics (Senda 1980).

B) There were other forms of planning. Many of them were borne of the necessity to develop crop rotation systems in areas where properties or farms were too small to easily organise a harmonious combination of fields and pastures. Such programmes could not be realised without an initial land consolidation. Most of the openfields which are – or were – present in the world were born in this way. In Western Europe, feudal lords and farming communities were responsible for most of these operations.

Since the organisation of openfields often dates back from the first part of the Middle Ages, documents on these operations have often disappeared. The land consolidation required a land survey and geometric operations of division. We generally have only scarce information on the way such operations were performed, but enough was left for us to be sure that some form of land planning was present. The schemes responsible for the openfield landscape were drawn at the scale of the elementary rural communities – the parishes. This means that the process of decision making and the ensuing planning operations remained in many cases local ones.

When the problem of landscape organisation is to combine efficiently the private cultivation of fields and the collective management of cattle rearing, there is no possibility of imagining many field patterns. The best solutions – i.e. the division of the land into two or three fields, with compulsory crop rotation and the collective use of fallow for grazing – appear as logical outcomes of the problem as soon as it is clearly analysed. As a result, these solutions could certainly have been invented independently in several places – Xavier de Planhol provided a good example of such a situation when he observed, in the 1950s, the creation of openfield systems in Anatolia by farmers who were not acquainted with European experiences (de Planhol 1958). Locally, a diffusion process occurred around the places where these solutions had been invented or reinvented, but what was fundamental for the theoretical interpretation of rural landscape was that a possibility existed to reach, through trial and error, the same result in different places.

In a way, the genesis of the openfield system evokes the genesis of American myths, as analysed by Claude Lévi-Strauss – the combination of elements was a natural outcome of their structural properties, which was the main idea of Roman Jakobson's structural linguistics. The analysis of openfield systems as developed by geographers from the 1920s offers some similarity to the structural analysis of phonemes, as initiated by Ferdinand de Saussure: in order to function, a system has to be made of different and complementary units. The difference between landscape analysis and structural linguistics evidently comes from the fact that rural landscapes cannot be easily decomposed into sets of minimal units.

We have more information on land planning operations as developed on the eastern margins of medieval Europe, and in some colonial areas. The progression of farming settlements did not result from a decision at the state level. It was the outcome of the initiatives of local entrepreneurs or feudal lords. They often chose to organise *Waldhuffendörfer*: they opened roads, divided lands in long plots on either side and gave them to the settlers. The Eastern Europe road settlements had their equivalent in the French Canadian *rang,* which allowed for the organisation of the rural landscape along the Saint-Laurent River and its affluents in Eastern Canada, in French Louisiana along the Mississippi River, and in some other localities. German settlers in Southern Brazil often chose the same solution.

In the case of *Waldhuffendörfer,* planning was restricted to the drawing up of farmsteads. It did not deal with systems of crop rotation and the combination of fields and pastures. Before the invention of the barbed wire and other modern types of fence, another form of planning operation was necessary for dissolving openfields systems; in order to break with the crop rotation they involved, it was impossible to grow hedges around their overly-long and narrow fields. The only solution was to proceed to a land consolidation. In Britain, where this transformation appeared as early as the late Middle Ages and gained momentum in the 18th century, enclosures had to be voted on by the Parliament. The land consolidation scheme itself was prepared and applied by local authorities. The situation was different in the countries where the old field systems were destroyed all at once by a state decision, as in Scotland, Sweden and/or Denmark (Johnston 1976; Pred 1986).

2.1.2   The Genesis of Unplanned Rural Landscapes and Noam Chomsky's Generative Grammar

A French historian, Annie Antoine, recently presented an original analysis of Western France rural landscapes (Antoine 2000). She used two types of documents: one, the old land maps prepared for the big landowners, the local nobility, bishoprics, and monasteries; and two, land law as it was expressed through the customs of three provinces – Anjou, Maine and Bretagne – and used in the local courts.

She studied more particularly the role of hedges in the rural landscape and the way fields were enclosed. She covered modern times, mainly the 18th and early 19th centuries.

It was the time when physiocratism was very popular in France; for this conception of economics, agricultural production was considered as the only source of wealth, which meant that the essential responsibility of economists and agronomists was to promote more rational ways of farming. As a consequence:

> For the observer unfamiliar with the countryside, what was beautiful is what bears wealth, the (cultivated) nature is beautiful because it is a promise of crops. And in order to be truly beautiful, it had to be completely cultivated (Antoine 2000: 226).

For a traveler or an agronomist with such a physiocratic perspective, the landscape of Western France was certainly not beautiful, since a wide part of the land was covered by rough pastures, gorse and broom, and remained unploughed. For them, all the land had to be enclosed. Each farm had to be located within the tract of land it used. All the fields had to be incorporated within the system of crop rotation in order to produce every year.

When analysing the evidence of cadastral maps, local land law and the judgments of courts concerning property rights, it appeared that farmers and the big landowners who had not been contaminated by the physiocratic vision had a different conception. A good example is offered by the great agricultural enquiries of the end of the 18th and the beginning of the 19th centuries – when consulting the summaries prepared at the scale of intendances and later departments, the perspective is totally physiocratic. When consulting the local reports on which these syntheses were built, the reader discovers another conception of what was farming, of what was a functional landscape.

For the physiocrats, hedges had to lock up herds. For local farmers, the hedge was "a bulwark, a cage, which helped much more often to protect a cultivated plot of land, to forbid an access to cattle, rather than to contain it"

(Antoine 2000: 169). Annie Antoine makes her analysis more explicit in the following quotation:

> The protest is general: those who wish to see a beautiful countryside, those who desire that roads do not look like ruts, all of them denounce the practices of farmers who persist in not ploughing the uncultivated land and using it in a disastrous way. Here is the crux of the problem: in the peasant practice, the cultivated land and the uncultivated land are the two components of the same system (Antoine 2000: 228).

The last sentence is essential – in the perspective of the local farmers, landscape is conceived as the materialisation of a farming system. It is made up of parts which are at the same time different and complementary. A farm needs rough pastures – in which there is no need for hedges, since they are open to all the farmers of the community – and hedges, which protect the cultivated fields from the teeth of wandering cattle. It means that according to densities, the materialisation of the farming system was conducive to a variety of landscapes:

> The scarcity of hedges, the fact that they are essentially present near the farms, or for enclosing large open spaces, bears witness for a very extensive use of land; on the reverse, a denser and above all more regular *'bocage'* is the *traduction* of a less 'savage' and more organized style of cattle rearing (Antoine 2000: 172).

The judgments of the local courts concerning land conflicts are particularly interesting, since they prove that all the members of local societies, farmers, landowners and judges, were perfectly aware of the nature of the landscape organisation and its finalities.

This study is fascinating for many reasons. In a way, Annie Antoine provides a confirmation for the theses developed by the geographers and historians working on *bocages* of Western France since the late 1930s – this type of functional organisation is linked with the role of cattle rearing in local farming systems. In another way, the study of Annie Antoine offers a completely new perspective on the way rural landscape were born and evolved – there were no rigid spatial models of land-use. Landscapes simply reflected dynamic farming system – farmers created more hedges and enclosed more land when their systems became more intensive.

Rural landscapes apparently conform to fixed spatial patterns, but this is a superficial view of the real processes: they are transitory forms, pertaining to certain families, adapted to the dynamics of farming systems. The functional view developed by the geographers of the 1930s and 1940s was not wrong, but it was incomplete. Traditional rural landscapes meet effectively functional needs, but they did not result from the imposition on the land of some kind of super-organic pattern by some mysterious authority. Instead, they were the result of conflicting interests in societies based on evolving farming systems.

The example chosen by Annie Antoine of Western France is a fascinating one, since it analyzes the genesis of rural landscapes that no authority had planned. In the *bocages* of Western France, the only actors were the landowners and the courts, which applied the rules of the local customs. Planning was only possible within the limits of private properties – i.e. big landowners dividing their land into farms. The rural landscapes were, in such a legal and social setting, a dynamic reality; hence their plasticity and the slow genesis of the *bocage* as it existed at the end of the 19th century.

The functional approach as used by Annie Antoine is certainly the most useful one for understanding the genesis, forms and dynamics of rural landscapes in areas where farming is the only activity, and landscapes evolve without consolidation and planning. The processes analysed by Annie Antoine differ from those used for the understanding of openfield systems. They are closer to those analysed by Noam Chomsky – they belong to the family of generative grammars.

The modern analyses as developed by Annie Antoine introduce something completely new in this field; instead of starting from the observer, they start from the perceptions, images and habits of the people who live and work in an area and shape it. They prove that the functional language of oppositions and complementarities is naturally used by all those who have to organise complex and complementary activities.

The language of functionality is not the only one the people who live in rural areas are using. As a result, research is now exploring other new paths.

2.2 Traditional Rural Landscapes and the Expressions of Collective Social Realities

2.2.1 Landscapes as an Expression of the Political and Cultural Unity of Social Groups

Kenneth Olwig has developed, during the last ten years, a fascinating analysis of the early significance of the term *landscape*. He presented his main results in his last book (Olwig 2002). Everyone knows that the term *landscape* was coined in the Netherlands in the 15th century to designate a new type of painting; thanks to the invention of modern perspective, it had become possible to propose realist representations of rural areas, cities, forests, hills, and mountains behind the scenes, which remained for a long time the central motifs. In van Eyck's paintings, it was the perspectives offered through the windows which introduced the landscapes.

In fact, the word *landscape* was widely used in the Netherlands and in other countries along the North Sea during the last centuries of the Middle Ages. Kenneth Olwig found it, with a slightly different spelling – *landskip* – in the border zone between Danish and German settlement in Schleswig and Holstein. In this area, it served to designate a small territory. A *landskip* was a physionomical unit, but it was also a social and cultural unit – the area for which the name was used was perceived as the home of a specific group, with its own customs, its own institutions. It was also a unit of political management. Kenneth Olwig notes similar ways of perceiving the human environment in Switzerland.

The thesis of Kenneth Olwig is based on this idea – for the farmers and dispersed population of rural areas, landscape was not basically perceived as scenery. It was a social and political construction embodied in a territory (Olwig 2002). Olwig proposes to renew landscape studies by restoring the political significance they had in traditional Western society, a significance that was lost during modern times, when the term was increasingly associated with art and painting.

Historians have reached similar conclusions elsewhere. Yves Durand, in his book *Vivre au pays au XVIIIe siècle* (Durand 1988), emphasised the way local populations conceived the region in which they lived – for them, the *pays* (which served as the root for *paysage,* landscape) was the fundamental social and political unit. It was the home country. The perception of the geographical reality was in a way a morphological (as it was based on the landscapes people were familiar with) and socio-political one (as it was shaped by the limits of local solidarity and political and administrative networks). In her book *Pays ou circonscriptions. Territoires de la France du Sud-Ouest,* Anne Zink explored the expression of the feeling of otherness in Ancient Regime France, and discovered that it resulted from the dialectics between the traditional small social and cultural units named *pays,* and the administrative circumscriptions imposed on the area by religious or political authorities (Zink 2000).

Geographers have focused, in their studies of traditional rural landscapes, on the fact that they were shaped according to the logic of crop rotation and farming systems. More recent research like the one developed by Annie Antoine shows that the local populations were perfectly aware of the spatial imperatives behind the farming system they used. For them, landscapes were not static morphological constructions – they were evolving structures based on the farming systems.

What Kenneth Olwig discovered was similar in the field of social, cultural and political systems – landscapes were not, for local populations, only based on the visual perception of a special combination of fields, hedges, walls, farms, villages etc. They were the embodiment of basic social, cultural and political systems.

2.2.2   Traditional Rural Landscapes and the Expression of Social Differentiation

The interpretation local populations placed on landscapes was a global one, when focusing on farming practices or social and cultural life. There were also piecemeal readings – each element was given a specific meaning. It happens often for houses and the gardens which surrounded them.

In Finland and other Nordic countries, farms were painted either in yellow or red: the well-to-do farmers used the yellow colour, which was expensive. The other components of the population had to content themselves with the red one, since this painting was a cheap by-product of the Falun copper mines in Sweden.

The social reading of farmhouses and other rural buildings was present in most societies, but it took on more significance during the 19th century, at a time when modernisation started in European rural areas, cities began to grow rapidly and the relations between rural and urban areas changed. Rural communities ceased to be self-enclosed realities. They began to be more open to urban attitudes. They more readily adopted urban styles.

When visiting the rural areas of Western Europe, most of the beautiful farms or handsome buildings in villages date back to the late 18th or 19th centuries. By the beginning of the 20th century, the permeability of rural areas to urban influences had become so great that the techniques of rural construction and built forms were directly copied from urban models, and had lost their previous originality. But there was a period, mainly before the construction of the railways, when urban influences began to transform the sensibilities in rural areas, without already outdating most of the techniques locally used. The people living in rural areas were increasingly aware of the existence of other rural communities and cities. They reacted to this new level of transparency by asserting their identities. Each rural cell wished to be the first to demonstrate how much it differed from its neighbours.

An example of this can be found in the beginning of the 19th century. At this time, there were about ten types of headgears for women in Brittany; by the end of the 19th century, there were more than 100. Each *pays,* and, in some parts of Brittany each parish, had its own (de Planhol 1988). Architectural forms began to be used as an indicator of the rural diversity of a country. It was in 1873 that Dr. Hazelius created the open-air museum Skansen in Stockholm – each part of the kingdom was represented by one of its farms, cautiously taken apart in the place where it stood and reconstructed in the capital.

The inhabitants were increasingly aware of this transformation – they began to consider the forms of their houses and farms as a testimony about what and who they were. Rural landscapes were transformed into sceneries by the people who inhabited them in order to display to the travelers and tourists who they were and how they wished to be seen.

The consequences of these new attitudes were spectacular. In provinces of southwestern France such as Quercy or Périgord, most farmers decided to add a pigeon loft to their houses as a way of stressing their social promotion through the French Revolution – in Ancient Regime France, only the nobility had the privilege to build pigeon lofts.

Farm houses ceased to be conceived only in functional terms. Their main building had to be built of stones or bricks, with beautiful door or window frames, and details of architecture were chosen for their elegance. A part of them pertained to the local know-how of masons or carpenters, and another part had been borrowed from urban architecture; hence the character of many rural areas in France, with their beautiful farms, mixing old local traditions and urban features. In the regions where sharecropping was widespread, landowners invested more in their farms than in the past – this explained the quality of rural houses in countries where sharecroppers were generally relatively poor, like Quercy or Périgord.

The population of rural areas used landscape to express the social status of its members and increasingly, during the late 18th and 19th centuries, to express its specificities when compared with other rural areas or cities. Landscapes were not totally transformed by the will to give them this kind of imprint – the result was achieved through the use of a few markers.

2.3 Traditional Rural Landscapes and the Relations of Human Groups with Nature and the Other World

2.3.1 Traditional Rural Landscapes as an Expression of Environmental Specificities

For the people living in rural areas, landscapes appeared both as an expression of the farming systems and as the material basis of social, cultural and political units. Other readings were also practiced – they stressed the natural dimension of landscape and the relations local groups developed with their environments.

Landscapes ceased to be seized upon as totalities. Observers focused on some of their features, such as slope, exposition to sun, vegetation, soils etc. They were used for their value as signs of interesting properties of the environment. The farmers of *terreforts,* in southwestern France, were able to differentiate between some fundamental types of soil: the *terreforts* themselves, being heavy, difficult to plough after the rains or when too dry, but with an important capacity for retaining water; *boulbènes*, being lighter, easier to plough, more acidic and prone to drying up rapidly in the summertime; or *rougets,* derived from clay on slopes oriented to the north. For maize, *terreforts* were more convenient, since they are cropped late, in September or October. *Boulbènes* were best for barley and wheat, which are harvested in early July (Toujas-Pinède 1978).

In this naturalist perspective, a landscape ceased to be read as a whole. It was made up of a combination of smaller areas or features, each one endowed with original characteristics. Geographers often rely on the analysis of toponymy to understand the way local people perceived and classified the natural features they took notice of – for example, what were the species of trees used as topographical indicators? How did farmers name the different types of soils they distinguished? What were the names they gave to the local topographic forms? Linguistic analysis is all the more interesting in that it often allows a person to know the period when each toponym was introduced – the forms and pronunciations of the names of trees, soils, rocks etc., differ from those now in use in the same area. Local people often remain conscious of the significance of these old terms, which means that they are still relevant today.

In these readings, landscape is considered by local peoples as a reality where the properties and characteristics of nature are made understandable through the presence of specific signs or markers. In the Brazilian *mata atlantica,* for instance, the presence of the whitish foliage of *ubumbas* proves that it is a secondary forest; more inland, a forest, a *mato grosso,* reflects the good quality of soils – in the State of Goyaz, rural settlement started in relation to the *mato grosso* characterises the area around Goiania.

2.3.2   Traditional Rural Landscapes and the Expression of Other Worlds

The elements of a landscape may appear as signs of a deeper reality. There are springs where naiads or the geniuses of water dwell, forests haunted by trolls or other spirits, trees inhabited by a god etc. Rocks, cliffs or other spectacular landforms are often the places where realities beyond what can be normally seen are present. Even in a plain with uniform vegetation, some spatial differentiation may occur in this way – areas of profanity are opposed to places loaded with sacredness. The ethnographers and geographers who study traditional societies in the Pacific Islands, Australia, the mountains of Indonesia or southeastern Asia, Africa or Indian America always stress the significance of these beliefs, and the fact that they contribute much towards structuring space (Raison 1977; Bonnemaison 1986).

Similar interpretations existed in European societies. Christianity struggled hard to destroy them, but did not succeed completely. In many cases, it had to compose itself using the previous religious systems – as a consequence, the Church often used the sacred places of the previous systems to build its new churches, chapels or crosses (Saintyves 1907). For the local people, the Christian cult practiced there was generally loaded with pagan reminiscences.

The great modern religions have developed universal systems for giving significance to human destiny and the whole cosmos. These systems do not speak of local realities: the scarcity of rains and the threat of drought it creates, the violent storms able to destroy crops just before the harvest, the floods and their destructive effects in alluvial plains etc. Paganism did not speak in the same way about salvation, damnation and sin. It did not offer the same possibilities of hope in a better life after death. It was, however, much more efficient for local concerns. Each of the gods or spirits it honoured was in charge of a specific element of nature: wind, rain, lightning or storms; with a specific tract of land: a source, a river; and with the cycle of life: birth, growth and death. In order to prevent local catastrophes, traditional beliefs offered precise and specific answers.

When the dangers were too great and threatened a whole social group and the totality of the landscape where it lived, specific measures were taken: people met to pray, dance, and make sacrifices in specific places; they also participated in processions which moved across the whole area in order to restore its purity. The Christian feast and rituals of Rogations took over much of this type of ceremony.

The analysis of the religious interpretations of landscapes, or of some of their elements, started in the 19th century (Mannhardt 1875-1877; Hahn 1896) and developed further in the 20th century (Sébillot 1908; Rantasalo 1919-1925; Lautman 1979). Mircea Eliade (1949) offered a good account of this theme. The studies of historians and ethnographers provided evidence of the persistence of these traditional beliefs in modern rural societies – an example is Carlo Ginsburg's work (1966) regarding northern Italy.

Simon Schama published a fascinating book on landscapes in 1995. Many geographers liked it, but did not succeed in using it. It was not an analysis of the material aspects of landscapes, but instead a reflection on the ways its elements – woods and forests, rocks or water – have been used in legends, in classical literature or in other literary productions; as guides, for instance. Instead of focusing on religious beliefs, the work was centered on mythologies. Such a reading stresses more the exceptional features of the environment than the trivial ones: it often speaks of very old and big oaks or firs, precipitous cliffs, curious rocks, lakes or ponds with deep and black waters, and/or impressive water falls. The universe such elements signal is that of the deep forces at work in nature before the intervention of man – it is the reason for which the landscape of Schama is never a cultivated one. Civilisation puts a ban on the forces which played such a role in the oral traditions of many societies.

Augustin Berque has recently wondered about the lack of an aesthetic dimension in the perception of landscape in most societies (Berque 1995, 1999). For him, before the invention of landscape in 4th century AD in China and the 15th century in Western Europe:

> ... the aesthetic feeling remains inserted into an ethnic more global perception, grasping directly a certain cosmology (i.e. the reasons of being a World) in the *geograms* of a given environment. Here are thus cosmophanic societies (Berque 1999: 53-54).

2.3.3   Partial Conclusion – A Wide Variety of Readings

We did not try to cover all the possible readings of traditional rural landscapes. We did not speak about, for instance, the particular perceptions concerning women, children and old persons. We mainly delved into the conceptions shared by adult and working men.

Until 30 years ago, geographers had mainly developed a specific type of landscape interpretation – they seized upon it as a totality, and explained its subdivisions as resulting from the combination of cultivation and cattle rearing in a particular farming system.

There are other forms of landscape readings. Sometimes, landscape is analysed as the spatial translation of the social, cultural and political structures of a group, as shown by Kenneth Olwig's work on the early *landskips* of the North Sea area, and now illustrated by many other studies all over Europe. The landscape as the home of a group still rests, however, on the idea that a landscape has to be seized as a whole.

Most of the other readings of landscape are piecemeal. Each place, each line, each elementary area appears as a sign. Its significance depends on its situation. Whenever markers are created by men, they are conceived in order to convey a clear meaning; such as the social situation of the owner of a house, or the originality of the population settled in a place.

When interpreting landscape as the visible forms of nature, men attribute a meaning to the presence of specific elements or features, such as big trees, ponds, caves or cliffs. Such markers have not been designed by men. Their meaning results from their supposed capacity to express hidden messages behind such and such element or feature.

The diversity of readings of the landscapes of rural areas by the populations who inhabit, shape and use them shows that the relations between visible forms and the significance men confer upon them are neither necessary, nor always pertaining to the same types. Landscapes may be read as global systems, each of their components being tied to the whole through functional links (as is the case for farming landscapes), or as global homes for specific social, cultural and political units. Landscapes may also be read as collections of independent signs, either created by men for conveying their ideas, or discovered by them as the expressions of the inner order of nature, or a mythical and religious Other World.

The analogies between landscape interpretation and linguistics are numerous: in both fields, people deal with wholes which may be split into discrete components; in both cases, a meaning is conferred upon these elements. In some cases, it results from the division of the reality itself and the articulation of its parts (an idea central to the first forms of structural linguistics, but equally present in the classical interpretations of rural landscapes). This analogy appears, however, as a superficial one, since a language is merely a juxtaposition of sounds, while a rural landscape could not exist and function if an image of the farming system was not present in the minds of the farmers who run it.

In other cases, landscape is made up of the juxtaposition of discrete signs. It offers no global property. The significance of those signs comes from the existence of signifiers, which are either built by men in order to communicate, or deciphered by them as testimonies to the real structure of the natural world, or to the presence of another world, a beyond which rules over our world.

Geographers tried to apply the recipes of semiology towards modernising their analyses of landscapes. It appears, however, that the differences between landscapes and languages are such that this analogy could not work very well – language is made of arbitrary signs, while a landscape is either structured by functional divisions, or doted by markers which reflect the clear intention of men or the existence of natural or supernatural systems. Some geographers, like Roger Brunet, tried to split the geographical space into sets of discrete units, i.e. chorems, just as language is divided into sets of phonemes, morphemes, semes etc. The chorems of Brunet may, however, always be divided a step further, to when the phonemes, morphemes, and semes are really the smallest units into which the continuum of the language can be split (Brunet 1980).

The analogy between linguistics and the geographical analysis of rural landscapes is certainly an interesting one, but it is not easy to take advantage of it in order to develop new insights on landscapes.

3. THE RISE OF URBAN PERSPECTIVES ON RURAL LANDSCAPES: THEIR AESTHETIC DIMENSIONS

3.1 Landscape Architecture at the Landscape Scale

The analogies between linguistics and landscape analysis are different when dealing with the creations of artists, designers, and landscape architects. In landscapes produced and experienced by traditional rural societies, there was certainly an aesthetic dimension in the way people were conceiving their homes in order to prove their belonging to a specific group or class, or to demonstrate their capacity to meet the criteria of excellence as imagined by urban dwellers. The impact of these aesthetic concerns was, however, limited by its scale – it modified points rather than lines or areas. The areas it transformed were too limited to modify the global character of landscapes.

These conditions changed when emperors, kings or princes began to organise huge gardens, parks and forests around their castles. We know of examples of these huge enterprises in many civilizations – the tradition of "paradises", as it evolved from antic Persia to Moslem Iran or Mughol India, was based on the drawing of rectangular gardens which covered tens or hundreds of hectares (Moynihan 1979). Their impact on the landscape was always important. Landscape architects did not try, however, to alter the totality of a landscape – parks appeared more as islands in the countryside than as attempts to change the outlook of a whole region. They were always enclosed by walls, in order to show clearly their nature of being enclaves inside rural areas.

The status of the landscape was different in Roma. The mosaics of the aristocratic villas displayed large tracts of rural areas designed for pleasure. They more or less intermingled with farmland. We only know of Plinus the Young's villa through his letters, and are unable to measure its real dimensions and its impact on the landscape. With the ruins of *villa Hadriana* close to Tivoli, we have direct evidence of the scale to which landscape planning was conceived of at that time – the gardens covered an important area, but in a deep valley, which limited their visual impact.

Feudal lords were more interested in the defensive quality of their castles than in landscape design. Since they often chose to fortify their defences using hills, the dungeon, towers and ramparts they built were seen from a wide area; but this did not change its global character. As exemplified by the *Allegoria del buono e cattivo governo* of Lorenzetti, the situation began to change in the later Middle Ages – the urban elites who chose to spend summertime in their villas were fond of gardens. Because of the accumulation of villas close to the cities, the character of the suburban regions changed – they became the *bel paesi* described by Emilio Sereni (1961).

The art of the suburban villa throve in the 16th century. In Venice, the aristocracy discovered an interest in farm investment in *Tierra firma* at a time when the profit-earning of trade was certainly lesser than in the past. The construction of a villa in a newly bought farm achieved two aims: it offered the landowner a means of controlling efficiently the management of his farms and the payment of rents – often a share of all the crops; it also provided urban families with a second home for summertime. Thanks to architects like Palladio, the design of the house was combined with that of a garden or a park. The idea of relying on the rules of linear perspectives to compose a majestic environment had become popular by the 1570s, with Cardinal Montalto, the future Sixte-Quint, drawing a beautiful park around his villa, close to the Roman forums.

In 16th and 17th-century Venice, as well as in 18th and 19th-century England, big landowners launched bold initiatives in order to transform the landscapes of the areas they owned. They had impressive gardens and parks designed around their villas or castles. Through the invention of the *ha-ha,* they discovered that it was possible to integrate the whole rural landscape into the views that can be discovered from the reception rooms of their residences, or when walking in the park. Even if they were not able to pay for as huge projects as the Versailles of Louis XIV or the Blenheim of the Duke of Marlborough, the impact of the landscape gardening they initiated was often as huge. They did not hesitate in relocating whole villages in order to obtain the kind of long range views they cherished (Jacques 1983; Hunt 1986).

Denis Cosgrove (1984) provided us, nearly 20 years ago, with a fascinating book on the history of this venture. He reminded us of the social role played by the literary and pictorial representations of landscape in the transformation of rural areas during the 17th, 18th and 19th centuries. The starting points were provided by the geometries that can be built out of the rules of linear perspectives (Le Nôtre's landscape architecture, for instance), or the images conveyed by the myth of Arcadia, and its literary and pictorial treatment during the Renaissance and the 17th century. The linear perspective was conducive to the drawing of long alleys or geometrical canals, which united the elements of the landscape around a central point of observation. The most important event, in the history of the Arcadian myth, was the interpretation, in the 17th century, by Claude Lorrain and other landscape painters. Instead of relying on linear perspective to suggest the depth of a scene, they played with the light – its reflection on the waters of the seas, ponds or lakes, the paler and bluer colors of backgrounds – and developed the art of the aerial perspective.

What was really new with Cosgrove was, however, his social and ideological interpretation of the landscape revolutions of the 17th and 18th centuries. The history he wrote deals at the same time with the aesthetics of the landscape designers of the time – Palladio in Italy, Lord Burlington, William Kent, Capability Brown and their imitators in Britain – the history of an emerging class of landowners, and the ideologies which motivated them.

The aim of the rich landowners was to offer a testimony of their taste. For them, transforming and beautifying a landscape was a means of showing their aptitude in aesthetic design and the knowledge they had gained of the most recent achievements in this field. To succeed in such an enterprise was certainly important for the society in which those wealthy landowners lived – this ensured for them the admiration of their friends and competitors. A first wave of landscape gardening had developed in Britain at the end of the 17th century, with many impressive parks designed along esthetic lines copied from Versailles and other French models. The new Whig aristocracy developed a critical view of this type of landscape architecture: the centrality of composition, the rigidity of lanes, and the way trees were trimmed – all appeared as signs of Louis XIV's absolutism. The aesthetics for which Pope and Shaftesbury pleaded in the 1710s or 1720s materialised, thanks to Lord Burlington and William Kent, as a new style of landscape architecture: the lack of rigidity of its lines, its winding paths, and the way long perspectives were animated by isolated trees or small groves – everything spoke of a society where liberty has triumphed over absolutism.

What proves the competitive character of landscape architecture in 18th century England was the total replacement of the aesthetics of the French garden with that of the landscape park, considered as being genuinely English. All the gardens of the 17th century and early 18th century were redrawn within less than a generation, from 1730 to 1760.

In their big landscape ventures, landowners sought not only to be acknowledged and praised by the other members of their social class – they also wished to reaffirm the legitimacy of their social power. Their capacity to beautify a landscape justified intellectually the right they had acquired through inheritance or purchase to rule over wide tracts of land. It was impossible to criticise them, since they imposed upon the landscape which they controlled the most admirable organisation it was possible to imagine.

3.2   Landscape as a Text

In the example chosen by Cosgrove, parks were designed by artists and transformed into landscapes thanks to the initiative of wealthy aristocrats. The message conveyed by the landscape was clear. It was partly an aesthetic one, and translated for a particular setting the general principles of composition which dominated at that time. It was also partly social – landowners tried to improve their status in the aristocratic circles they frequented, and to reaffirm, through the cultural achievements best exemplified by the huge tracts of land they had beautified, the economic and social responsibilities they had.

Regarding this type of landscape, the problems landowners and landscape architects had to resolve are perfectly known, thanks to the documents they left behind. Their aim was not to discover the best land system for crop rotation and the combination of cultivation and cattle rearing. They had an interest in these problems, since they were essential for the economic performance of their farms, but that did not interfere with their aesthetic ambitions. Landscapes ceased to appear as systems of well-adjusted components, or as mosaics of features receiving a meaning through a few spectacular markers or symbols. The interest shifted to the message landowners and landscape architects tried to convey.

Landscapes were messages. They had to be studied as texts. They tried to translate into a visual form abstract ideas, philosophical meditations, or religious convictions. The parks designed at the end of the 18th century were full of false Gothic ruins, mock temples, and pseudo-graves. They reminded visitors that all culture is mortal, decadence always follows prosperity, and that the only lasting pleasures are associated with wisdom, the study and contemplation of nature, reflection on the human condition, and a sincere admiration for the great men who were able to transcend the prevailing mediocrity of humankind.

The grammar of the 18th-century garden displaced an older one, which was based on the taste for Arcadia as a shared sojourn of shepherds, gods and goddesses. There, the main elements had been the sources, the basins where naiads were settled, and caverns and grottos, which gave a glimpse of the chthonian forces at work in the universe.

The grammar of Western gardens was generally a simple one. Landscapes were more difficult to interpret when they had to translate into visible forms more complex messages. James Duncan (1990) analysed in this way the Buddhist cityscape of Kandy, in Sri Lanka. The kings who reigned in this inland capital at the beginning of the 19th century knew that their power was a fragile one – they were threatened by the British coastal settlements. In order to survive, they had to preserve the unity of the society they ruled over. For them, the only way to achieve such a result was to diffuse the message of the Buddhist faith, in its most favorable version for the civil power they represented.

The messages embodied in rural landscapes are generally much less complex than those orchestrated by the Kandian dynasty and the artists who worked for it in the 19th century.

3.3   The Impact of Arcadian Dreams

For a long time, the areas transformed by the Arcadian visions of landscape remained relatively scarce and small, except in the suburban belt of Italian or southern France *bel paesi,* or in the areas of the English countryside where the residences of the landed aristocracy were the more numerous – the Home Counties. Even there, they generally covered less than one-tenth of the total area.

Changes began to occur at the end of the 18th century. The Arcadian dream took on a new form: instead of being based on Greek and Roman mythology and a romantic vision of the lives of shepherds, it was from then on associated with the pure valleys of the Alps, or other more or less archaic rural communities. The language of this new form of dream was less academic: instead of temples, ruins of castles and churches and mock tombs, landscapes had to be decorated by just a few thatched cottages, a milk shed and a sheep barn. During the last decades of the 18th century, parks were invaded, just as in the case of the Trianon in Versailles, by these expressions of the new Arcadian sensibility.

As long as the taste for rural scenes was deeply intertwined with the knowledge of Greek or Latin mythologies, its impact was limited to the aristocracy and the learned fractions of the society. In its new form, however, Arcadia became accessible to middle class tastes. The economic transformations associated with the Industrial Revolution and the building of railroads popularised tourism – an increased number of urbanites could for the first time visit the small peasant communities along the coasts or in the mountainous areas, which served as models for the new sensibilities. Because of the railway and later the tramway, to live in a suburban setting ceased to the dream of only a minority of wealthy persons – the suburbanisation of the middle classes was partly the outcome of this new vision of Arcadia as a form of paradise on Earth. Cynthia Gorrha-Gobin has, for instance, shown how the American feminism of the 19th century was responsible for the genesis of this new form of urban environment (Ghorra-Gobin 1987, 1997).

If the dream of city-dwellers was to opt for a more rural form of life, the result was quite different from what has been expected. All rural life rapidly disappeared from most suburbs. Instead of a ruralisation of the city, what was observed was an accelerated form of urban sprawl.

4. RURAL LANDSCAPES IN THE POSTMODERN AGE

4.1 New Types of Activities and the Social Composition of Postmodern Rural Areas

The social and professional composition of rural areas has completely changed during the last few decades. The first effect of the industrial revolution had been a simplification of the professional activities locally present – most of the craftsmen had disappeared.

With a continuously growing agricultural productivity, farming densities are continuously declining. In the areas where farming is still the major activity, it means a drastic decline in overall densities, with a spiral of desertion – young families leaving localities where it is impossible to find the services they need for raising their children.

The other possibility is a change in the professional composition of rural communities – their populations are stable, or growing, even if the land which was cultivated 50 years ago by five, 10 or more farmers is operated today by only one. Either the local population has opted for other professional activities, or people coming from other rural areas or cities have settled and compensate for the out-migration of indigenous families. In either case, rural areas have ceased to be – socially – farming communities, even if fields and pastures still cover most of the land.

Rural areas are in this way transformed into new suburban or rurban ones. Their population is diverse. It is composed of: a minority of farmers, often native from the area, but with highly differentiated forms of production (truck farming, cattle raising or a combination of farming and touristic activities); groups of local origin but with urban-type activities in the sector of industry or services; newcomers migrating from cities or abroad with the same type of activities but a different background; retirees, etc. It means that rural populations are culturally more diverse than in the past. They are bearers of new forms of dreams.

4.2 Cultural Evolution and the Postmodern Dreamed Landscape – Another Conception of Nature

As long as modernity lasted, the idea of rural areas as a form of Arcadia, a kind of paradise on Earth, survived. Even with the modernisation of farming activities, it was still possible, 50 years ago, to find farms in which cows were milked by hand and where people sold their production to their neighbours. In the farm courtyards, there were chickens or ducks wandering freely. Even if farmers used fertilisers, these chemical inputs were small and not too toxic for human beings.

Except for a few old style farms still preserved for pedagogic reasons and some dude ranches created for entertainment, farming in the postmodern era has lost its friendly and amateur-like character. It has become a highly technical activity, and its practitioners have to specialise in order to master the new technologies they have to use. It is increasingly difficult for them to open their exploitations to visitors. Farming has ceased to appear as an activity of the public domain, with skills that everyone was able to understand. In such a context, it becomes increasingly difficult to assimilate rural areas to the pastures, flocks and shepherds of Arcadia. Farming has ceased to be an activity able to fuel the dreams of urban dwellers.

Postmodern societies have torn the links to the civilization of Antiquity – either Greece or Rome – which had subsisted well into the 20th century. Latin and Greek have practically disappeared from the secondary school curricula, even in Latin-speaking countries (even in Italy). Greece itself should be considered apart from this, since Modern Greek has so many similarities with the Greek spoken at the time of Pericles, Plato or Aristotle.

The perception people have of rural areas has changed for other reasons. Until a generation ago, the contrast between the areas used by agriculture and wild areas did not seem very sharp. Farming involved a modification of nature, the destruction of wooded area and the creation of meadows or pastures which differed at least partly from the natural moors which they replaced. Farming activities did not appear, however, as a threat to nature.

The situation is completely different today. It is up to a point a consequence of the modernisation we just described of contemporary farming: fields or pastures are increasingly similar to chemical laboratories, using huge quantities of fertilisers in order to improve yields, and a wide variety of pesticides in order to eliminate weeds and diseases. Cows, sheep and pigs are increasingly born of carefully selected parents. The farming environment is increasingly an artificial one. Its management endangers natural equilibriums: soils are polluted by the overuse of fertilisers and pesticides; the water of ponds and rivers is transformed through eutrophication because of their overly high fertiliser content; the destruction of insects threatens the health of birds and deprives them of most of their food, etc.

The conjunction of the break with the Arcadian tradition and new forms of farming is conducive to a new conception of nature – nature exists only where there is no human activity. Hunting has to be prohibited, since it interferes with the composition of the pyramids of wild fauna – angling has to disappear from the rivers for the same reason. Woods were often used as rough pastures by cattle, sheep or pigs. These activities have to be banned since they compete with wild species, and destroy some plants because of their edible qualities.

In contemporary perspectives, nature exists only when it is severely preserved from any human interference. The managers of natural parks know how difficult a problem this is – tourists wish to have access to these protected areas in order to discover wild fauna in its natural setting, but they then disturb the animals, by modifying their eating habits through the food they have offered them or have left behind in camping or picnic sites. The grizzlies of the Yellowstone Park are dependent on the food they find in these areas, a situation which is not without danger for visitors.

The environment of ecologists is completely different from that of classical or romantic poetry or literature. It means that at least a part of the former rural areas had to be transformed back into "genuine" nature. Everybody knows that such a change requires a long time, and that the parks subsequently produced will need very careful human management in order to look really "natural". The fauna of great mammals has to be constantly checked in order to know whether it does not outnumber the local resources, and programs of selective killing have to be introduced etc.

4.3 Cultural Evolution and the Postmodern Dreamed Rural Landscapes – New Forms of Communities

The fascination for rural villages, especially those located in the remote parts of mountainous areas or along the rocky seashores of the Atlantic Ocean or the Mediterranean Sea, stemmed from the image of closely-tied communities they gave – the idea was born with the *Letters of a Savoyard Vicar* by Jean-Jacques Rousseau. Its success was general all over Europe.

Kenneth Olwig has shown that the *landskips* of Schleswig-Holstein were good examples of this form of social organisation. Ferdinand Tönnies, the great theoretician of the idea of *Gemeinschaft*, was born in the German part of Schleswig. Olwig suggests that his conceptualisation of rural communities was implicitly based on his early experience in this part of the German-Danish border.

The ideal of a community is not dead, but it has ceased to be associated with rural groups devoted to farming. At the time of postmodernity, the idea that we had to wait for building the utopias we dreamed of had completely vanished. Utopias had to be built now, and if it was possible, where we were living. It was the motor behind the American hippies' communes in the late 1960s and early 1970s, and their European equivalents. It survived this period of contestation. Much of the people who settle in rural areas wish to participate in some form of community.

Local traditional communities are sometimes still used as references. They had a strong identity, clearly exhibited through the shapes they had given to landscapes and the markers they had scattered over them. Why not rely on the same forms to build an identity for the newborn communities of the present? Hence the passion with which many new rural dwellers try to preserve all the details of the farms they have bought. Hence the emphasis given to forms which have lost their functional meaning, and have been transformed into pure markers: the haystacks in Slovenia, as shown 30 years ago by David Sopher (1979), the pigeon lofts of Quercy or Périgord in France, the huge chimney stack of former "*tués*" in the Jura mountains, the use of small brown tiles or Roman tiles depending on the regions of Western and Mediterranean Europe etc.

The new communities that try to use the folklore inherited from the past also take over their feasts, their rituals and their ceremonies. More numerous, however, are those who rely on other models. Some groups are looking for Oriental examples, and try to build their identities on some form of Zen Buddhism, transcendental meditation or other fashionable philosophy imported from eastern or southern Asia. The idea of a socialist community did not disappear with the popular democracies of Eastern Europe and the Soviet Union. The main change is that *kolkhozes* (collective farms) have ceased to be considered as valuable models. This ideal is still alive among youths of both North and South America.

4.4   Cultural Evolution and the Postmodern Dreamed Landscape – The Countryside as a Playground

Most of the newcomers do not settle in rural areas because they offer tracts of "genuine" (in fact, artificially reconstructed) nature, or because they wish to participate in new forms of communities. What they appreciate in the relatively low-density areas of the contemporary countryside is the possibility they offer to practice activities which involve the use of large tracts of land. New "ruralites" wish to jog in pleasant environments, practice golf on the perfect greenery of links, play tennis on well-trimmed courts, climb up the closest cliffs, raft down the rapid streams of nearby mountains, practice deltaplaning etc.

The areas covered by the different sport equipments are often important – much more than for the parks of 18th-century Britain, for instance. Golf links, the areas devoted to equestrian sports, football, handball, basketball and other playgrounds are smaller, but with the parking lots they need, the land they cover is also extensive.

Between fields and the "genuine" nature of ecological parks, there is room for an intermediary category: a humanised form of nature, made up of a mixture of woods and clearings, with paths for jogging, walking, hiking, etc. The management of these areas is a difficult one, since they support high densities of visitors during a part of the year, which means that it is necessary to care for the replacement of trees and the survival of many plant or animal species. In some rural regions, the cultivated part of the landscape now covers only a small proportion of the total area.

The rural landscapes of postmodern societies are, in this way, utterly different from those which preceded them.

4.5   The Countryside as a Palimpsest of Unrelated and Utterly Different Land-Uses

All these forms of activities need special equipments and well-designed environments in order to be performed under the best conditions. It means that they have to be managed by professionals. The countryside has ceased to be a domain for intuition or improvisation. The decisions which concern it have to be carefully prepared. Each land-use has its own specificities, which means that it is difficult to use the same tract of land for different activities.

The different land-uses are not only incompatible. Tensions appear between them. The people who seek to experience new forms of communitarian life are generally glad to be close to natural parks, but are afraid of the risks inherent in the presence of dangerous mammals: wolves, bears, boars, etc. They do not like the style of most of those who settle in rural areas simply in order to practice their preferred sports, since they are generally noisy and uncultivated young persons. Ecologists try to forbid all human activities within natural preservation areas, even simple visits. The parkland designed for leisure is sometimes squat on by nomads. Because it is partly wooded, it may attract all types of petty offenders.

Postmodern landscapes are, in this way, shaped by conflicting interests. It is not a new situation – the problems which arose between ploughmen and cattle raisers hold a central place in the development of traditional crop rotation systems. What is new is that the conflicting groups do not always live in the contested areas – ecologists fight for the preservation of all the natural parks, whatever their location. The members of the new communities share an ideal which was often elaborated in faraway locations; they get some support from all of those who have the same beliefs as them. The industry of sport equipment involves heavy investment. It is backed by important firms, which try to strongly defend their interests wherever they are threatened.

It means that the natural outcome of the decisions and conflicts which appear in such locations is generally untidy and chaotic. Each piece of land is well managed, but the whole landscape is not a harmonious one. In order to create more pleasant environments, some measure of planning is needed. Unfortunately, there is no option other than passing zoning regulations –

each land-user is very jealous of their right to organise the tracts they own or rent in the best of their interests.

5.   CONCLUSION

Landscapes are the outcomes of human interests and activities. Their forms may be interpreted as a language, but it is not a universal one. In this paper, we proposed a classification of the languages of rural landscapes based on the economic, social and cultural position of the groups which were or are responsible for their genesis: one, as the languages of function, the generative grammars of landscape elements and the semiotics of religious signification for traditional farming groups; two, as the rhetoric of harmony, purity and social status and power for upper or middle class urbanites from the Middle Ages to the 20th century; and three, as the languages of genuine nature, amenities and open air activities for the new rural population which resulted from the 20th century revolutions in mobility and activity.

Linguistic models are helpful for geographers who try to decipher and interpret specific forms of rural landscapes. They may rely on: firstly, the dialectical relations between words and things at all stages of evolution (the naming of soil, plants, environments), and the naming of the countryside itself; secondly, the models of structural linguistics and generative grammars for classical forms of agrarian landscapes; and thirdly, semiotics for the aesthetic and social readings which were so important in the religious fields of purely rural societies, or in the ideological ones for modern urban societies.

Geographers have an obvious interest in borrowing tools developed by linguists, but they have to know that none of these tools would be able to provide them with a universal key for reading and interpreting landscapes.

REFERENCES

Antoine, A. (2002). *Le Paysage de l'historien. Archéologie des bocages de l'Ouest de la France à l'époque moderne.* Rennes: Presses Universitaires de Rennes.

Berque, A. (1995). *Les Raisons du paysage. De la Chine antique aux envionnements de synthèse.* Paris: Hazan.

Berque, A., Conan, M., Donadieu, P., Lassus, B. & Roger A. (1999). Mouvande. Cinquante mots pour le paysage, Paris: Editions de la Villette.

Bonnemaison, J. (1986). *Les Fondements d'une identité. Territoire, histoire et société dans l'archipel du Vanuato,* Paris: ORSTOM, 2 vol.

Brunet, R. (1980) La composition des modèles en analyse spatiale, *L'Espace géographique,* 9, 4, 253-265.

Cosgrove, D. (1984). *Social Formation and Symbolic Landscape.* London: Croom Helm.

Duncan, J. (1990). The City as a Text: The Politics of Landscape Interpretation in the Kandyan Kingdom. Cambridge: Cambridge University Press.

Durand, Y. (1988). *Vivre au pays au XVIIIe siècle*. Paris: PUF.

Eliade, M. (1949). *Traité d'histoire des religions*. Paris: Payot.

Ginsburg, C. (1966). *I Benandanti: Richerche sulla stregoneria e sui culti agrari tra cinquecento e seicento*, Turin.

Ghorra-Gobin, C. (1987). Les Américains et leurs territoires. Mythes et réalités. Paris: La Documentation française.

Ghorra-Gobin, C. (1997). *Los Angeles. Le mythe américain inachevé*. Paris: CNRS.

Greimas, A.-J., Renier, A., Castex, J., Panerai, Ph., Petitot-Cocorda, J., Ostrowetsky, S. & Guilbaud, G.Th. (Eds.) (1979). Sémiotique de l'espace - Architecture, Urbanisme, sortir de l'impasse. Introduction de Jean Zeitoun. Paris: Denoël-Gonthier.

Hahn, E. (1896). *Demeter und Baubo. Versuch einer Theorie der Entstehung unseres Ackerbau*. Lübeck.

Hunt, J.D. (1986) *Garden and Grove. The Italian Renaissance Garden in the English Imagination: 1600-1750*. London: Dent.

Jacques, D. (1983). *Georgian Gardens. The Reign of Nature*. London: Batsford.

Johnson, H.B. (1976). *Order Upon the Land*. New York: Oxford University Press.

Lautman, F. (Ed.) (1979). *Religions et traditions populaires*. Paris: Editions de la réunion des Musées nationaux.

Mannhardt, W. (1875-1877). *Wald- und Feldkulte*. 2 vol. Berlin.

Moynihan, E.B. (1979). *Paradise as a Garden. In Persia and Moghul India*. New York: Braziller.

Olwig, K. (2002). *Landscape, Nature and the Body Politic: From Britain's Renaissance to America's New World*. Madison: University of Wisconsin Press.

Planhol, X. (1958). *De la Plaine pamphylienne aux lacs pisidiens, nomadisme et vie pastorale*. Paris: Maisonneuve.

Planhol, X. (1988). *Géographie historique de la France*. Paris: Fayard.

Pred, A. (1986). *Place, Practice and Structure*. Cambridge: Polity Press.

Raison, J.-P. (1977). Perception et réalisation de l'espace dans la société Mérina. *Annales E. S.C.*, n 3, 412-432.

Rantasalo, A.V. (1919-1925). *Der Ackerbau im Volksaberglauben der Finnen und Esten mit entsprechenden Gebräuchten der Germanen vergliche*, 5 vol. Helsinki: Sortavala.

Saintyves, P. (1907). *Les Saints successeurs des dieux. Essais de mythologie chrétienne*. Paris: Nourry.

Schama, S. (1995) *Landscape and Memory*. London: Fontana Press.

Sébillot, P. (1908). *Le Paganisme contemporain chez les peuples celto-latins*. Paris: Doin.

Senda, M. (1980). Territorial Possession in Ancient Japan: The Real and the Perceived. In Association of Japanese Geographers (Eds.), *Geography of Japan* (pp. 101-120). Tokyo: Teikoku Shoin.

Sereni, E. (1961). *Storia del Paesaggio agrario italiano*. Bari: Laterza.

Sopher, D. (1979). The Landscape of Home. Myth, Experience, Social Meaning. In D. Meinig (Ed.), *The Interpretation of Ordinary Landscapes* (pp. 129-149). New York: Oxford University Press.

Toujas-Pinède, C. (1978). Les collines et les plateaux du Quercy. In F. Taillefer (Ed.), *Atlas et géographie du Midi toulousain*. Paris: Nathan.

Zink, A. (2000). *Pays ou circonscriptions. Territoires du Sud-Ouest sous l'Ancien Régime. Concevoir l'autre*. Publications de la Sorbonne. Paris.

Chapter 3

"THIS IS NOT A LANDSCAPE": CIRCULATING REFERENCE AND LAND SHAPING

Kenneth R. Olwig
Department of Landscape Planning, SLU-Alnarp, Sweden

1. LANDSCAPE AS REPRESENTATION AND THE REPRESENTATION OF LANDSCAPE

A look at a definition of landscape in an ordinary dictionary reveals that landscape can mean both a representation of a scene (pictorial scenery) and that which might be thus represented, as when "a portion of territory that can be viewed at one time from one place" is represented in a painting of, for example, "natural inland scenery" or as "a particular area of activity: SCENE":

---

land·scape
Etymology: Dutch landschap, from land + -schap -ship
Date: 1598
1 a : a picture representing a view of natural inland scenery **b** : the art of depicting such scenery
2 a : the landforms of a region in the aggregate **b** : a portion of territory that can be viewed at one time from one place **c** : a particular area of activity : SCENE (political landscape).
3 obsolete: VISTA, PROSPECT
(Merriam-Webster 1995: *landscape*; Olwig 1993)

Landscape is thus both a form of representation and something that is represented. This raises issues, I will argue, concerning the historical and present day character of the interrelationship between the representation and that represented. I will suggest in the following that the relationship is

H. Palang et al. (eds.), European Rural Landscapes:
Persistence and Change in a Globalising Environment, 41-65.
© *2004 Kluwer Academic Publishers. Printed in the Netherlands.*

circular. The particular form of representation can shape the landscape represented, and the landscape thus represented can shape its representation. This circularity, furthermore, can end in a form of self-referential circulating reference in which the landscape is shaped in its own representational image, and the distinction between representation, and that which is represented, is lost. I will further suggest that the notion of landscape as representation, when seen from an historical perspective, is broader than the simple representation of landscape as scenery, and that it also includes that which is here termed the *political landscape* (see also Mitchell 1996, 2000). This essay is thus concerned with the way history can help us to understand the ongoing process of interaction between changing representations of landscape and the landscape that is thus represented. This understanding can suggest future strategies for the way we (as landscape researchers, architects and planners) represent landscape and, thereby, come to influence the way the future political and material landscape is shaped.

1.1   Real and Surreal Landscape Representation

We cannot comprehend the world directly, as it *presents* itself to us, because the information we receive through our senses is so vast and complex that it is, in and of itself, incomprehensible. It is for this reason we create *representations* of the world that enable us to reflect upon it and give it order, structure and meaning.[1] These representations can be expressed in the form of spoken or written language, by graphic and pictorial means, or by a combination of the graphic and the written, as in a theorem in geometry. Representations can even take more visceral forms, as in the case of sculpture, dance and ritual, models or architecture, that can be pictured graphically, and described in words, but which must be used, experienced and lived to be fully grasped (Langer 1953). If these representations seem to work, and to help us create a world that functions and makes sense, then these representations will be taken for granted as being essentially equivalent to the world they represent. We then tend to forget that they are representations, and see them rather as a direct presentation of reality. Most educated people, for example, have learned to think of the world in terms of a Newtonian physics that is rooted in the theorems of Euclidean geometry in which a straight line is the shortest distance between two points; and we drive across bridges and work in buildings that not only have been

---

[1] I am here taking what Denis Cosgrove (2003) has termed "a *semiotic* approach to landscape", which "is skeptical of scientific claims to represent mimetically real processes shaping the world around us. It lays scholarly emphasis more on the context and processes through which cultural meanings are invested into and shape a world whose 'nature' is known only through human cognition and representation, and is thus always symbolically mediated".

constructed according to those theorems, but which also function as bridges and buildings. Einstein's atomic physics was initially viewed with suspicion because he questioned these solid foundations when he sought to develop a model of the universe in which there are no straight lines because it is really curved. It would be easy to dismiss such "nonsense" if it were not for the atom bomb. Perhaps the real power of that bomb was epistemological and metaphysical rather than physical – it exploded worldviews, making it clear that there is more than one physics, and that they can be used to shape very different realities. Artists, like scientists, are capable of such epistemological bombs. The Belgian painter, René Magritte, I will suggest, created works of art that could do much to explode established conceptions of landscape and help us to grasp that people may not only perceive the same landscape differently, but that they may actually shape and dwell within ontologically different landscapes.

It is because René Magritte's paintings confront established notions of what is real that they have a powerful impact as art. This is why these works are termed "surreal," a term that, in ordinary language, suggests the bizarre, but which literally means "above the real." Thus, rather than dismiss Magritte's work as bizarre, we should see it as a door to a meta-theoretical understanding of the way representations construct differing notions of what is real. The surreal can thereby help us to transcend frozen conceptions of the real that may no longer serve our best interests. I am going to use it as a door to the interpretation of our representations of landscape.

1.2   This is <u>not</u> a ~~Pipe~~ Landscape

Magritte's most famous artwork might well be his 1929 painting of a pipe with the text, *This is not a pipe* (Figure 1). It has even inspired a small book by Michel Foucault (Foucault 1998). The painting is titled *The Treason of Images*, and an important dimension of the painting's message is the artwork's jarring juxtaposition of a realistic looking picture of a pipe and a written sub-text that contradicts the visual image. The painting thus, at a basic level, reveals the "treasonous" way that visual images convey a convincing, yet inexplicit and insubstantial message, when compared to that of the sub-text. The picture is indeed *not* a pipe, even if it looks like a pipe. The picture contains, however, a *representation* of a pipe, or, to be more precise, it is a representation of the idea of a pipe; an ideal, archetypal pipe. This point is brought out in a second version of this painting in which the image of the pipe, and the text that tells us that this is not a pipe, is placed upon the image of a canvas lying on an easel. Above the easel another image of the same pipe floats freely in space like an idea in a cartoon. The point of the painting, when read in this way, is that we should be careful not to confound our representations with the ideas and objects that they represent.

The potentially treacherous relationship between graphic image and sub-text ought to be of importance to students of landscape because the juxtaposition of graphic image (be it picture or map) and written sub-text is at the heart of landscape as a subject of study and a medium of education. A lecture on landscape is not complete without slides, yet we often overlook the seductive polysemy of the landscape image (Barthes 1972; Tuan 1979). Where one scholar sees nature in a landscape scene with the sub-text *The Alheath* (Denmark), another may see culture, and neither is likely to understand why their respective interpretations are so different. It requires lengthy study to disentangle the meanings of nature, culture and place embedded in the national perception of the landscape of the heath (Olwig 1984; Sack 1997; Spirn 1998). The problem, however, is more than the polysemy of the visual image, it is also explicitly related to the visual representation of landscape, which is explicitly treated in other Magritte paintings.

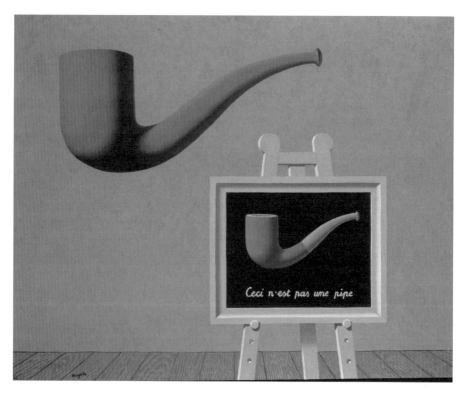

Figure 3-1. Les deux mystères, 1966, by René Magritte. © Photothèque R. Magritte – ADAGP, Paris 2004; EAÜ.

Figure 3-2. Where Euclid walked, by René Magritte, 1955. © Photothèque R. Magritte –
ADAGP, Paris 2004; EAÜ.

1.3   Through the Window's Looking Glass

In another artwork by Magritte (Figure 2) we see an easel in a room upon
which stands a painting of an urban landscape. The same landscape, which is
the subject of the painting on the easel, is simultaneously visible through a
window in the room in which the painting stands. The painting is positioned

in such a way that the representation on the canvas overlaps the scene seen through the window, so that one blends into the other. In this case Magritte's artwork seems to be making the point that both the landscape on the easel and the landscape seen from the window in Magritte's painting are, like the pipe, not landscapes, but representations of landscape. The painting has a sub-text, which reads: *Where Euclid walked*, and if we look closely we see that the tower of a building has the same Euclidean shape as the boulevard that stretches from the window out to the horizon. The painting is thus not just a representation of landscape, but a representation of the way a perspective representation of landscape is created through the use of Euclidian geometry as framed by a square window.

Figure 3-3. Woodcut from 1531 book on drawing technique by Johann II of Bavaria and Hieronymus Rodler.

Art historians will know that Magritte's window cannot be just any window, but must be the famous window described in Leon Battista Alberti's Renaissance classic, *On Painting* (Alberti 1956), which provided the first written instruction on the geometric laws of perspective drawing (see Figure 3). Alberti was, among other things, a cartographer, and there is much to indicate that the techniques of perspective drawing were extrapolated from the sciences of surveying and cartography – the landscape drawing is thus basically a map projected from an oblique angle, rather than from the top down (Edgerton 1987; Cosgrove 1988). Central to Alberti's construction of the perspective illusion was the frame: "I inscribe a quadrangle of right angles, as large as I wish, which is considered to be an open window through which I see what I want to paint" (Alberti 1956: 56).

Some scholars, such as Denis Cosgrove and Stephen Daniels, have contributed to our understanding of the representation of landscape by focusing largely upon the way we represent landscape pictorially, in the tradition of Alberti's window. For these scholars landscape tends to become one with the representation, rather than with that which is represented outside the window: "A landscape is a cultural image, a pictorial way of representing or symbolising surroundings" (Daniels & Cosgrove 1988: 1).[2] Magritte's painting, *Where Euclid walked*, when seen from the perspective of Daniels and Cosgrove, could read: "This *is* a landscape," because landscape in their view is fundamentally a form of representation, albeit a form of representation that can have a profound effect on the perception and design of the material world (see, for example, Daniels 1989, 1993, 1999; Cosgrove 1998). The assumption that landscape was a form of artistic representation eventually led Daniels and Cosgrove to suggest a "post-modern perspective" in which the:

> … landscape seems less like a palimpsest whose "real" or "authentic" meanings can somehow be recovered with the correct techniques, theories or ideologies, than a flickering text displayed on the word-processor screen whose meaning can be created, extended, altered, elaborated and finally obliterated by the merest touch of a button (Daniels & Cosgrove 1988: 8).

A palimpsest is a parchment paper that can be overwritten, but which still leaves a trace of what has previously been written, so that it is possible to

---

[2] Magritte's painting can explain differences amongst landscape researchers. Two landscape scholars may both commonly study landscape by means of what are essentially maps and framed representations of landscape scenes, whether they be graphics, photos or LANDSAT images. The one scholar, however, focuses on the morphology of the objects seen in the landscape – the mountains, rivers, fields, barns, cities etc., whereas the other is more interested in surveying the spatial geometries that frame the landscape – the distance between mountain and river, field and farm, and farm and city. The two may study the same landscape, yet they somehow find themselves in fundamental disagreement about the object and methods of landscape study. They think that they are studying the landscape beyond the window, but actually, as Magritte's painting suggests, they are studying differing aspects of the landscape as framed by Alberti's window. If one wishes to transcend such differences it is necessary to re-examine the contradictions embodied in the way we represent landscape in our studies, rather than assume that these contradictions simply reflect differences that are out there, beyond the window.

reconstruct earlier texts.[3] To penetrate the layers imprinted upon the land and thereby gain a deeper understanding of the landscape as a kind of palimpsest has arguably been the traditional goal of landscape research. Cosgrove and Daniels' statement thus presents a direct challenge to what many landscape researchers have seen to be their scholarly mission. Hence, if Cosgrove and Daniels are right, they raise serious epistemological questions about the scientific value of much that has passed for landscape research. I would argue that they are right, but only when landscape researchers conflate a particular mode of landscape representation, with that which is represented, and thereby effectively reduce landscape to its representation.

1.4   Shaping the Land

A basic problem with the reduction of landscape to a form of representation is the question of how a representation can exist unless it represents something beyond itself? The philosopher Ralph Waldo Emerson makes a statement about landscape that superficially resembles Cosgrove and Daniels' emphasis on landscape as a form of artistic representation, but there is a difference. In Emerson's view there is a land that is present prior to its representation in art, but it first takes on the shape of a land – a lands(c)hape – when it is unified by the eye of the artist (in this case a poet):

> The charming landscape which I saw this morning, is indubitably made up of some twenty or thirty farms. Miller owns this field, Locke that, and Manning the woodland beyond. But none of them owns the landscape. There is a property in the horizon which no man has but he whose eye can integrate all the parts, that is, the poet. This is the best part of these men's farms, yet to this their warranty-deeds give no title (Emerson 1991: 7).

Artists, as Emerson suggests, have the power to integrate the parts that make up our environment and shape them into a landscape whole. The artist is here shaping the land in a sense that is quite in agreement with the fact that the derivation of the *-scape* in land*scape* is closely tied to that of *shape* (Olwig 1993). Artists, however, are not alone in giving the land the shape of a meaningful totality. Landscape architects and planners, communities of citizens and representative political bodies, can all both represent and shape the land as a whole. There is thus a process of circulation involved in the relationship between the particular form of representation, the concrete things represented and the way those things are shaped as a landscape whole.

---

[3] Actually, a word-processor is not really all that different from a palimpsest, because previous texts remain layered on the hard disk of the computer, and are recoverable with the correct techniques.

In the following I will first examine some of the problems with the Albertian window on landscape, as a particular mode of representation before moving on to a consideration of alternative forms of representation and their relationship to that that is represented.

1.5   The World Transformed Through the Frame of the Looking Glass

Magritte's *Where Euclid Walked* shows, as noted, a painting of a landscape on an easel and the same landscape as seen through a window. The landscape seen through the window seems highly familiar, even though the artist has done everything possible to make us aware that this is not a real city, but an ideal city constructed in the imagination of the artist. Why then does the city seem familiar and real, or even surreal? The reason is that the geometric/cartographic principles outlined in Alberti's book have subsequently been applied in the shaping of the land outside the window by surveyors and architects in the design of buildings and city plans (Söderström 1996) – a plan is literally "a large-scale map of a small area" (Merriam-Webster 1968, *plan*).

Alberti was not just a cartographer, he was also a painter and architect (like his friend Fillippo Brunelleschi from whom he learned the technique of perspective drawing), and he was deeply concerned with the importance of geometry. "It would please me," he wrote, "if the painter were as learned as possible in all the liberal arts, but first of all I desire that he knows geometry" (Alberti 1956: 90). Buildings have geometric shapes and boulevards are often designed to appear to taper into a distant vanishing point. This is not only true of the obviously man-made constructions of the urban landscape, but also of the less apparently man-made constructions of the countryside, where fields have been surveyed, enclosed and given aesthetic form according to the same geometric laws (Barrell 1972; Cosgrove 1988, 1993). This makes it is possible for the landscape scholar to confuse the pictorial landscape representations that they study inside their offices, with the landscape seen through the office window. There is a good fit between the landscape representation inside the window and that outside because the same geometric principles have often shaped both.

The work of Daniels and Cosgrove is deeply indebted to a British literary tradition of landscape study exemplified by the work of Raymond Williams and John Barrell. In this tradition the rise of interest in landscape was linked largely to the process of 18th century British enclosure (Barrell 1972, 1980, 1987; Cosgrove 1984; Williams 1973). Cosgrove, however, showed that the link between enclosure and landscape could be traced back beyond British enclosure to the physical and social agricultural transformations that occurred on the *terraferma* of 16th-century Venice (Cosgrove 1988, 1993). Cosgrove showed that surveying, cartography, and the related conceptualization of the land as scenery, were all tied to the forceful appropriation, as private property, of common lands and un-enclosed farmland by a powerful elite. Cosgrove's work was preceded by that of art historians, such as Samuel Edgerton, who focused on the link between the pioneering of perspective drawing in Renaissance Florence and the rise in Florence of a mercantile economy and the related capitalization of land as property (Edgerton 1975). I would add, to Cosgrove and Edgerton's line of argument, that the development of this scenic conception of landscape was also tied to the Renaissance emergence of the nascent nation state – which in turn played an important role in promoting urban and regional planning, as well as enclosure (Olwig 2002b).

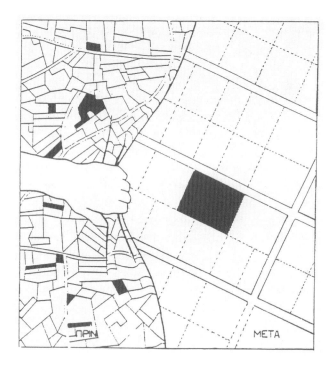

Figure 3-4. Greek poster intended to convince farmers to enclose their land. My thanks to Hans Sevatdal for this illustration.

Surveying and cartography made it possible to take a complex mosaic of fields and grazing lands, that was closely attuned to an equally complex and varied ecological pattern of biotopes, and reduce it to geometric, privately owned fields (see Figure 4) whose ecology was simplified and made uniform by draining, clearance of woods, and the development of new (mono)crops such as potatoes, turnips and clover. The way in which the mapped representation became one with the land is illustrated by John Norden's statement in his 1607 *Surveyor's Dialogue*: "Is not the fielde itselfe a goodly map for the lord to looke upon, better than painted paper?" It is such a goodly map because: "A plot rightly drawne by true information, discribeth so the lively image of a manor, and every branch and member of the same, as the lord sitting in his chayre, may see what he hath, and where and how he lyeth, and in whole use and occupation of every particular is upon suddaine view" (quoted in Kain & Baigent 1992: 5).

The landscapes of production created through enclosure were complemented by the recreative (though still productive) landscapes created by the landscape garden architect – but underlying them both was the same process of enclosure, the same surveying and cartographic impulse, and the same re-engineering of the land. It is thus possible to view the rise of interest in landscape as part and parcel of a largely positive process of modernization, beginning in the Renaissance, that has brought the Western World to its current level of agricultural productivity and welfare, while stimulating the aesthetic creation of landscapes of great beauty – albeit at a sad but necessary cost to "traditional society" and the rural poor. It is also possible, however, to view these developments more negatively as the expression of an imperial appropriation of both human and physical landscape, in the interest of a privileged minority (Mitchell 1994; Olwig 2001a, 2002b). To this critique can be added that of scholars like James C. Scott who, in his book, *Seeing Like a State: How Certain Schemes to Improve the Human Condition Have Failed*, has cogently argued that these map driven landscape "improvements" have laid the groundwork for environmental and social deterioration (Scott 1998; see also Kjærgaard 1994).

The landscape is not simply a form of representation, but rather an expression of a circular, dialectical, interaction between differing modes of representation and processes of social and environmental change that transform both. The recovery of the history of the interaction between landscape representation and transformation simultaneously raises serious questions about the naïve realism that characterizes much landscape study, particularly in the natural sciences, where the representation of landscape as map or graphic pictorial representation is, I will argue, not infrequently confounded with the landscape represented (see also Linehan & Gross 1998).

1.6   The Genie in the Map

The map is not the territory, as A. Korzybski warned long ago in an argument, later further developed by Gregory Bateson, that calls to mind Magritte's painting of a pipe that is not a pipe, but a representation of a pipe (Korzybski 1948; Bateson 1972). The tendency to confuse the representation with that which is represented applies to both perspective drawing and the map because both derive from surveying. The map, and the perspective drawing, often appear to present a precise representation of a segment of territory outside the window, but it does so by subtly inscribing the topology of our rounded, curvilinear world within a flat, static, Euclidian gridded space.[4] The single point perspective landscape drawing makes use of cartography's gridded space to create an illusion of three-dimensional bodily space – but the illusion remains a form of ocular deception. The objects appear to have a "smooth" curvilinear, bodily topologic shape, but they have in fact been frozen in a two dimensional Euclidean gridded, or "striated," space[5].

To some, the fit between our landscape representations and the empirical reality out there might be seen to corroborate the scientific veracity of the landscape discipline. However, for some students of science, such as Bruno Latour, this fit might suggest a questionable, self-referential, scientific syllogism (Latour 1999). Are we studying a reality out there, or are we basically corroborating a particular construction of the world? Latour argues that between the cartographic representation and that which is represented "there is neither correspondence, nor gaps, nor even two distinct ontological domains, but an entirely different phenomenon: circulating reference" (Latour 1999: 24). Latour bases his conclusions, in part, on anthropological fieldwork that he undertook amongst field scientists in Brazil. His study of the scientists' research process led him to observe:

> Remove both maps, confuse cartographic conventions, erase the tens of thousands of hours invested in … [their] atlas, interfere with the radar of planes, and our … scientists would be lost in the landscape and obliged once more to begin all the work of exploration, reference marking,

---

[4] One cannot create a flat map of a round world without some form of distortion. We have learned to accept certain map projections as giving the conventional shape of, for example, the continents. But if the projection is changed the shape of these bodies will change radically. By the same token, when we use maps to measure, for example, a coastline, we conventionally ignore the fact that the coastline on the gridded space of the map is, strictly speaking, made up of a myriad of angles. As fractal geometricians will tell you, if you calculate these angles into your measure, the length of the coastline will vary wildly according to the scale of your calculations.

[5] On smooth vs. striated space see Deleuze & Guattari (1988: 474-500); on Euclidian vs. topological space see Olwig (2001a).

triangulation, and squaring performed by their hundreds of predecessors (Latour 1999: 29).

This prompts the observation:

Lost in the forest, the researchers rely on one of the oldest and most primitive techniques for organizing space, claiming a place with stakes driven into the ground to delineate geometric shapes against the background noise, or at least to permit the possibility of their recognition. Submerged in the forest again, they are forced to count on the oldest of the sciences, the measure of angles, a geometry whose mythical origin has been recounted by Michel Serres (Latour 1999: 41-42).

"Yes," he concludes, "scientists master the world, but only if the world comes to them in the form of two-dimensional, superposable, combinable inscriptions" (Latour 1999: 29).

Latour does not deny that the field scientists he has studied may have made an important contribution to science, but he is struck by the fact that this is only a partial knowledge that is highly governed by the scientist's geometric tools, much like, I would add, Magritte's landscape painting of where Euclid walked. As Latour puts it:

We have taken science for realist painting, imagining that it made an exact copy of the world. The sciences do something else entirely – paintings too, for that matter. Through successive stages they link us to an aligned, transformed, constructed world (Latour 1999: 78-79).

The map (and hence also the perspective landscape drawing) gives a grand overview of the forest, which doubtless enables the scientists to see patterns that otherwise would not be visible, but it also, Latour suggests, can cause the scientists to overlook other salient information. "In a beautiful contradiction," he writes:

the word 'oversight' exactly captures the two meanings of this domination by sight, since it means at once looking at something from above and ignoring it (Latour 1999: 38).

Enclosure involved the use of the techniques of surveying and cartography to transform complex ecological environments into neatly divided spatial mosaics of forest, meadow and field that could be easily measured, valued, sold as real estate to private owners as well as taxed and regulated by government (Figure 5). It is for this reason that Latour's analysis comes to mind when one sees how landscape ecologists use more or less the same cartographic tools – though they may be digitally beefed up as GIS, LIS or Landsat images – in order to similarly segment the environment into areal biotopes and ecological communities. The map is the perfect tool upon which to inscribe two dimensional, geometrically measurable, areas of space. And for this reason one cannot help but question whether "nature" really is organized, as many landscape ecologists suggest, into biographical islands linked by corridors, along which animals and insects travel using, one presumes, carefully hidden mental maps (Linehan and Gross 1998; on mental maps see Tuan 1975). Is the landscape ecologist's zoning of nature into biotopes a reflection of a reality out there, or is it more a reflection of the epistemology of the map that is both the scientist's tool and the tool of the surveyor who has walked the land, arm in arm with Euclid, transforming it into a world that appears to be pre-zoned? Do scientists thereby help maintain a framework of knowledge that, as Latour argues, "does not reflect a real external world that it resembles via mimesis, but rather a real interior world, the coherence and continuity of which it helps to ensure" (Latour 1999: 58)?

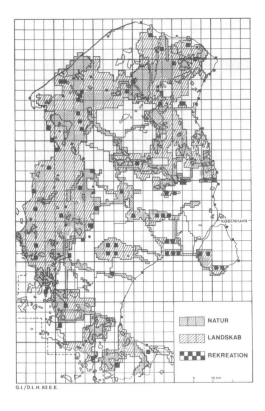

Figure 3-5. Danish planning map showing areas designated as nature, landscape and recreation. The white areas fall into none of these categories.

1.7 Do we Preserve Nature, or its Representation?

For the field scientists studied by Latour their landscape representations are subtitled *nature*, and these representations thereby not only provide an overview of that nature, they may also tend to "overlook" the human content of the landscape, as a number of Nordic researchers have recently documented (Lehtinen 1991; Hanssen 1998; Mels 1999). The representations of scientists and administrators play, in turn, an important role in environmental planning and the "preservation" of nature in natural parks and elsewhere (Jones & Daugstad 1996, 1997). In an analysis that draws on arguments that are quite parallel to those of Latour, Mels (2002: 147-149) argues that: "By trading on what Lefebvre called 'the power of landscape' the texts and maps decouple human practices from nature conservation policies. The power of maps and descriptions is also enhanced by and enhances the authority of the photographs in the plan." According to Mels this is because:

> The rationale behind the images, in contrast to the maps, is nowhere discussed. Apparently they speak for themselves, being "not only a natural scene, and not just a representation of a natural scene, but a *natural* representation of a natural scene, a trace or icon of nature *in* nature itself, as if nature were imprinting and encoding its essential structures on our perceptual apparatus". ... In a way, the park plan is a special case of a 'pictorial turn' where the burden of proof rests on maps and images rather than on work and material experiences (Mels 2002: 147-49; quotation from Mitchell 1994: 15).

In Mels' analysis this situation is quite similar to the circular referentiality described by Latour:

> Taken together, these modes of representation are extraordinarily effective at communicating a sense of spatial transparency, pure presence, and unity, unrelated to practices of framing nature. In one word, the plan operates through the *repetition* of space. Instead of offering an insight into the complexity of social space, the maps, images, and texts take on the shape of a green tautology (Mels 2002: 149).

The problem would appear to be that some natural scientists and park planners are confusing the graphic representation of landscape in maps and pictures, with the landscape beyond the Albertian window. It might thus be useful to try and break this ring of representation by going out the window in search of an alternative understanding of the landscape we have learned to know largely through graphic representations.

2. AN ALTERNATIVE LANDSCAPE ONTOLOGY OF HABITUS, PRACTICE AND CUSTOM

The English word for landscape allows for a confusion of meaning that is not found in its equivalents in the Romance languages, such as the French word for landscape, *paysage*. This confusion, in turn, leads to confusion with regard to the representation of landscape. The *pays* in *paysage* refers to the polity or country of a people, and is equivalent in meaning to the land in *land*scape in the earliest uses of the word in the Germanic languages, including English (Olwig 2002a; Claval this volume).[6] The word land in a statement such as "the land rose in rebellion," is thus a social phenomenon, a country or a people (Merriam-Webster 1968, *land*). It is in this sense that a land is constituted by the people that have rights in that land, such as the right to representation in Parliament (England is the land of the English). The *land* of a people was historically constituted by smaller lands – for example the common lands belonging to the village community, as represented by the commoners with a "landright" to representation in a village representative body, such as a village *thing* or *moot*.[7] The English term land, as used here, is applied not just to the people of a land in the social sense, but also to the material substance (e.g., dry land, meadowland) that is encompassed by such a land. This would not be the case in French where a separate word, *terre*, would be used for land in the physical sense.

It is the tendency for the root *land* in *land*scape to slide in meaning from a social entity (land = *pays*) to the physical substance that can lead to confusion with regard to the representation of the land in landscape. Is the landscape represented in the painting primarily the social entity (land = *pays*), or is it something primarily physical (land = *terre*). The earliest Dutch Renaissance *landschap* paintings were arguably representations of the land of a people, in which material phenomena represented on the canvas functioned as the symbolic referent for the nature or –*schap* of the land as a polity (Olwig 2002a). The Dutch suffix –*schap* is spelled in differing ways in different Germanic languages in different eras (e.g., –*scipe* in Old English and –*scape* in Modern English), but all are equivalent to the modern English suffix –*ship*. This suffix generates an abstraction. Thus, there might be two friends, citizens or fellows in a room, all concrete beings, but between them they share something abstract and difficult to define, *friendship, citizenship*

---

[6] *Paysage* appends the suffix *age* to pays in much the way as *schaft* is appended to *Land* in the Germanic languages, or *ship* to *town* in English. *Pays* carried essentially the same connotations of areal community and people as country and land. The equivalent Italian terms, *paése* and *paesàggio* carry the same meaning (Gamillscheg 1969).

[7] *Landright* – "Law of the land; legal rights of natives of the country; legal obligation connected with land or estate" (O.E.D. 1971, *landright*).

or *fellowship*. The suffix *–ship* designates "something showing, exhibiting, or embodying a quality or state" of being, in this case, that of a *friend*, *citizen* or *fellow*. *–Ship* thus designates the abstract nature, state or constitution of something. *–Ship*, however, is also akin to the word *shape*, in the sense of "shape, form, create" (Merriam-Webster 1968, *-ship*). The connection between *–ship* and *shape* is apparent in the sense that a phenomenon is seen to have been shaped according to its underlying "nature," "state" or "constitution". A *landschap* painting would thus provide a concrete way of representing the abstract "nature", "state" or "constitution" of the land through the medium of iconographic reference to the material phenomena that are characteristic of the land as a polity. The painting might depict scenes of people ice-skating together on a pond surrounded by a dike. The people ice-skating together could here symbolize the land (*pays*) as community shaped according to the nature of the laws made by its *landschap* representative body. The dike and pond would then symbolize the ability of this community to shape the place of their dwelling. Landscape paintings, in this sense were thus a representation of the *shape* or condition of "**c** : a particular area of activity : SCENE (political landscape)." The meaning of this landscape representation changes subtly, but radically, when landscape is defined not as a symbolic representation of the nature of a polity, but as "**1 a** : a picture representing a view of natural inland scenery" (Merriam-Webster 1995, *landscape*). It is this scenic notion of landscape that, as Cosgrove tells us, "emerged around the turn of the sixteenth century to denote a painting whose primary subject matter was natural scenery" (Cosgrove 1984: 9). Now, we are dealing not with the representation of the nature of a land as a polity, but with nature reified as the physical land that constitutes "in*land* scenery."

2.1 Landscape as the Theater of State

The "substantive" legal and political sense of landscape as a place and polity existed well before the sixteenth century, but the specifically scenic sense of landscape essentially developed as the 16th century was drawing nigh. This mode of representation was created, I have argued, as a means by which the emergent Renaissance central state could set the stage for itself as the power most competent to represent the land as a natural entity defined by its physical boundaries. It was developed as part of the then ongoing clash between, on the one hand, the Renaissance princes and kings, who wanted to centralize power under the royal (e)state, and, on the other, representative bodies such as parliament, which sought to maintain the customary rights of their constituents. Landscape thus provided, the scenic backdrop for the theatrical representation of what became known as the theater of state (in which the proscenium arch fills the role of the Albertian window) (Olwig 2002a). As the French philosopher, Régis Debray, has remarked, because nobody has ever either seen or heard a state, a state must, at any price, make itself visible and let itself be heard: "It is the theater of the state which creates the state, just as the monument creates memory" (Debray 1994: 66).

The map and the perspective drawing provided a means whereby the state could represent the land as *shaped* by the state's commanding gaze, and oversight.[8] The map and the landscape scene put us in a position above and outside the landscape, from which we "overlook" it, and structure its space, through the geometric framework of the Albertian window. This, however, is only one means of representing and organizing the shape of the land. Representative legal bodies also, as has been seen, provide a means of representing and shaping the nature of the land. The constitutional basis for such bodies was characteristically rooted in the precedence of custom based upon the inherited practices of its members. This was the *-ship* of the land, the source of its "nature," "state" or "constitution." Custom is commonly reduced to timeless tradition by modernists (Giddens 1990). This is a misunderstanding, however, even if custom is ostensibly rooted in "time out of mind" precedence. Custom is, in fact, progressively brought up to date through the reinterpretation of precedence in light of present circumstance and the present sense of justice. The historian Eric Hobsbawm argues that tradition "must be distinguished clearly from 'custom'" because whereas tradition denies change, custom reaffirms it (Hobsbawm 1983: 2-4).

Whereas tradition is commonly linked to relatively nebulous phenomena such as *identity* and *culture*, custom provides the foundation for an important dimension of the laws that govern society – particularly in countries that have inherited the English system of common law.[9] Common law is thus extrapolated from customary practices through a legal process of trial by judge and jury in which unwritten custom becomes formalized as written law. Common law thus works by creating a body of law that is analogous to that generated by custom. It differs in this respect from written statutory law, which, unlike common law, is imposed from above by state institutions. Statutory law tends to be inspired by the rationality of geometry – hence its identification with a *justice* that is "*right*" under the "*rule* of law." "Custom lies upon the land," according to Edward Coke, the 17th century jurist, who vehemently defended the rights of parliament against the prerogatives of the monarch. Custom had "'two pillars'": "common usage", and "time out of mind". It is on the basis of these pillars that customs "are defined as a law or right not written; which, being established by long use and the consent of our

---

[8] This was well before the era of the surveiliant gaze incorporated into Jeremy Bentham's *panoptican*, as so vividly described by Foucault (Foucault 1979).

[9] On the nebulous character of the culture in cultural studies see Mitchell (1995).

ancestors, hath been and is daily practiced" (quoted in Thompson 1993: 97, 128, 129).[10]

Custom generates a "habitus", according to the historian E.P. Thompson, drawing upon the theory of the French anthropologist cum sociologist, Pierre Bourdieu. Habitus is defined by Thompson as "a lived environment comprised of practices, inherited expectations, rules which both determined limits to usages and disclosed possibilities, norms and sanctions both of law and neighbourhood pressures" (Thompson 1993: 102). The habitus literally means something like *embodied habit*. The habitus is the setting in which its inhabitatants develop a corporal and multiple sense of place which is distinguishable from the absolute space of the visitor's (or bureaucrat's) mapped landscape. Bourdieu gives the following explanation of the difference between the structured space of the map, analogous to the analytical framework of structuralist theory, and lived realm of practice:

> It is significant that "culture" is sometimes described as a *map*; it is the analogy which occurs to an outsider who has to find his way around in a foreign landscape and who compensates for his lack of practical mastery, the prerogative of the native, by the use of a model of all possible routes. The gulf between this potential, abstract space, devoid of landmarks or any privileged centre ... and the practical space of journeys actually made, or rather of journeys actually being made, can be seen from the difficulty we have in recognizing familiar routes on a map or town-plan until we are able to bring together the axes of the field of potentialities and the "system of axes linked unalterably to our bodies, and carried about with us wherever we go", as Poincaré puts it, which structures practical space into right and left, up and down, in front and behind ... (Bourdieu 1977: 2).

The map, as Bourdieu notes, not only provides a mode of de-centered orientation to the environment that is radically different from that of the local inhabitant, it also generates a particular representation of reality that is taken for granted, like the pipe that is really not a pipe in Magritte's artwork:

> Knowledge does not merely depend, as an elementary relativism teaches, on the particular standpoint an observer "situated in space and time" takes up on the object. The "knowing subject", as the idealist tradition rightly calls him, inflicts on practice a much more fundamental and pernicious alteration which, being a constituent condition of the cognitive operation, is bound to pass unnoticed: in taking up a point of view on the

---

[10] Though Coke did a great deal to illuminate the role of custom as the foundation for the common law of England, his advocacy of common law helped lay the basis for a generalized set of laws that furthered the interests of the propertied against the interests of those whose use rights to the land were founded upon custom (Blomley 1994).

action, withdrawing from it in order to observe it from above and from a distance, he constitutes practical activity as an *object of observation and analysis, a representation* (Bourdieu 1977: 2).

The perniciousness of the map's unnoticed cognitive positioning is that it becomes the basis for a mode of thought in which the structure of the representation is privileged over the practices that produced the phenomena that it seeks to represent. The map, for Bourdieu, exemplifies the structuralist mode of thought that he is seeking to critique. Structuralism thus elucidates the structure in the representation (in this case a map) and then reproduces that structure in the representations used to analyze that which has been mapped, thereby creating a form of circulating reference in which the practices that produced the mapped phenomena have a tendency to disappear from view.[11] This same mode of analysis might be applied to the relationship between *social formation* (a French structural Marxist term) and the spatial structure of central perspective pictorial representation (Cosgrove 1984). The point here is not to deny the insights of the structural analysis of the means by which the Renaissance development of central point perspective created the basis for a significant approach to landscape representation. The point, rather, is to critique the way in which the structural logic of this mode of analysis has a tendency to become self-referential, whereby landscape not only comes to be defined in terms of its representation, i.e. as "a pictorial way of representing or symbolising surroundings", but it also, through a process of circular representation, comes to define that which is represented in terms of its own geometric logic. Landscape, hereby, ceases to be a representation of the nature of the land, it becomes one with nature; naturalizing all that it is shaped in its image.

Concepts such as habitus and practice, when seen in relation to custom, provide a means of breaking the circle of circulating reference, and thereby open the window to an understanding of practices and spatial understandings that do not fit the binaries of structuralist thought and the geometries of the map.[12] These concepts may thus provide tools for the understanding of how, for example, generations of farmers and pastoralists have sustainably managed common lands without segmenting the land into a mapped and

---

[11] I have attempted a similar critique of the structuralism of the Swiss-French psychologist and historian of science, Jean Piaget, who privileges Euclidian space over topological space in his analysis of childhood cognitive development (Olwig 2001a).

[12] Bourdieu's notion of habitus is somewhat analogous to the notion of "anthropological" (or "existential") space as the "*paysage*" (landscape) created through the practice of "*lieu*" (location in geometrical space) (Certeau 1990: 174), that has been developed by Bourdieu's countryman, the philosopher and anthropologist Michel de Certeau (Certeau 1984).

parceled landscape – thus defying the scientific logic that assumes that the use of such a landscape necessarily must end in a tragedy of the commons (Hardin 1977; McCay & Acheson 1987). An awareness of the process of circulating reference might help prevent the specter of landscape ecologists seeking to preserve the "biodiversity" of unenclosed lands using the same cartographically driven, top-down, methods that were used to enclose and simplify the surrounding enclosed lands (Olwig 2001b; Scott 1988). It might also help environmentalists and state administrators to comprehend, and take account of, the differences between top-town, scientific approaches to landscape, and that of, for example, the pastoralists and agriculturalists whose practice shapes that landscape (Hanssen 1998; Setten 2004).

3. CONCLUSION – PERSPECTIVES FOR APPLIED LANDSCAPE RESEARCH

The recognition that landscape is both a representation and that which is represented – and that both exist in a relationship of *circulating reference* in which each works to shape the other – opens the door to a clearer understanding of the way differing forms of representation have historically helped generate differing landscape transformations. The recognition that landscape is not a pre-given structure, designed by a celestial architect or Nature, but something that humans have actively shaped – in their imaginations, in their representations, in their polities, and in their surroundings – opens the window to the realization that landscape is not monolithic. There is, thus, not simply the landscape as represented through the graticule of the map or the Albertian window, but also the landscape as represented by habitus and custom, by common law, and by representative bodies. It is not my suggestion that landscape researchers should privilege one or the other form of landscape. The overview and the map provide useful tools for both research and administration, but they also can overlook salient information and it must be recognized that they represent, literally, a particular *perspective* on landscape that is not, or should not, be hegemonic. There should be room for other, subaltern, representations, even if they may exist in a tensive and contradictory relationship to the dominant scenic and cartographic representation of landscape in the sciences and in state planning (Olwig 2001a).

We live in a world of property rights that have been carefully plotted on the graticule of the map and proudly viewed, as landscape scenery, from the windows of its proprietors. Individual ownership and state surveillance and control are part and parcel of daily existence. We also live in a world of custom in which we actively share community use rights in common environments according to moral codes controlled by social inclusion or exclusion. The room I am writing this in, like the windows of the computer screen upon which I write, and the plot of land on which I live, is full of rectangles. But in a little while, I will leave the room and beat a winding ancient path to the sea across a square field acquired by the mercantile family of a neighboring navy captain that appropriated the land from the local farming community. I will thereby use my feet to lay my own representative mark on the land, shaping the path in accordance with a time out of mind law that itself lies on the land, and is apparent to anyone who knows the landscape. When I reach the sea I will be able to collect mussels and rest on the sand, up to the high water mark, because this is my *land right*. I will infuriate my militant nouveau riche neighbor by crossing his land in full view of his window. This will not bother me, however, because his fury gave me the idea for the conclusion to this essay.

When studying or shaping the landscape, I would conclude, one should not just look through the window; one should also look at it, and consider how it frames one's view. Unless one is a god, or a charlatan, one will never entirely escape the encirclement of the circulating reference between our representations and that which is represented. But one can ponder the question of what it might mean to be "on the right side of the window".[13]

REFERENCES

Alberti, L.B. (1956 [1435-36]). *On Painting*. London: RKP.

Barrell, J. (1972). *The Idea of Landscape and the Sense of Place*. Cambridge: Cambridge University Press.

Barrell, J. (1980). *The Dark Side of the Landscape: The Rural Poor in English Painting 1730-1840*. Cambridge: Cambridge University Press.

Barrell, J. (1987). The Public Prospect and the Private View: The Politics of Taste in Eighteenth-Century Britain. In J.C. Eade (Ed.), *Projecting Landscape* (pp. 15-35). Australia: Humanities Research Centre Australian National University.

Barthes, R. (1972 [1957]). *Mythologies*. New York: Hill and Wang.

Bateson, G. (1972). *A Theory of Play and Fantasy. Steps to an Ecology of Mind*. New York: Ballantine.

Battisti, C. & Alessio, G. (1975). *Dizionario Etimologico Italiano*. Firenze: Instituto Di Glottologia, G. Barbèra.

Blomley, N.K. (1994). *Law, Space, and the Geographies of Power*. New York, Guilford Press.

Bourdieu, P. (1977). *Outline of a Theory of Practice*. Cambridge: Cambridge University Press.

---

[13] This question was raised by my colleague in the Department of Landscape Planning at SLU-Alnarp, Roland Gustavsson, in response to the version of this essay that I presented in my key-note address to the PECSRL Conference, August 2002, in Tartu, Estonia.

Certeau, M. de (1984). *The Practice of Everyday Life*. Berkeley: University of California Press.

Certeau, M. de (1990). *L'invention du quotidien: Arts de faire*. New Edition. Vol. I. Paris: Gallimard.

Claval, P. (2004). The Languages of Rural Landscapes. *This Volume*.

Cosgrove, D. (1984). *Social Formation and Symbolic Landscape*. London: Croom Helm.

Cosgrove, D. (1988). The Geometry of Landscape: Practical and Speculative Arts in Sixteenth-Century Venetian Land Territories. In D. Cosgrove and S. Daniels (Eds.), *The Iconography of Landscape* (pp. 254-276). Cambridge: Cambridge University Press.

Cosgrove, D. (1993). *The Palladian Landscape: Geographical Change and Its Cultural Representations in Sixteenth-Century Italy*. Pennsylvania: Pennsylvania State University Press.

Cosgrove, D. (1998). Cultural Landscapes. In T. Unwin (Ed.), *A European Geography* (pp. 65-81). London: Longman.

Cosgrove, D. (2003). Landscape: Ecology and Semiosis. In H. Palang & G. Fry (Eds.), *Landscape Interfaces: Cultural Heritage in Changing Landscapes* (pp. 15-20). Dordrecht: Kluwer.

Daniels, S. (1989). Marxism, Culture and the Duplicity of Landscape. In R. Peet & N. Thrift (Eds.), *New Models in Geography* (pp. 196-220). Vol. II. London: Unwin and Hyman.

Daniels, S. (1993). *Fields of Vision: Landscape Imagery and National Identity in England and the United States*. Cambridge: Polity Press.

Daniels, S. (1999). *Humphry Repton: Landscape Gardening and the Geography of Georgian England*. New Haven: Yale University Press.

Daniels, S. & Cosgrove, D. (1988). Introduction: Iconography and Landscape. In D. Cosgrove & S. Daniels (Eds.), *The Iconography of Landscape* (pp. 1-10), Cambridge: Cambridge University Press.

Debray, R. (1994). *L'Etat séducteur: Les révolutions médiologiques du pouvoir*. Paris: Gallimard.

Deleuze, G. & Guattari, F. (1988). *A Thousand Plateaus: Capitalism and Schizophrenia*. London: Continuum.

Edgerton, S. (1975). *The Renaissance Rediscovery of Linear Perspective*. New York: Basic Books.

Edgerton, S. (1987). From Mental Matrix to Mappa Mundi to Christian Empire: The Heritage of Ptolemaic Cartography in the Renaissance. In D. Woodward (Ed.), *Art and Cartography: Six Historical Essays* (pp. 10-50). Chicago: University of Chicago Press.

Emerson, R.W. (1991 [1836]). *Nature. Nature/Walking*. Boston: Beacon Press.

Foucault, M., (1979 [1975]). *Discipline and Punish: The Birth of the Prison*. Harmondsworth: Penguin.

Foucault, M. (1998 [1973]). *Dette er ikke en pibe*. Copenhagen: Hans Reitzels Forlag.

Fritzner, J. (1886-1896). *Ordbog om Det Gamle Norske Sprog*. Kristiania: Ny Norske forlagsforening.

Gamillscheg, E. (1969). *Etymologisches Wörterbuch Der Französischen Sprache*. Heidelberg: Winter.

Giddens, A. (1990). *The Consequences of Modernity*. London: Polity Press.

Gove, P.B. (Ed.) (1968). *Merriam-Webster. Webster's Third New International Dictionary of the English language, Unabridged*. Springfield, Mass.: G. & C. Merriam.

Hanssen, B.L. (1998). *Values, Ideology and Power Relations in Cultural Landscape Evaluations*. Bergen: Dept. of Geography, University of Bergen.

Hardin, G. & Baden, J. (1977). *Managing the Commons*. San Francisco: W.H. Freeman.

Hobsbawm, E. (1983). Introduction. In T. Ranger & E. Hobsbawm (Eds.), *The Invention of Tradition* (pp 1-14). Cambridge: Cambridge University Press.

Jones, M. & Daugstad, K. (1996). Cultural Landscape under Administration – a Conceptual Analysis. In M. Ihse (Ed.), *Landscape Analysis in the Nordic Countries* (pp. 162-188). Stockholm: FRN.

Jones, M. & Daugstad, K. (1997). Usages of the 'Cultural Landscape' Concept in Norwegian and Nordic Landscape Administration. *Landscape Research*, 22, 3, 267-281.

Kain, R.J.P. & Baigent, E. (1992). *The Cadastral Map in the Service of the State: A History of Property Mapping.* Chicago: University of Chicago Press.

Kjærgaard, T. (1994). *The Danish Revolution, 1500-1800: An Ecohistorical Interpretation.* Cambridge: Cambridge University Press.

Korzybski, A. (1948). *Science and Sanity: An Introduction to Non-Aristotelian Systems and General Semantics.* Lakeville, Conn: The International Non-Aristotelian Library Publishing Company.

Langer, S.K. (1953). *Feeling and Form: Theory of Art.* New York: Charles Scribner's Sons.

Latour, B. (1999). *Pandora's Hope. Essays on the Reality of Science Studies.* Cambridge: Harvard University Press.

Lehtinen, A.A. (1991). Northern Natures. *Fennia* 169:1, 57-169.

Linehan, J.R. & Gross, M. (1998). Back to the Future, Back to Basics: The Social Ecology of Landscapes and the Future of Landscape Planning. *Landscape and Urban Planning,* 42, 207-23.

McCay, B.J. & Acheson, J.M. (Eds) (1987). *The Question of the Commons: Culture and Ecology of Communal Resources.* Tuscon: The University of Arizona Press.

Mels, T. (1999). *Wild Landscapes: The Cultural Nature of Swedish Natural Parks.* Lund: Lund University Press.

Mels, T. (2002). Nature, Home, and Scenery: the Official Spatialities of Swedish National Parks. *Environment and Planning D: Society and Space,* 20, 135-154.

Merriam-Webster (1995). *Collegiate Dictionary.* Springfield, MA.: Merriam-Webster.

Mitchell, D. (1995). There's no Such Thing as Culture: Towards a Reconceptualization of the Idea of Culture in Cultural Geography. *Transactions of the Institute of British Geographers* N.S. 20, 102-116.

Mitchell, D. (1996). *The Lie of the Land: Migrant Workers in the Californian Landscape.* Minneapolis: University of Minnesota Press.

Mitchell, D. (2000). *Cultural Geography: A Critical Introduction.* Oxford: Blackwell.

Mitchell, W.J.T. (1994). Imperial Landscape. In W.J.T. Mitchell (Ed.), *Landscape and Power* (pp. 5-34). Chicago: University of Chicago Press.

Olwig, K.R. (1984). *Nature's Ideological Landscape: A Literary and Geographic Perspective on its Development and Preservation on Denmark's Jutland Heath.* London: George Allen & Unwin.

Olwig, K.R. (1993). Sexual Cosmology: Nation and Landscape at the Conceptual Interstices of Nature and Culture, or: What does Landscape Really Mean? In B. Bender (Ed.), *Landscape: Politics and Perspectives* (pp. 307-343). Oxford: Berg.

Olwig, K.R. (2001a). Landscape as a Contested Topos of Place, Community and Self. In S. Hoelscher, P.C. Adams & K.E. Till (Eds.), *Textures of Place: Exploring Humanist Geographies* (pp. 95-117). Minneapolis: The University of Minnesota Press.

Olwig, K.R. (2001b). Time out of Mind – Mind out of Time: Custom vs. Tradition in Environmental Heritage Research and Interpretation. *International Journal of Heritage Studies*, 7, 4:,339-354.

Olwig, K.R. (2002a). *Landscape, Nature and the Body Politic: From Britain's Renaissance to America's New World.* Madison: University of Wisconsin Press.

Olwig, K.R. (2002b). Landscape, Place and the State of Progress. In R.D. Sack (Ed.), *Progress: Geographical Essays* (pp. 22-60). Baltimore: Johns Hopkins University Press.

O.E.D. (1971). *Oxford English Dictionary*. Oxford: Oxford University Press.

Sack, R.D. (1997). *Homo Geographicus: A Framework for Action, Awareness, and Moral Concern*. Baltimore: Johns Hopkins University Press.

Scott, J.C. (1998). *Seeing Like a State: How Certain Schemes to Improve the Human Condition Have Failed*. New Haven: Yale University Press.

Setten, G. (2004). The *Habitus*, the Rule and the Moral Landscape. *Cultural Geographies*, in press.

Spirn, A.W. (1998). *The Language of Landscape*. New Haven: Yale University Press.

Söderström, O. (1996). Paper Cities: Visual Thinking in Urban Planning. *Ecumene, 33*, 249-281.

Thompson, E.P. (1993). *Customs in Common*. London: Penguin.

Tuan, Y.-F. (1975). Images and Mental Maps. *Annals, Association of American Geographers* 65, 2, 205-213.

Tuan, Y.-F. (1979). Sight and Pictures. *Geographical Review, 69*, 4, 413-422.

Williams, R. (1973). *The Country and the City*. New York: Oxford University Press.

Chapter 4

NAMING AND CLAIMING DISCOURSE
The "Practice" of Landscape and Place in Human Geography

Gunhild Setten
Department of Geography, Norwegian University of Science and Technology, Trondheim, Norway

1. SETTING THE SCENE

This is a paper about concepts, classification and the ordering of knowledge. Conceptualising the world consists of labelling knowledge by another name – we employ suitable concepts using our worldly experiences and knowledges. Withers (1996: 275) claims that "classification is intrinsic to knowledge", hence, "we label knowledge as an inevitable consequence of ordering the world". The relationship between concepts and categories and the world is thus dialectical – concepts and categories "are contexts and subjects of geographical experiences" (Relph 1985: 21). Therefore, there is a need to always "be sensitive to the reciprocal relationships between geographical 'texts' and the epistemological contexts of their production and use" (Withers 1996: 275). Consequently, the dialectics of language (i.e. its concepts and classifications) and their relationship to the world provide meaning and direction to the world. Even more importantly, dialectics of language annex the world.

In order to develop and further illustrate such a claim, this paper will focus on how the world becomes conceptualised, classified and ordered into *landscape* and *place* respectively, and how this can ultimately be seen as a process of annexation. Landscape and place are well-established concepts within human geography, and have been debated for decades (e.g., Tuan 1974; Relph 1976; Entrikin 1991; Cresswell 1996, 2003; Smith 1996a; Cosgrove 2000; Olwig 2002). Related disciplines such as philosophy (e.g., Casey 1993; 2001), anthropology (e.g., Basso 1996), and art history (e.g.,

H. Palang et al. (eds.), European Rural Landscapes:
Persistence and Change in a Globalising Environment, 67-81.
© 2004 *Kluwer Academic Publishers. Printed in the Netherlands.*

Lippard 1997) are also increasingly taking an interest in a world conceptualised as landscape and place. Why, then, might it be necessary to examine these concepts' histories and workings again? Given that concepts are essential analytical tools employed within the social sciences and the humanities, there is every reason to support ongoing contemplation in regards to how they work (e.g., Tuan 1991). More specifically, scientific debates concerning landscape and place often prove to be complicated, confusing and elusive (e.g., Casey 2001; Cloke & Jones 2001). Both concepts are used interchangeably and as separate categories. The overwhelming research activity done contributes to the maintenance of the multifarious natures of these concepts – "each can be used in several senses and at many levels of theoretical sophistication" (Smith 1996a: 189).

As well, the concepts of landscape and place are not only of relevance within the confines of academia. They are to a great extent employed within the areas of physical planning and environmental management (e.g., Geelmuyden 1993; Jones & Daugstad 1997; Widgren 1997). One of the most crucial dimensions in regards to the conceptual practice of landscape and place, both within scientific and applied discourses, is their role in debates over what interests are to be represented in the shaping of our surroundings. The already-noted dialectical relationship between conceptual practice and the world is thus pertinent, in regards to the power of the concepts themselves and their abilities in creating reality. This means that not only does language provide meaning and direction to the concepts of landscape and place, it (more importantly) also orders landscapes and places in a concrete fashion – for example, classifying landscapes into *everyday landscapes* or *landscapes of high scenic value* can be seen as strategies aimed at creating both a conceptual order and an ordering of physical landscapes in a concrete manner. There is a relative lack of reflection over how notions of ownership and the annexation of the world in a concrete manner are involved in such an "ordering process". On the one hand, ownership in this context refers to how our differing conceptual practices are strategies for writing oneself into and annexing a discourse, or parts of a discourse, or at best creating new discourses. On the other hand, it refers to how such an exercise contributes to an annexing of what discourse ultimately refers to, i.e. on a concrete level. I argue that ownership in regards to concepts, discourses and realities is a crucial ingredient in any scientific endeavour. What therefore concerns me is how the practice or employment of landscape and place within a geographical discourse *structures* our concepts, discourses and realities? Consequently, I conceptualise *ownership* as being outside of the traditional territories of legal and financial systems. Ownership is rather "considered in terms of space engaged in action and ideology in a process of empowerment" (Crouch & Parker 2003: 398). Thus conceptual practice is ultimately social, and likely to be contested.

This paper will proceed based on an idea concerning how the concepts of landscape and place are employed in order to annex *scientific* discourse. I will be looking at how these two core concepts of human geography are powerful classificatory devices for directing, ordering and annexing a discipline, and thus the practitioners within that discipline. The empirical material is consequently not directly derived from landscapes and places in the concrete, material form; which represents only one way of framing and classifying knowledge. We also frame and classify our knowledge within written texts. A text that seems particularly suited to serving the purpose of this discussion is *The Dictionary of Human Geography* (Johnston et al. 1981a, 1986, 1994a, 2000a). The four editions of this dictionary are powerful classificatory and ordering frameworks in regards to the conceptual employment of landscape and place – the dictionaries can be seen as shaping and ordering the reality of a number of scholars. This is particularly important because the successive editions make subtle claims to providing a certain temporary and spatial discursive frame for the discipline of human geography as a whole. Because they are widely read, they represent major sources of information within the discipline. This is due to the fact that they also often provide good and condensed introductions into very complex discussions. In addition, dictionaries such as this one are important sources concerning change in conceptual practice.

Before looking at the *landscapes* and *places* of the dictionaries, I find it useful to briefly reflect upon how the dialectics of concepts, categories and realities are made, transformed and ordered.

2. CONCEPTUAL WORKINGS

"To have a concept is to understand a category", says American geographer Jonathan Smith (1996a: 190). To have a concept thus means that one understands what the category in question includes and excludes. One also understands "the nature of the included items so that one can relate to them in an appropriate manner" (ibid). For example, having a concept of "horse" allows me to be able to discriminate between a horse and a cow. Concepts are, however, not only of use in terms of classification: "they also advise those who hold them of an appropriate deportment" (ibid). Thus, my concept of horses is that some are friendly and some are not, which tells me that I should consequently approach an unfamiliar horse warily. My concept of *horse* thus affects my attitude towards a horse. It is, in other words, an appropriate concept of horses – it advises my behaviour and helps me in deciding as to how to act in the world. Hence, concepts *do* things, they have consequences:

Concepts name experiences. Foucault argues that the process of naming does not depend on what you see, but on elements that have already been introduced into discourse by structure (Foucault 2002; see also Gerber 1997). In other words, what is already categorized and named structures how we are able to comprehend what we see. This does not imply that concepts are definite – concepts such as landscape, place and horse come in many shapes and sizes. What this does mean is that good concepts are flexible in the sense that they are able to claim a "default position" in regards to any particular debate. Hence, there are claims to ownership inherent in conceptual practice.

Concepts create realities. Having a conceptualised vision of the world is having a notion of what is real.[1] Conceptualising the world – i.e. giving it some kind of existence – is a social process. Concepts are thus meaningless when seen apart from alternative understandings of the same concept – or apart from alternative concepts.

Thus, *concepts aim at truth-making.* Concerning concepts, "… metaphor and image are conceived not as surface representations of a deeper truth but as creative intervention in making truth" (Cosgrove 1990: 344). "Debates about words and categories are [thus] always moral endeavours in that they attempt to bring about a vision of the world" (Gerber 1997: 2). Any conceptual practice is therefore aimed at attitudinal improvement. Such moral endeavours are both conceptual and concrete, as Gerber asserts above. They are also ultimately contested.

Therefore, it follows that *concepts aim at persuasion.* Concepts aim at persuading us that what we experience, find truthful and real actually *is* what we experience, find truthful and real. In other words, what we know *is* what we know. Arriving at knowledge is a process of conceptual persuasion: "The result is knowledge. But the process is persuasion" (Myerson & Rydin 1996: 15).

It follows from the above that the workings of concepts are a continuous process of social contestation because:

Concepts draw borders between categories – i.e. we draw borders between us. Because conceptualising and categorising is a mental exercise, concepts and categories are not something found "out there" – they are rather, something mental "in here". Why and how people experience and thus conceptualise and categorise the world differently is therefore a crucial question. Equally important is, however, how we choose to *practice* (i.e. employ) our concepts once we have categorised them and arrived at certain

---

[1] Note, however, that not everything conceptualised needs to be materialised. A notion of God could exemplify that.

knowledge. In order to understand the mechanisms involved in a rather blurred and elusive conceptual debate about landscape and place, it is thus necessary to dismantle the rhetoric involved. There is a need to determine how realities are advocated, "by presenting the partial as complete, the artificial as natural, the contingent as essential, and the optional as necessary" (Smith 1996b: 4). In other words, how are we conceptually persuaded into making it appear that there is no *practical* choice, i.e. how do we come to be persuaded into certain conceptual practices? It is thus time to move into the main body of this paper.

3. LANDSCAPE AND PLACE – "THE SAME BUT DIFFERENT"[2]

My interest in the relationship between landscape and place is very much the result of having worked with landscapes for quite a number of years, both conceptually and concretely. Through that work, I have observed that the concept of place surfaces frequently in what might be termed "landscape literature", both within geography and outside of the discipline. Place often plays a subordinate or casual role in this literature – it is placed, so to speak, on the fringes of what appears to be the subject matter. The concepts are also often used interchangeably, as if landscape and place is the same thing. The fact that they are sometimes used interchangeably indicates that they must be of relevance for one another. The relevance is, however, sometimes hard to identify. There is another fact that is equally important – that landscape and place are two different concepts.

These rather confusing observations led me to start reflecting more systematically on what the relationship between landscape and place might consist of. Research regarding the relationship between landscape and place needs to be grounded in two crucial issues. Firstly, the conceptual practice needs to be examined within scientific discourse. We need to have a sense and understanding of how scientists organise information and observations. Conceptualisation is an obvious way to organise – we frame our experiences by employing suitable words for them. Secondly, how conceptual practice is part and parcel of how to be in the world needs to be looked at.[3] In the

---

[2] This section has the character of a personal account, and will not to any extent make reference to other scholarly work. I also wish to emphasise that my reference to the discussion with Prof. Robert Sack must be seen only as an illustration of the point I wish to make in this paper, and is thus not a personal reply to Prof. Sack, a speech of defense or an attempted "attack" on him.

[3] Researching the conceptual practice of landscape and place therefore involves working on two different levels – a discursive and a non-discursive level. A discourse can simply be understood as that which people say, and all concepts are thus part of discourse. In

following, I will draw on these issues when looking at how geographers have employed landscape and place in order to categorise and classify their information and observations.

My professional identity is very much based on being a landscape geographer. I therefore consider a concept of landscape to be "mine" in the sense that I believe I am entitled to employ *landscape* when framing and giving direction to my research and objects of research. I do not employ the landscape concept randomly. I attempt to inscribe myself and my writing within a landscape discourse so that it allows for "my" concept to be acknowledged, so that I can annex part of a discourse and also in some ways even the concrete landscape it might refer to. Consequently, I am distancing myself from other landscape concepts and understandings that belong to other people (or other scientists). This became very clear at a research symposium in September 2000, as my landscape concept was subjected to explicit questioning. I was presenting a paper that dealt with how landscape can be a lived phenomenon – always in the making – and how we through our lives in the landscape are producers of rules concerning appropriate behaviour in the landscape (Setten 2001, forthcoming 2004). I was pointing out how a dialectical relationship between people and land might work, and how differing landscape practices are both a beginning and an end in such a relationship. My main concern was to demonstrate that this use of dialectics is of a moral kind, because we often tend to judge other people's practices as good or bad based on how we ourselves experience and conceptualise the landscape. I was arguing that passing moral judgements is very often the cause of conflicts within the landscape, and that we need to acknowledge the power of conceptual practice in order to understand such conflicts.

American geographer Robert Sack, who was commenting on the paper, asked me whether I could have employed the concept of *place* rather than that of *landscape*.[4] I was both surprised and provoked by this. I was surprised, because converting *landscape geography* to *place geography* was

---

discourse what we say (in the broad sense) is framed: it is given structure, meaning and direction. Importantly, we then act upon this frame. Even more importantly, discourse is therefore practice, but not all practice is discourse. This is due to the fact that some practices are not reducible to "speech" acts; e.g. being deprived of one's home – the home being something that often provides a sense of place – can be a violently-felt experience not reducible to a speech act (or discourse). There is hence a thin, but still important, line that needs to be drawn between these levels; not in the least with regard to how we are able to understand questions of ownership inherent in conceptual practice and concepts' workings. Undoubtedly, the way we conceptualise the world, based on our information and observation, affects the way we "do" the world. I therefore acknowledge that one can, with difficulty, distinguish between these dimensions. See for example Ricoeur (1972), Livingstone (1992), and Setten (1999) for a more thorough discussion.

[4] I have received a similar comment in a reviewer's report: "I did wonder why the author did not simply replace 'landscape' with 'place' as her main category."

not something that I had considered. And I felt provoked because "swapping concepts" was not as simple and straightforward as it might appear. According to Professor Sack, I was arguing for an understanding of landscape that equalled his understanding of place. He thus challenged me to start reflecting on the degree to which landscape and place are: "the same things, things that differ only in degree or that are very different kinds of things, and which one is the key in the mutually constitutive relationship" (Sack 2000: 136). If we were arguing for the same thing, how come we chose to order, classify and frame our observations and information into landscape and place respectively? Given that there are two concepts involved, we could not possibly be talking about the same thing – could we? In order to find possible answers to these questions, I chose to turn to *The Dictionary of Human Geography* (Johnston et al. 1981a, 1986, 1994a, 2000a) – a text claiming to be a guide to the concepts, terms and theories used in human geography.

4.   THE "PRACTICE" OF LANDSCAPE AND PLACE

The Dictionary of Human Geography was initially published in 1981 (Johnston et al. 1981a), with a second edition in 1986 (Johnston et al. 1986), a third in 1994 (Johnston et al. 1994a), and a fourth in 2000 (Johnston et al. 2000a). The successive editions guide you through what are claimed to be the chief concerns of human geographers, the themes and approaches within the discipline presented through keywords and written text. Within each main entry, cross references are made to other entries so the reader or the user may follow different directions if they choose to do so within the edition's system of cross-referencing. Given that my main concern dealt with the concepts of landscape and place, I set out to go where the dictionaries directed me. It is clear that what I set out to do was not an innocent mapping of how practitioners of these concepts have classified their research objects into landscape and place respectively. Rather, I was tracing a situated order of academic knowledge as it is expressed through complex writing strategies. There are two issues in particular that help to explain this situated order.

First, the dictionaries are firmly placed within the bounds of Anglo-American geography. In line with post-colonial trends influencing the social sciences, this aspect is subject to increasing concern in the prefaces to the third (Johnston et al. 1994b) and fourth (Johnston et al. 2000b) editions. The editors express their awareness of the dangers involved in being "concerned almost entirely with English-language words, terms and literatures" (Johnston et al. 2000b: viii). What is meant by: "English-language words, terms and literatures" is, however, not made clear – are they literatures written in the English language or literatures written by scholars having the English language as their first language? The editors have employed a narrow definition of the English language, meaning that writings not published in Great Britain, the United States, Canada or Australia are mostly not included. Partialities and limitations linked to a geography of language are therefore inherent, because the perspectives present leave out a tremendous amount of important work. The editors of the fourth edition are hence stating that the dictionary does not actually constitute the "authorised" version of the discipline. This has led them to display a second concern, that of "the slipperiness of our geographical 'keywords': of the claims they silently make, the privileges they surreptitiously install, and of the wider web of meaning and practice within which they do their work" (ibid: vii). English-language words, terms and literatures impact heavily on what, for example, Nordic, Southern European or African human geography might be about. Therefore, my notions of landscape and place cannot be thought of as being removed from the conceptualisations found within English-language geography. This is, however, not the main issue here. The point is that in order to understand why Professor Sack was questioning my use of concepts, there is a need to have a sense of the workings of concepts in general; and more specifically, concerning how his question might feed into an idea of a conceptual practice where notions of order and ownership are inherent.

Let us thus move on by looking more closely at what can be found in successive editions of the dictionary in regards to the concepts in question. Table 1 shows the main entries for *landscape* and *place* in addition to three related main entries: *placelessness, cultural landscape* and *Landschaft*, as they are presented in each edition. The vertical columns represent which references are made to other main entries. Entries that are underlined are cross-references between the five main entries chosen. Because my choice of entries is selective, I have left out important entries, for example those of *sense of place, space, region, territory,* and *morphology*. My concern – which is to demonstrate that the stories told in the dictionaries can be seen as attempts at creating a conceptual order that affects the discipline and its practitioners – is, however, still valid.

My reading and understanding of the conceptual histories related in the dictionary is just as partial and potentially limited as those of the authors and editors; i.e., it is situated. Thus, there are two things in particular that surprised and puzzled me. Firstly, given the fact that the concepts of landscape and place are often used interchangeably and therefore seem to be categorising the same thing, it is surprising that there is no cross-referencing between them. Secondly, landscape is and has been one of the core concepts within geography (both human and physical). It therefore puzzled and surprised me even more that there was no entry for landscape before the third edition (Johnston et al. 1994a). It is thus useful to look closer at the authors' and editors' choices for cross-referencing. We will find that both landscape and place are referred to under the entry for *placelessness* (Duncan 1994). It is stated that *placelessness* is: "The existence of relatively homogenous and standardised LANDSCAPES which diminish the local specificity and variety of PLACES that characterized pre-industrial societies" (Duncan 2000: 586). This is a highly visualised and scenic understanding of landscape. At the same time, it is an understanding of *place* that equals my understanding of *landscape*. *Landscape* is here cast as merely an object for interpretation, rather than as a context of experiences (Relph 1985). A similar understanding is displayed in the entry for *cultural landscape* (Cosgrove 2000). There is a lengthy presentation on Carl Sauer's Berkeley school of thought regarding landscape geography, and its today thoroughly-criticised perspective on landscape. The critique has on the whole been targeted against the Berkeley emphasis on materiality and morphology, and its lack of emphasis on meaning and social and cultural processes (see Duncan 1980). Particularly noteworthy is that what might be identified as a Nordic perspective on landscape (e.g., Olwig 2002) is found in the entry for *cultural landscape*. A well-established understanding of landscape among many Nordic scholars is thus erased from the entry on landscape.

Table 4-1. Schematic presentation of main entries and cross-referencing (Johnston et al. 1981a, 1986, 1994a, 2000a).

Main Entry Edition	Place	Placelessness	Cultural Landscape	Landschaft	Landscape
Johnston et al. (Eds.) (1981). The Dictionary of Human Geography. Oxford: Blackwell.	Humanist geography		Berkeley School; Cultural hearth; Genre de vie; Geopiety; Topophilia; Phenomenal environment; Environmental determinism; Behavioural environment; Culture	Cultural landscape, Berkeley School, Historical geography; Empiricism	
Johnston et al. (Eds.) 1986. The Dictionary of Human Geography. (2nd ed.). Oxford: Blackwell.	Humanist(ic) geography; Intentionality; Nature; Culture		Berkeley School; Cultural hearth; Geopiety; Topophilia; Human agency; Phenomenal environment; Culture; Environmental determinism; Behavioural environment	Cultural landscape; Human agency; Culture; Morphology; Berkeley School; Historical geo.; Phenomenalism; Empiricism	

Main Entry Edition	Place	Placelessness	Cultural Landscape	Landschaft	Landscape
Johnston et al. (Eds.) (1994). The Dictionary of Human Geography. (3rd ed.). Oxford: Blackwell.	Locale; Sense of Place; Humanistic Geography; Structuration; Political geography	Landscape; Place; Humanistic geography; 'Sense of place'; New towns; Suburbs; Conservation	Cultural geography; Culture; Berkeley School; Environmental determinism; Culture area; Landscape; Culture; Iconography; Nature	Landscape; Regions; Human agency; Morphology; Berkeley School; Historical geo.; Vertical themes	Territory; Landschaft; Environmental determinism; Cultural landscape; Berkeley School; Existentialism; Iconography; Text; Postmodernism
Johnston et al. (Eds.) (2000). The Dictionary of Human Geography. (4th ed.). Oxford: Blackwell.	Space; Identity; Territory; Territoriality; Globalization; Placelessness; Sense of Place; Humanistic Geo.; Positivism; Law; Phenomenology; Locality; Production of Space; Structuration theory; Political geography; Locale; Time-space compression; Convergence; Communities; Nationalisms; Identity politics; Apartheid; Ethnic cleansing; Gender and geography; Patriarchy; Restructuring; Economic geography; Difference; Postmodernism; Postmodernity; Boundaries; Internet; Critical human geography; Resistance; Transgression; "Nature"; Social exclusion; Racism; Nationalism; Essentialism	Landscape; Place; Humanistic geography; Culture; Sense of place; New towns; Suburbs; Postmodernism; Postmodernity; Globalization; Time-space compression; Community; Location	Cultural geography; Environmental determinism; Human agency; Culture; Berkeley School; Culture area; Geography of animals; Nature; Culture; 'Morphology'; Landscape; Colonization; Modernization; Cultural ecology; Post-colonial; Political ecology; Hybrid; Social construction; Representation; Power; Iconography; Theory; Ecological; Text; Lifeworlds; Communities; Territorialities; Imperialism; Sense of place	Landscape; Regionalism; Cultural landscape; Human agency; Morphology; Berkeley School; Historical geography; Vertical themes	Representation; Territory; Landschaft; Environmental determinism; Cultural landscape; Berkeley School; New Cultural geography; Theory; Image; Ideological; Geography and art; Text; Post-structuralism; Feminist geographies; Marxist geography

If the lack of cross-referencing between landscape and place and the fact that landscape was only introduced in the third edition (Johnston et al. 1994) was puzzling and surprising, it is no less striking that the focus on place has – indicated simply by looking at Table 1 – become overwhelming. The nature of the references indicates that it is the so-called "linguistic turn" identified with post-structural currents within the social sciences over the last 15 to 20 years that seems to have led – at least in terms of an Anglo-American viewpoint – the discipline of human geography into a state of "place craze". Thus we ultimately need to get a sense of why the use of *place* during this period has gained such academic ground (e.g., Cresswell 1996, 2003; Cloke & Jones 2001). Let us therefore revisit Robert Sack's question. Professor Sack was implying that landscape and place is the same thing but that he still preferred place over landscape – therefore suggesting that they are also different. What bearing might that preference have on the discipline in more general terms? Is landscape becoming a less interesting and less relevant, maybe even redundant, way of categorising the world? Along these lines, what can the concept of place provide that landscape can not? Given that within Nordic human geography there has been a reluctance to take on the debate concerning place, is landscape then a more relevant concept for human geographers working within a Nordic context?[5] How could such relevance possibly be demonstrated?

An attempt to answer these complex questions is outside the scope of this paper. I will instead end with some reflections regarding how an encyclopaedic ordering of knowledge can inform a more general debate on the workings of concepts and classifications.

5.   NAMING AND CLAIMING DISCOURSE

Encyclopaedias have: "long sought to combine summation of universal knowledge with a rational *order* for that knowledge" (Withers 1996: 275-276, my emphasis). During the Enlightenment, encyclopaedias became means by which one could bring the world to order. As stated in the introduction to this paper, order rests upon classification, and: "classification was central to contemporary intellectual enterprise and to advance profoundly useful knowledges" (ibid: 276). There was hence an emphasis and value given to such ordering of knowledge – i.e. practicing encyclopaedism – that was based upon social, political and philosophical reasons. Encyclopaedism became part of a progressive European project, and eventually nationalism. Notions of utility, reason and modernity were therefore part and parcel of an encyclopaedic ordering of the world. Encyclopaedic ordering was therefore a moral endeavour – bringing the world into encyclopaedic order became a means by which one could

---

[5] As noted earlier, these are not questions of importance to the academic world only. Landscape architect Anne Katrine Geelmuyden (1993) picked up on this, with reference to how the concepts of landscape and place were employed within physical planning and environmental management in Norway. Ten years ago, she stated something that has bearings on what also concerns me; "Now it is place, understanding places and analysing places that are at the core of the rhetoric, as if 30 years of developing a landscape analysis has been a waste of time. Is this something new or merely the emperor's new clothes?" she asked (1993: 152, my translation).

improve the human condition in general, and an individual's deportment in particular.

Geography played an important part in this modernist project of disciplining the world. Withers (1996) demonstrates that geography was part of the *scientific* order, i.e. geography was situated in relation to other classificatory knowledges of 18th-century encyclopaedias (for example, mathematics, astronomy and mechanics). Consequently, encyclopaedic geography was also part of the order of the *geography* at the time, not least in connection to the mapping and discoveries of the non-European world. Geography was thus in the service of a grand scientific and political project. This indirectly answers the questions concerning what sort of geography was being produced, and for whom such geography and ordering was useful. Withers (1996: 290) also claims that the encyclopaedic ordering tradition of the 18th century was "a marker of national intellectual capacity", and "the peaking of national encyclopaedic projects in Europe in the nineteenth century is an expression of imperial political authority and engagement, through empirical science, with the empire of natural knowledge".

Ordered knowledge was thus about power; "intellectually, over the objects ordered; socially, over those without that knowledge but for whom encyclopaedias, grammars and dictionaries were produced by elite groups in society; and philosophically, in terms of human mental control over the perceived disorder of the natural world" (Withers 1996: 291-293).

The question is: What bearing do such historical claims have on the role of the contemporary editions of *The Dictionary of Human Geography*? The powerful conclusion drawn by Withers is relevant to my claims and questions regarding the classification of landscape and place, as I have presented them throughout this paper. *Intellectually*, the dictionaries exercise power over the objects ordered – the dictionaries, or rather the authors of the entries, claim a certain right (and hence social ownership) to define places and landscapes by means of including and excluding scholarly work. This is supported by the choice of cross-references. Similarly, an absence of cross-referencing must also be seen as a strategy for creating order. *Socially*, the dictionaries are the product of those who are considered to be leading experts within the various fields. The dictionaries are hence produced by certain experts in order to teach students of geography the necessary precision of terminology based on the intellectual power over objects. *Philosophically*, as was stated by Withers, the power is reflected in terms of human mental control over the perceived disorder of the natural world. Today, this last claim to power is better seen as the disorder of the *social* world. The dictionaries are social and cultural products, and they need to be understood as such. The point is, however, that there are *naturalising* claims made through the choice of entries and cross-referencing. The successive editions of the dictionary might therefore be seen as temporal attempts at *disciplining* a discipline. Looking at the prefaces and the editors' introductions, this seems to be the case with the first edition published in 1981 (Johnston et al. 1981b). The editorial team states that the dictionary: "is presented as a basic reference work for human geographers; we hope it will prove to be as indispensable to them as an atlas" (Johnston et al. 1981b: xii). By the publishing of the fourth edition in 2000, the editors' introduction reflects a sobriety on behalf of Anglo-American geography that was not found 20 years earlier. This reflects not only an intradisciplinary change, but also, as stated earlier, postcolonial concerns within the social sciences and the humanities as a whole. *The Dictionary* cannot, however, escape its ordering and disciplining role. There is thus always a need to reflect on the power of language, the power of naming, the complicated claims we make through the language we use, and the ways we choose to frame our knowledge – be it either in encyclopaedic or other forms.

REFERENCES

Basso, K.H. (1996). *Wisdom Sits in Places. Landscape and Language Among the Western Apache*. Albuquerque: University of New Mexico Press.

Casey, E. (1993). *Getting Back Into Place. Toward a Renewed Understanding of the Place-world*. Bloomington & Indianapolis: Indiana University Press.

Casey, E. (2001). Body, Self and Landscape, a Geophilosophical Inquiry Into the Place-world. In P.C. Adams, S. Hoelscher & K.E. Till (Eds.), *Textures of Place. Exploring Humanist Geographies* (pp. 403-425). Minneapolis: University of Minnesota Press.

Cosgrove, D. (1990). Environmental Thought and Action: Pre-modern and Post-modern. *Transactions of the Institute of British Geographers NS*, 5, 344-358.

Cosgrove, D. (2000). Cultural Landscape. In R.J. Johnston, D. Gregory, G. Pratt & M. Watts (Eds.), *The Dictionary of Human Geograph.y* 4th Ed. (pp. 138-141). Oxford: Blackwell.

Cloke, P. & Jones, O. (2001). Dwelling, Place and Landscape: An Orchard in Somerset. *Environment and Planning A*, 33 (4), 649-666.

Cresswell, T. (1996). *In Place/Out of Place Geography, Ideology, and Transgression*. Minneapolis: University of Minnesota Press.

Cresswell, T. (2003). Landscape and the Obliteration of Practice. In K. Anderson, M. Domosh, S. Pile & N. Thrift (Eds.), *Handbook of Cultural Geography* (pp. 269-281). London: Sage.

Crouch, D. & Parker, G. (2003). 'Digging-up' Utopia? Space, Practice and Land Use Heritage. *Geoforum*, 34, 395-408.

Duncan, J.D. (1980). The Superorganic in American Cultural Geography. *Annals of the Association of American Geographers*, 70 (2), 181-198.

Duncan, J.D. (1994). Placelessness. R.J Johnston, D. Gregory & D.M Smith (Eds.), *The Dictionary of Human Geography*. 3rd Ed. (p. 444). Oxford: Blackwell.

Duncan, J.D. (2000). Placelessness. In R.J Johnston, D. Gregory, G. Pratt & M. Watts (Eds.), *The Dictionary of Human Geography*. 4th Ed. (pp. 585-586). Oxford: Blackwell.

Entrikin, JN. (1991). *The Betweenness of Place: Towards a Geography of Modernity*. London: Macmillan.

Foucault, M. (2002). *The Archaeology of Knowledge*. London: Routledge Classics.

Geelmuyden, A.K. (1993). Landskapsanalyse – planredskap og erkjennelsesvei. *Byggekunst*, 3, 152-157.

Gerber, J. (1997). Beyond Dualism – The Social Construction of Nature and the Natural and Social Construction of Human Beings. *Progress in Human Geography*, 21 (1), 1-17.

Johnston, R.J., Gregory, D., Haggett, P., Smith, D., & Stoddart, D.R. (Eds.) (1981a). *The Dictionary of Human Geography*. Oxford: Blackwell.

Johnston, R.J., Gregory, D., Haggett, P., Smith, D., & Stoddart, D.R. (1981b). Editor's Introduction. In R.J. Johnston, D. Gregory, P. Haggett, D. Smith & D.R. Stoddart (Eds.), *The Dictionary of Human Geography* (pp. x-xii). Oxford: Blackwell.

Johnston, R.J., Gregory, D. & Smith, D.M. (Eds.) (1986). *The Dictionary of Human Geography*. 2nd Ed. Oxford: Blackwell.

Johnston, R.J., Gregory, D. & Smith, D.M. (Eds.) (1994a). *The Dictionary of Human Geography*. 3rd Ed. Oxford: Blackwell.

Johnston, R.J., Gregory, D. & Smith, D.M. (1994b). Preface to the Third Edition. In R.J. Johnston, D. Gregory & D.M. Smith (Eds.), *The Dictionary of Human Geography*. 3rd Ed. (pp. vii-viii). Oxford: Blackwell.

Johnston, R.J., Gregory, D., Pratt, G. & Watts, M. (Eds.) (2000a). *The Dictionary of Human Geography*. 4th Ed. Oxford: Blackwell.

Johnston, R.J., Gregory, D., Pratt, G., & Watts, M. (2000b). Preface to the Fourth Edition. In R.J. Johnston, D. Gregory, G. Pratt & M. Watts (Eds.), *The Dictionary of Human Geography*. 4th Ed. (pp. vi-ix). Oxford: Blackwell.

Jones, M. & Daugstad, K. (1997). Usages of the "Cultural Landscape" Concept in Norwegian and Nordic Landscape Administration. *Landscape Research*, 22, 267-281.

Lippard, L. (1997). *The Lure of the Local. Senses of Place in a Multicentered Society*. New York: The New Press.

Livingstone, D. (1992). *The Geographical Tradition*. Cambridge, MA: Blackwell Publishers.

Myerson, G. & Rydin, Y. (1996). *The Language of Environment. A New Rhetoric*. London: UCL Press.

Olwig, K.R. (2002). *Landscape, Nature and the Body Politic. From Britain's Renaissance to America's New World*. Madison: The University of Wisconsin Press.

Relph, E. (1976). *Place and Placelessness*. London: Pion.

Relph, E. (1985). Geographical Experiences and Being-in-the-World: The Phenomenological Origins of Geography. In D. Seamon & R. Mugerauer (Eds.), *Dwelling, Place and Environment. Towards a Phenomenology of Person and World* (pp. 15-31). Dordrecht: Martinus Nijhoff Publishers.

Ricoeur, P. (1971). The Model of the Text: Meaningful Action Considered as Text. *Social Research. An International Quarterly of the Social Sciences*, 38 (3), 529-562.

Sack, R.D. (2000). Casey, E.S. The Fate of Place: A Philosophical History. Book review. *Progress in Human Geography*, 24, 136-137.

Setten, G. (1999). Den 'nye' kulturgeografiens landskapsbegrep. *Nordisk Samhällsgeografisk Tidskrift*, 29, 55-71.

Setten, G. (2001). Farmers, Planners, and the Moral Message of Landscape and Nature. *Ethics, Place and Environment*, 4 (3), 220-225.

Setten, G. (2004). The *Habitus*, the Rule and the Moral Landscape. *Cultural Geographies,* in press.

Smith, J.M. (1996a). Ramifications of Region and Senses of Place. In C. Earle, K. Mathewson & M. Kenzer (Eds.), *Concepts in Human Geography* (pp. 189-211). Lanham: Rowman & Littlefield.

Smith, J.M. (1996b). Geographical Rhetoric. Modes and Tropes of Appeal. *Annals of the Association of American Geographers*, 86 (1), 1-20.

Tuan, Y.F. (1974). *Topophilia. A Study of Environmental Perception, Attitudes and Values.* Englewood Cliffs, New Jersey: Prentice-Hall.

Tuan, Y.F. (1991). Language and the Making of Place: A Narrative-descriptive Approach. *Annals of the Association of American Geographers*, 81, 684-696.

Widgren, M. (1997). Landskap eller objekt: Kring kulturminnesvårdens problem att hantera landskapets historia. In J. Brendalsmo, M. Jones, K. Olwig & M. Widgren (Eds.), *Landskapet som historie* (pp. 5-16). NIKU Temahefte 4. Oslo: Norsk Institutt for Kulturminneforskning.

Withers, C.W.J. (1996). Encyclopaedism, Modernism and the Classification of Geographical Knowledge. *Transactions of the Institute of British Geographers* NS, 21, 275-298.

Chapter 5

BETWEEN INSIDENESS AND OUTSIDENESS – STUDYING LOCALS' PERCEPTIONS OF LANDSCAPE

Katriina Soini
MTT, Agrifood Research Finland, Environmental Research, Finland

1. POST-PRODUCTIONISM AND THE CHALLENGES FOR RURAL LANDSCAPE RESEARCH

In European countries, the environmental and landscape management of rural areas has faced several new challenges along with the changing role of agriculture. The quality of food, landscape values, and the vitality of rural areas have increasingly been emphasised in the latest agricultural and rural policies. At the same time, the consumers are increasingly taking part in food production and in rural – especially landscape – development; for example, demanding the preservation of landscapes and maintenance of biodiversity (Marsden 1995; Pierce 1996; Macnaghten & Urry 1998). Society is thus reasserting its hold on rural nature, land and culture.

This development, which has been called *post-productionism* has introduced a new complexity into rural space (Marsden 1995, Yliskylä-Peuralahti 2003). Rural landscape is no longer a natural product of rural activities, agriculture and forestry, but rather an arena of different and conflicting interests, and therefore a contested object of planning. The agrarian population is smaller, whereas the share of other rural residents and part-time residents is increasing. The farmers are no longer alone in their involvement in rural areas, although their production systems and behaviour still have the most impact on them.

In addition, symbolic values have taken precedent over the material values – the rural areas have become more consumptive (Shucksmith 1994;

H. Palang et al. (eds.), European Rural Landscapes:
Persistence and Change in a Globalising Environment, 83-97.
© 2004 *Kluwer Academic Publishers. Printed in the Netherlands.*

Pierce 1996), as people who are separated from the process of production encounter the rural environment more and more as a commodity. For them, rural landscape may appear as a marginal place that evokes nostalgia and fascination (Shields 1991). As taste and fashion are shaping landscape, both the biological and the cultural preservation of landscape incur the danger of invented natures and traditions of "landscape" or "heritage industry", and a transformation of the form and meaning of landscape (Macnaghten & Urry 1998; Hopkins 1998).

Culturally-oriented landscape research, which considers landscape as socially and culturally constructed, has extended our understanding of rural landscape and its development. At the same time, however, it has dissociated the practical needs of landscape management and planning. It does not necessarily provide landscape management and planning with applicable knowledge about the landscape perceptions of different interest groups. Furthermore, the perceptions of local residents have received relatively little attention, probably due to the many new interest groups that have emerged in the field, such as NGOs and citizens. In addition, local peoples' insights into landscape have been considered methodologically much more difficult to deal with (Mitchell 2001).

The methodological questions are related to the language and concepts that are chosen for the study. In the social sciences, the concepts used, as well as the object of the study, are socially constructed (Berger & Luckmann 1992). Thus, the researcher has to be at least aware of the existence of the various knowledge systems and meanings that relate to the landscape. Landscape as an academic concept has been widely discussed, and a partial consensus on the ambiguity of the concept has been reached. In fieldwork, the complexity of the concept emerges from a different perspective, as the academic definitions of landscape differ from the knowledge system of local people. This implies very practical problems when an empirical study on landscape perceptions is employed. How is it possible to deal with landscape issues in the in-depth interviews or in the survey, if the locals conceptualise the landscape in their own and personal ways?

This chapter is an attempt to describe briefly a conceptual framework for exploring locals' perceptions of rural landscapes, and to reveal some practical implications of a case study. First, the concepts of landscape and place are discussed in an academic context, and then brought into the empirical study. Afterwards, the experiences of a case study, in which the concepts were applied, are presented. In brief, the article is about the fact that what an image is depends on how it is studied (Smith 1985).

2.   FROM LANDSCAPE BACK TO PLACE

Landscape is a very central concept in contemporary human geography. The history of the concept is long and varied, and will not be discussed here in detail. To summarise, in the discipline of geography landscape has been defined as an area, a scene, a way of seeing or a representation, depending upon the chosen approach. In addition, landscape is widely and loosely used beyond geography, and even as a part of our everyday language. Despite the wide scale of the definitions – or rather due to this – it is not necessarily the most appropriate concept for empirical purposes.

Place is also an elusive concept. Although place had already existed in the geographical literature in earlier decades, it first became a real object of interest and active study during the 1970s (Buttimer 1976; Relph 1976, Tuan 1976; Gold et al. 2000). Place implies a sense of place, an ability to recognise different places and different identities of a place (Relph 1976). The philosophical foundation for the humanistic study of place is found in phenomenology, which takes as its starting point the phenomena of the lived world of immediate experience and seeks clarification through careful observation and description.

Place and landscape are in many ways intertwined with each other, but they do have certain differences. In academic geography, landscapes are seen more as objects of interpretation than as contexts of the experience – as place is typically seen as (Relph 1989). Places are lived in: they permeate everyday life and provide meaning in people's lives, whereas landscape is predominantly thought of as a visual concept, as an image. Visual appearance is an important feature of all places, too. However, a scene may be of a place, but the scene itself is not a place, as Tuan (1974) has put it. Accordingly, all place experiences are not necessarily landscape experiences (Relph 1976).

Utilising a phenomenological approach, it becomes possible to think of landscapes not as external objects, but as sites of *dwelling*. Dwelling is for Heidegger (1971) the process by which a place, in which we exist, becomes a personal world and a home. Dwelling involves a lack of distance between people and things, and an engagement that arises primarily through using the world, rather than through scrutiny (Thomas 1993). Despite the subjective and practical character of the concept, it is important, as Hincliff (2003) and Mitchell (2001) have noted, to include spatiality and temporality regarding dwelling. Otherwise, the study is in danger of remaining ahistorical and ageographical.

Dwelling is closely related to the question of *insideness* and *outsideness* (Lowentahl 1962; Relph 1976; Cosgrove 1998). This pair of concepts has been used, although the necessity and validity of the concepts have also been challenged.[1] Nevertheless, the insider-outsider approach provides a systematic way of dealing with different perceptions of landscape and place – the insiders are considered a part of the place; and it is considered that the place is a part of them. Landscape is for them partly invisible, because the experiences, senses and knowledge that construct the place give content to the landscape. By contrast an outsider, such as a traveller, is much more able to experience the qualities of landscape, although from a distance. The outsiders are able to project and prospect the landscape, while the insiders live within the scene. Relph (1976) has classified different types of insiders and outsiders[2], but it is hardly possible to be purely an insider or an outsider of a place, because the self can never be separated from the place. It is exactly for this reason that landscape seems an inappropriate concept when focusing on the insiders' perceptions of landscape.

The phenomenological approach and the concept of place have been revived in the field of environmental management and research (Harvey 1996; Macnaghten & Urry 1998). Capturing the rich variety of human relationships with resources, lands, landscapes and ecosystems, place has been considered to have a great potential for bridging the gap between the science of ecosystems and their management, which many other approaches fail to achieve (Spretnak 1997; Butz & Eyles 1997; Williams & Stewart 1998).

The idea of understanding environment as a lived phenomenon, as a place, is also in line with the current rural policy of the European Union. The policy is being integrated into environmental and landscape management, and the implementation increasingly leans on local communities and their initiatives. A sense of place becomes of crucial importance regarding the principles of rural environment and landscape management, the local people, and their values with regard to the environment. As Sack has put it, "We need place to alter nature, to protect and restore it; we need place to separate ourselves from nature and to re-engage with it" (Sack 1997: 254).

---

[1] Olwig (2001) for example recently introduced the *place ecology of landscape* to bridge the dichotomy of nature and man, insiders and outsiders. He sought to focus on the process of dwelling in place as the key to understanding the physical and social nature of landscape. Place ecology thus shifts the emphasis towards the activity of those who *create places* through dwelling.

[2] Relph (1976) distinguished different types of insideness and outsideness. In existential outsideness, all places assume the same meaningless identity and are distinguishable only by their superficial qualities. Objective outsideness (a planner's perspective) exists when people separate themselves emotionally from the places which they are planning and restructure them according to principles of logic, reason and efficiency. Incidental outsideness applies only to those places in which we are visitors and towards which our intentions are limited and partial.

3.   TOWARDS EMPIRICAL WORK

The complexity of the concepts of landscape and place appears to increase when empirical research is employed. In this respect, some differences between languages may appear. In Finnish, *maisema*, for *landscape*, means scenery, a view, or picturesque. In a plural form, it refers to a region. The Finnish term *paikka*, for place, also has several meanings. In the dictionary of the Finnish language (1992), *paikka* has been defined as: 1) a restricted *or* 2) an unrestricted space; 3) a location, a position, a site; 4) a spot or a patch in a landscape or in an area; 5) a situation or a moment; 6) a part of something. *Paikka* originates from the German word *spaik*, which means a spoke of a wheel. It has come to mean a place of a different colour, or a patch or a hand gear. In Finnish, *paikka* seems to refer to something which is different from its surroundings. In addition, many activities are typically combined with the word *paikka* (place), contrary to *maisema* (landscape). Landscape may be beautiful or ugly, varied, wooden or agricultural, but it does not tell about the things that take place there. Landscape is thus thought of as a kind of framing, not as a process (see also Cresswell 2003: 275).

In the Finnish language, landscape is restricted to the visual appearance, and place is predominantly conceived as a site or location. Neither of the concepts turns out to be appropriate in an empirical study, and a need for medium concepts emerges. The classical concept, *genres de vie*, introduced by the regional geographer Paul Vidal de la Blanche, provides a possibility. It has been translated as *lifeway*, *patterns of living*, or *ways of life*, referring to the connections between livelihood, social organisation and cultural traditions within particular bio-physical settings (Claval 1998; Buttimer 2002). Each lifeway include a symbolic world that is closely tied to group identity and characteristic behaviour routines, and interaction networks linked to the socio-spatial ordering of everyday life. Lifeway emphasises place as a setting for everyday routines and social interaction, and the physical dimensions of the place are in danger of remaining secondary – especially when employed in field work.

On the contrary, the concepts *everyday world*, *everyday landscape* or *everyday environment* seem to include physical framing. They refer to the world that we experience directly, and in which we are located *both* physically and socially (Katz & Kirby 1991). This spatial dimension is essential when a certain physically defined area is to be studied.

In geographical literature, the main focus has been on distinguished – rather than everyday – landscapes (Jones & Daugstad 1997). The latter have often been characterised as insignificant, against which valued landscapes can be contrasted. Recently, a new interest in the everyday landscape – or *landscape of practice* (Cresswell 2003) – has emerged. Looking at the issue from the perspective of rural people, to consider landscape as lived and practiced seems especially relevant. The home yard, open fields, farmhouses, forests and lakes might be a part of their everyday landscape, as they comprise a frame for rural people's everyday routines, or their *Place-ballet*, as Seamon (1980) has called them – thus generating a strong sense of place. If everyday environment is significant, it is probably worth taking care of. For this reason, the everyday environment of rural people is also of high value from the point of view of rural landscape management.

If the concept of everyday environment could be considered as applicable, how should a researcher apply it to fieldwork? According to Relph, the role of researcher is something between that of an insider and an outsider. He calls the researcher, or essayist of a place, an *empathetic* insider, whereas the locals are *existential* insiders. An essayist is, as Ryden (1993: 249) has put it: "both a sensitive wanderer and a conscientious cartographer, able both to feel the life and meaning of a place and to inscribe its patterns on paper." An essayist necessarily combines and balances the perspectives of the insider and the outsider; intimacy and closeness, which enable exact and sympathetic depiction of a place on the one hand, and on the other hand detachment and distance, which bring larger patterns and connections into focus. He must be aware of and capable of describing both the invisible and the visible landscape.

When studying human dimensions of landscapes, each field of science using various kinds of research methods has its own advantages and disadvantages. Methodological triangulation, which has been favoured by social scientists, involves using more than one method, and may consist of within-method or between-method strategies. It can be considered either as a process of cumulative validation, or as a means of producing a more complete picture of the investigated phenomena (Brennen 1995; Kelle 2001). The idea is to find the most appropriate method for each research question.

The best way to achieve an understanding of another's subjective world is through participatory observation; for example, by living among the local people, the insiders. However, this method is not always possible, and even not appropriate depending on the aims of the study. In addition to the in-depth interviews, surveys and participatory observation, exploitation of visual material, and the exploitation of maps have been found as interesting, due to the visual character of landscape (Lilley 2001; Soini 2001; Toupal et al. 2001). Both maps and landscape reduce the multi-sensual experience to visually-encoded features, and then organise and synthesise these into a meaningful whole. According to Macnaghten & Urry (1998), maps capture aspects of nature and society through visual abstraction and representation:

they provide distance and objectivity in regards to what is being sensed; they organise and articulate control over what is being viewed. Maps, like all artistic works, are rich in possibilities, and they certainly have their uses in human landscape research – the critical question deals with *how* they are used and interpreted.

The conceptual and methodological background briefly described above was applied to a study focusing on the perceptions of rural people regarding their surroundings, with natural values and agricultural biodiversity especially in focus. The research was carried out in Lammi, a small municipality located in southern Finland with a population of 5,600. Lammi is a rather typical Finnish rural municipality, with problems of depopulation and pressure to diversify economic life, including agriculture. However, Lammi has many strengths and possibilities compared to other rural municipalities of the same size. Lammi has a long and colourful history, an advantageous location rather near growing cities, and a varied landscape with many naturally and culturally attractive areas. Lammi also has an important summerhouse settlement concentrated on the shores of its lakes.

The study was done in two phases. First, a survey was sent to all the rural people of Lammi. Afterwards residents, both permanent and part-time, were interviewed in three villages that were chosen for the study.

4.   EVERYDAY ENVIRONMENT – SAME BUT DIFFERENT

In the beginning of the study, the definition of everyday environment as it applied to each respondent was determined, in order to find out the physical dimensions of their everyday environment. This was achieved using a survey that asked the respondents to describe and mark their everyday environment, which they knew and experienced daily. They were also asked to mark it on a map that was attached to the survey form. Later on, in the beginning of the interview, people were asked to draw their everyday environment on a blank paper; afterwards, a printed map of the neighborhood was used during the interview. Some respondents regarded the everyday environment as a difficult concept – they did not know whether it was a large or a small area, socially or physically defined. Some other people argued that their everyday environment varied according to the seasons. However, most of the respondents were intuitively able to define their everyday environment in one way or another.

The structure of a place is about its mixture of nature, meaning and social relations (Sack 1997). This was also the case in this research. Typically, the everyday environment included elements of the physical environment, such as buildings, ways and routes, elements of the natural and cultural landscape, fields, forests, lakes, etc. Although people, neighbors or communities were sometimes mentioned, the everyday environment was never determined by social factors alone. Rather, it was taken as a physically defined environment where the social relations take place (Agnew 1993). It was also striking that information regarding the quality and sensual experiences of the everyday environment was often added, although this was not asked for.

The size of the reported everyday environment varied considerably from the surroundings of the home yard to the village and working place, and to locations of spare-time activities. Thus, there was no single definition for everyday environment, but rather a wide range of spatially different life-ways. The everyday environment of the part-time residents appeared to cover a wider area than that of the locals. Additionally, they described the changes of the environment – such as eutrophication of the lake, the logging and cutting of the forest – in much more detail. Furthermore, the borders that they set for their everyday environment seemed to be wider and more permeable. These results refer to the *behavioral insideness*[3] of the part-time residents.

5.   PLEASANT PLACES – PLACES OF MY OWN

Pleasant places in this study were thought to be significant for the people in terms of *home*. To have a home is to *dwell* (Relph 1976). Home is a foundation of identity for people as individuals, and as members of a community. Home is not just the house or apartment one happens to live in, and it is not something anywhere that can be exchanged; but rather, it is an irreplaceable center of significance. This may seem very philosophical and obscure, but in fact it can be a common, everyday element of experience. Sack (1997) describes home as a locus of control, which encompasses both physical and cultural environments. One is at home when such control is possible. In addition, a sense of responsibility and a need to take care is associated with home, as well as a mutual dependence between human and home – we exercise control over home, but at the same time home empowers us. Although the meaning of home has been weakened as it has become

---

[3] Vicarious insideness refers to an experience of a place without actually visiting it, in contrast to behavioral insideness, which consists of being in a place and seeing it as a set of objects, views and activities arranged in certain ways and having certain observable qualities. Empathetic insideness demands a willingness to be open to the significances of a place, to feel it, to know and respect its symbols. Existential insideness is the most fundamental form of insideness – place is experienced without deliberate and self-conscious reflection, and yet is full of significances (Relph 1976).

marketable, exchangeable and sentimentalised, it is still a central point of existence and individual identity.

The places that people mentioned were classified as physical (natural or built environment), social, emotional, sensual, functional or institutional in character. Although they were predominantly related to a certain geographical place or to an area, emotional bonds or sensitive feelings related to the places were often reported. For example, the expression of "home or home yard" was preferred instead of "house". Additionally, ownership of the place was mentioned. Expressions like own fields, own forest, or continuity of farming in the next generation were those that expressed rootedness, care and concern for the place.

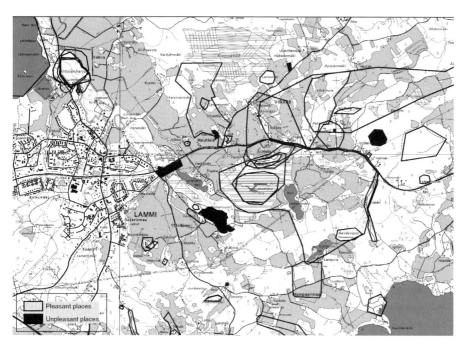

Figure 5-1. Perceptions of pleasant and unpleasant places. This figure presents the combined perceptions of pleasant and unpleasant places of the respondents from five villages. For most of the respondents, who made markings on the map (20 persons), "pleasant place" was a rather wide area around home. Four respondents marked the nature and culture conservation area at Untulanharju (NW corner). Markings were also concentrated in the surroundings of the mire of Lammi and the rock of Halila. Both of them are nature conservation areas. Unpleasant places in the same area were remarkably less common than pleasant places, and were reported by only 13 respondents. They were typically very precisely defined and outlined. © National Land Survey of Finland, permission number 110/VASU/04, chart: Niina Puronummi.

In general, the part-time residents described the pleasant places with much more versatility than did the permanent residents. They told about the facilities that the environment provided, and the quality of the services was evaluated, which told about their expectations of the everyday environment. They also described more openly the emotional bonds they had to those places, such as memories and the social meaning of the present village. The locals' senses of place appeared to be spatially and emotionally much more confined than those of the part-time residents.

6.   UNPLEASANT PLACES – PLACES OF OTHERS

When defining everyday environment, home, a place where one is rooted, people tend to find things that they would like to include there. For that reason, it is not probable that many unpleasant places are mentioned. In the survey, in contrast to pleasant places, unpleasant places appeared to be scarce (Figure 1). Additionally, they were more specifically defined, e.g., certain spots, such as cuttings, gravel pits, or dumping areas; even a poorly maintained nature conservation area was cited as an unpleasant place. The feelings associated with them included a clear sensatory load: smells, noises and visual character of the places were included in the depiction of these places. Whereas the status of the pleasant places was generally more or less static, unpleasant places were dynamic in character: negative changes in the quality of the environment were reported.

Very often unpleasant places were uprooted from the self. They were considered under the control of others, of neighbours or of the municipality. It is worth emphasising that almost half of all the respondents who replied to this question were of the opinion that there were actually no unpleasant places in their surroundings at all, which seems to refer either to the strong need to exclude unpleasant places from the everyday environment or to a good sense of place. The naming, placing and describing of unpleasant places is an interesting issue to be studied further. This brings up a question, among others, as regards to what the relation is between unpleasant places, topophobia, placelessness, and empty places in the rural context.[4]

---

[4] In addition to unpleasant places, placelessness, topophobia and empty places express the relationship between human and environment. *Placelessness*, introduced by Relph (1976) is the opposite of sense of place, referring to the rootedness people feel towards many "modern places". Tuan (1974) has described the negative bond between people and place as *topophobia*. *Empty places* are places that are simply empty of meaning (Bauman 2001; Sack 1997). They are not seen, although they exist, and not visited, because one might feel lost or insecure.

7.   CULTURALLY AND NATURALLY VALUABLE PLACES – PLACES OF THE OUTSIDERS?

The everyday environment of rural people may include pleasant or unpleasant places, regardless of the physical character of the environment. The subjective value of the place is not necessarily dependent on the more-objectively defined natural or cultural qualities. When people were asked to name culturally or naturally valuable places, they usefully named areas with historical and archaeological remains, or distinctive natural features. There were also some "ordinary", or even "poor" areas in terms of objective values. How do local and part-time residents see the natural and cultural values of their everyday environment? What are the natural and cultural values for them and why?

The survey pointed out that almost half of the respondents could not name or did not want to name any naturally valuable areas. Those who replied attached natural values to areas that were part of a nature conservation plan, to the places with which they had a very personal relationship, or to nature in general, such as lakes, a river, or eskers.

Compared with naturally valuable places, people more often had shared opinions about culturally valuable places. For example, the Church of Lammi, as well as the surroundings of the Mommila Estate, were appreciated most among the respondents, even by those who were living elsewhere. As there were again many respondents who could not answer the question at all, we may wonder whether the rural people have left the valuation of the environments to the authorities and experts.

8.   ADDED VALUE OF USING MAPS

In addition to the regular survey, interviews and participatory observation, mapping tasks were used. It was obvious that people were enthusiastic about the maps – in the survey, most of the respondents returned the map. They complained if they could not find certain places on the map, and some people wanted to keep the map, or asked to receive a copy of it. In the interview, the map acted as an intermediary device when discussing places and landscapes.

There were great differences between the markings people made on the maps: some people marked places in detail and with extra explanations, while others merely indicated their everyday environment with a single circle. However, the markings on the maps provided a general view of how people perceived their environment and its places, although they were not comparable as such. In addition to the markings, the maps also had other functions in the study. The process of reading and interpreting the map might be considered as a significant part of identifying self in the everyday environment. The maps encouraged people to look for, show and tell about places that they often visit, which they regarded as beautiful, and to tell of changes that had taken place. Although the map lacks the temporal dimension of the places, the verbal descriptions, both in a survey and interviews, supplemented the information found on the map.

9.   CONCLUSIONS

Traditionally, landscape has been for rural people, and especially for farmers, a part of their lifeway, as it has been closely connected with their livelihood. Recently, the agricultural, environmental and rural policy has intervened in this relationship and the environment has been in a way separated from the context of daily practices and experiences; first as a result of environmental protection, and later due to the increasing focus on environmental management, including a strong emphasis on landscape and biodiversity issues. Along with this development, rural people have been compelled to reassess their relationship with their everyday environment.

A proper understanding of the insiders' views of landscape has been considered as an important, though challenging, research task. The difficulties are partly derived from contemporary cultural geography, which has tended to ignore material spaces and to see landscape merely as ideology. Still, landscape, especially for rural people, is full of practices and material spaces, as well as meanings. By directing the focus of the research on the lived everyday environment, it might be possible to rediscover a part of the materiality of the landscape. This means that academic concepts should be set aside while conducting empirical research and local people should be approached through their everyday practices, using their own language, which is not necessarily an easy task for the researcher. As Relph (1989) put it when referring to Dardel's (1952) idea of *geographicality*:

> Though [geographical experiences] are not commonly known by name, they are experiences which everyone has and which require no textbooks or special methods to be appreciated. They go directly from place to person and from person to places … Geographicality is therefore unobtrusive, inconspicuously familiar, more lived than discussed. It is in fact a naming of the geographical forms of being-in-the-world (Relph 1989: 20-21).

The present case study demonstrated that it is not possible to ignore the place, even if the everyday environment was used as an intermediating concept – the everyday environment was lived in, sensed and experienced. It thus seems possible to approach landscape and place issues via the concept of everyday environment. The study also revealed that it turned out to be much more difficult than anticipated for the locals, the insiders, to define the pleasant and unpleasant, natural and cultural places. The place experience was a mixture of feelings, senses and memories which people were not necessarily able to locate physically.

In addition, a shared opinion of the values of the everyday environment did not exist, which is an interesting and important result from the point of view of landscape planning and management. This observation emphasises the subjective character of individual perceptions, and probably those of experts, which will be focused on next in the study.

At first glance, it might seem self-evident that there are different perceptions of landscape, and the classification of insideness and outsideness may appear as unnecessary or naive. However, this classification provides a framework for understanding the behavior of interest groups, which is needed for rural landscape management. To ignore the dichotomy between human and nature or insideness and outsideness does not necessarily provide any further knowledge of how to deal with the social issues of rural landscape management in practice. As the rural landscape is increasingly created and managed, the cultural processes related to the human/environment relationship have to be analysed in a systematic way as well.

REFERENCES

Agnew, J. (1993). Representing Space: Space, Scale and Sulture in Social Science. In J. Duncan & D. Ley (Eds.), *Place/Culture/ Representation* (pp. 251-271). 2nd Ed. London: Routledge.

Bauman, Z. (2001). *Notkea moderni*. Tampere: Vastapaino.

Berger, P. & Luckmann, T. (1966). *The Social Construction of Reality*. Finnish edition: Raiskila, V. (Ed.) (1994). *Todellisuuden sosiaalinen rakentuminen*. Helsinki: Gaudeamus.

Brannen, J. (Eds.) (1995). *Mixing Methods: Qualitative and Quantitative Research*. Avebury: Aldershot.

Buttimer, A. (1976). Grasping the Dynamism of Lifeworld. *Annals of the Association of American Geographers*, 66 (2), 277-291.

Buttimer, A. (2001). Sustainable Development: Issues of Scale and Appropriateness. In A. Buttimer (Eds.), *Sustainable Landscapes and Lifeways. Scale and Appropriateness* (pp. 7-34). Cork: Cork University Press.

Butz, D. & Eyles, J. (1997). Reconceptualizing Senses of Place: Social Relations, Ideology and Ecology. *Geografiska Annaler*, 79B:1, 1-25.

Claval, P. (1998). *Introduction to Regional Geography*. Malden: Blackwell Publisher.

Cosgrove, D. (1998). *Social Formation and Symbolic Landscape*. Madison: University of Winsconsin Press.

Dardel, E. (1952). *L'Homme et la Terre: Nature de la Realite Geographique*. Paris: Presses Universitaires de France.

Haarala, R., Lehtinen, M., Grillros, E-R., Kolehmainen, T., Nissinen, I., Eronen, R., Suorsa, M. (Eds.) (1992). *Dictionary of the Finnish Language. Suomen kielen perussanakirja.* Kotimaisten kielten tutkimuskeskus. Helsinki.

Gold, J., Stock, M. & Relph, T. (2000). Classics in Human Geography Revisited. *Progress in Human Geography*, 24, 4: 613-619.

Harvey, D. (1996). *Justice, Nature, and the Geography of Difference*. Oxford: Blackwell Publishers Ltd.

Heidegger, M. (1962). *Being and Time*. New York: Harper and Row.

Heidegger, M. (1971). *Poetry, Language, Thought*. New York: Harper and Row.

Hopkins, J. (1998). Signs of the Post-Rural: Marketing Myths of a Symbolic Countryside. *Geografiska Annaler*, 80B, 2, 65- 81.

Jones, M. & Daugstad, K. (1997). Usages of the 'Cultural Landscape' Concept in Norweigian and Nordic Landscape Adminstration. *Landscape Research*, 22 (3), 267-81.

Katz, C. & Kirby, A. (1991). In the Nature of Things: the Environment and Everyday Life. *Transaction Institute of British Geography. N. S.* 16, 259-271.

Kelle, U. (2001). Sociological Explanations between Micro and Macro and the Integration of Qualitative and Quantitative Methods. *Forum: Qualitative Social Research*, 2, 1.

Lilley, K. (2000). Landscape Mapping and Symbolic Form. Drawing as a Creative Medium in Cultural Geography. In I. Cook, D. Crouch, S. Naylor & J. Ryan (Eds.), *Cultural Turns, Geographical Turns: Perspectives on Cultural Geography* (pp. 370-438). Englewood Cliffs: NJ Prentice-Hall,

Lowenthal, D. (1962). Not Every Prospect Pleases. *Landscape*, 12, 2, 19-23.

Macnaghten, P. & Urry, J. (1998). *Contested Natures*. London: Sage Publications.

Marsden, T. (1995). Beyond Agriculture? Regulating the New Rural Spaces. *Journal of Rural Studies*, 11, 3, 285-296.

Mitchell, D. (2001). The Lure of the Local: Landscape Studies at the end of a Troubled Century. *Progress in Human Geography*, 25, 2, 269-281.

Pierce, J. (1996). The Conservation Challenge in Sustaining Rural Environments. *Journal of Rural Studies*, 12, 3:215-229.

Relph, E. (1976). *Place and Placelessness*. London: Pion.

Relph, E. (1989). Geographical Experiences and Being-in-the World: the Phenomenological Origins of Geography. In D. Seamon & R. Mugerauer (Eds.), *Dwelling, Place and Environment* (pp. 15-32). New York and Oxford: Columbia University Press Morningside Edition.

Ryden, K. (1993). *Mapping the Invisible Landscape. Folklore, Writing, and the Sense of Place*. Iowa City: University of Iowa Press.

Sack, R. D. (1997). *Homo Geographicus. A framework for Action, Awareness, and Moral Concern*. Baltimore and London: The Johns Hopkins University Press.

Seamon, D. (1980). Body-Subject, Time-Space Routines and Place-Ballets. In A. Buttimer & D. Seamon (Eds.), *The Human Experience Space and Place* (pp. 148-65). London: Croom Helm.

Shields, R. (1991). *Places on the Margin. Alternative Geographies of Modernity*. London and New York: Routledge.

Shucksmith, M. (1994). Conceptualising Post-Industiral Rurality. In J. Bryden (Ed.), *Towards Sustainable Rural Communities* (pp. 125-132). Ontario: University of Guelph Press.

Smith, S. (1995). Constructing Local Knowledge. The Analysis of Self in Everyday Life. In J. Eyles & D. Smith (Eds.), *Qualitative Methods in Human Geography* (pp. 17-37). Cambridge: Policy Press.

Soini, K. (2001). Exploring Human Dimensions of Multifunctional Landscapes through Mapping and Map-making. *Landscape and Urban Planning*, 57, 225-239.

Spretnak, C. (1997). *The Resurgence of the Real: Body, Nature and Place in a Hypermodern World*. New York: Addison-Wesley.

Thomas, J. (1993). The Politics of Vision and the Archaeology of Landscape. In B. Bender, (Ed.), *Landscape: Politics and Perspectives* (pp. 19-48). Oxford: Berg.

Toupal, R., Zendeno, M.l, Stoffle, R. & Barabe, P. (2001). Cultural Landscapes and Ethnographic Cartographies: Scandinavian-America and American-Indian Knowledge of the Lands. *Environment Science and Policy*, 4, 171-184.

Tuan, Y.F. (1974). Space and Place: Humanistic Perspective. *Progress in Geography*, 6, 211-252.

Tuan, Y.F. (1976). Humanistic Geography. *Annals of the Association of American Geographers*, 66 (2), 266-276.

Tuan, Y.F. (1989). Surface Phenomena and Aesthetic Experience. *Annals of the Association of American Geographers*, 79 (2), 233-241.

Williams, D. & Stewart, S. (1998). Sense of Place. An Elusive Concept that is Finding a Home in Ecosystem Management. *Journal of Forestry*, 18-23.

Yliskylä-Peuralahti, J. (2003). Biodiversity – a New Spatial Challenge for Finnish Agri-Environmental Policies? *Journal of Rural Studies*, 19, 215-231.

Chapter 6

LANDSCAPE CONSUMPTION IN OTEPÄÄ, ESTONIA

Tõnu Oja & Monika Prede
Institute of Geography, University of Tartu, Estonia

1. INTRODUCTION

Estonia's rural development has been much affected by the economic decline during the recent transition period, which has involved land reform and a process of privatisation in almost all of Estonia. However, the Otepää region forms a specific case that is largely different from other rural areas in Estonia – while most rural areas suffer from a severe shortage of financial opportunities, Otepää belongs to the few remarkably able to attract investments, and as a result it is experiencing fast development and elevated construction activity (for example, the number of residential houses in the region has increased by 9.6 percent within the last five years. See Remm & Oja 2003). The reason for this is based on the region's image of being "a nice place", based on landscape as a natural value. Exploring the region's fascinating end-moraine landscape in the interest of nature conservation resulted in a nature park, and the hilly relief makes Otepää the main winter sports center in Estonia – the landscapes have given the area a well-known recreational image. To some extent, different uses of landscape compete with each other for the same spatial resources. To some extent they are complimentary, and certain developments in infrastructure support all activities that are at the same time supported by the imagery of the region.The aim of this study is to assess the role of landscape as the main resource and basis of tourism and the recreation economy in the Otepää region; and to analyse perceptions held by the local people and visitors to the region, of changes in landscape that have occurred over the last decades.

H. Palang et al. (eds.), European Rural Landscapes:
Persistence and Change in a Globalising Environment, 99-112.

2.   LANDSCAPE CONSUMPTION

Whenever we use a resource, we consume it. Still, "consumption of landscape" as a resource has its particularities. These are partly related to the functions of landscape being distinguished slightly differently by many authors (e.g., Farina 2000; Mander & Oja 1999). Much of landscape consumption can be seen as (re)charging one's mind, which is a self-reflecting (recurrent) process related to an understanding of landscape. Landscape has also been considered as a holistic combination of different functions having a meaning, expressing a sense of harmonious, social and aesthetic unity, and reflecting collective memory (Cosgrove 1998). Also, ethics and value choices affect landscape evaluation, resulting in the image of a region having landscape as an essential part of that image (Hanssen 2003).

The functions of landscape seen from another angle imply different aspects of landscape consumption. The predominant ways of landscape consumption in the Otepää region are recreational (for both quieter pursuits and more active sports), educational and ecological/nature conservational. Related to these aspects is the aesthetic value of the landscape (in the case of Otepää, this is much related to the diversity of the landscape and its semi-openness). Also, for local people it is a place to live in (consuming the residential function), and it has been a production area (however, now with a decreasing importance in agriculture and with marginal remnants in forestry). As an ecologically sound area, Otepää to some extent serves also as a compensatory area for industrial parts of Estonia – in social terms, the compensation value is expressed in the recreational use of the region. Other aspects of landscape use in the area are not remarkable (Prede 1997; Sooväli et al. 2003).

Landscape as a resource has certain specific properties. Landscape is a partially renewable resource – being largely the result of natural processes shaping the landscape. Some aspects of landscape are renewed due to natural processes – still, most of them do this so slowly that in a human timeframe we cannot rely on the natural renewability of landscape. Landscape is a resource that can be built up: socio-economic activities have changed most landscapes, there are many elements in landscape that are of anthropogenic origin (both "nice" and "ugly"), and much of the losses in the natural side of the landscape can be "renewed" by human activities. The "ugly" elements in landscape show that landscape as a resource can be degradable – to an even larger extent, this degradability of a resource can be seen in many industrial settings as a leftover side-effect of the consumption of certain mineral resources, for example. Also, much of the landscape consumption – i.e. use of the aesthetic or recreational values by the end-user – is subjective, as is the understanding of what "nice" means, and it depends pretty much on the image one holds of the region. Another way of consuming the landscape is through the development of real estate. In the Otepää region this is an actively ongoing project, meaning a partial build-up of the landscape resource along with a degradation of its natural ecological value and, probably, enhanced residential value. The link between recreational value and real estate development is controversial – improved infrastructure increases the recreational value of the region, but the more developed the area becomes, the less it offers for nature tourism. Consuming the landscape as a resource means using the physical landscape, its representation and, the imagery – the portion of these compounds varies between different aspects of consumption. Furthermore, there is an ongoing process of interaction between changing representations of landscape and the landscape that is thus represented (Olwig 2004).

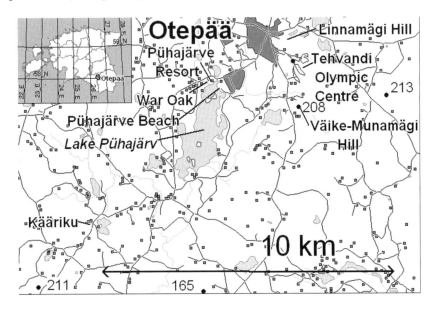

Figure 6-1. The study site.

The changes in the landscape may decompose stable image – the identity of the region. The maintenance of the identity depends to a great extent on the speed of change. We may assume that a critical threshold concerning the speed of change exists in excess of what pushes the reception of landscape into feeling the loss of identity. In turn, the critical threshold for speed of change (the addition of new elements into the landscape) depends much on the landscape type. In places such as Otepää, the hilly natural scenery provides the possibility of absorbing the visibility of changes, and to locate their visibility to a smaller area. The critical threshold for speed of changes is thus higher here than in the case of open and flat scenery, where any addition is seen much earlier and from further away. Maintaining traditional landscape with a cultural and historical heritage is an important prerequisite for sustainable tourism, and strengthens landscape identity values for locals (Sooväli et al. 2003).

3.   THE STUDY SITE

The study site – the Otepää region (58°04'N, 26°30'E) – is known as tourism and recreation area, a winter sports training center situated in South Estonia on the Otepää Upland (Figure 1). Four aspects characterising the Otepää region can be brought out. Firstly, being located on an end-morainic upland formed 15,000-10,000 years ago during the recurrent onset and recession of the glacial sheet covering Estonia has led to the formation of a hilly and diverse landscape with many lakes between the hills (there are 53 lakes in the Otepää Community). For a small group of specialists in geology the area is representative of a type – "Otepää" has become an adverbial to characterise certain geomorphologic formations (Karukäpp 1997). Secondly, it has been a recreation area since the 1920s, when poets and painters chose the shores of Lake Pühajärv (286 ha of water table with five little islands in it and surrounded by a picturesque winding shoreline) – the center of the region – for their summer resorts. The beach of Lake Pühajärv has been assigned the title of a Blue Flag Beach, and it is known for the different cultural events that attract a high number of visitors. The whole area is densely packed with pensions and farms providing tourism accommodation (over 1,600 beds altogether as counted from the web pages). An active vacation has been transformed from a privilege of the selected few into (almost) everyone's right in Estonia, and this has brought along mass tourism. In regards to tourism, the picturesque landscape is supported by historical and cultural values – written notes about Otepää date back to 1116, and most of the present-day 21 villages came into being in the beginning of the 12th century (Otepää 2003). Thirdly, Otepää has been an international cross-country skiing center since the 1950s, when the place with the hilly landscape and long-lasting snow cover became recognised as a perfect area for training. Moreover, the Olympic Center of Tehvandi with its perfect roller track was erected to host the Soviet ski teams in summer. Otepää has been added to the organisers' list of World Cup events in cross-country skiing. The summer activities include an international ski-roller competition (Saku Suverull) involving many of the world's leading ski athletes. An increasing availability of activities and competitive sports combined with general changes in peoples' attitude and lifestyles has led to an upsurge in the number of so-called "Sunday skiers" in the area. Downhill skiing became popular in Estonia in the 1990s, so ski lifts were established on several hills in the area. Among many hills, the previously popular nature tourism attraction – the Väike Munamägi Hill – has been partly changed into a skiing center. Fourthly, most of the area is protected as a nature park (taken under protection in 1957; since 1979 in its present size), Lake Pühajärv (with a park as a conservation object since 1929) and the Väike-Munamägi Hill have been actively exploited as tourism attractions. During the 1960s to 1970s, combined nature and cultural tourism became popular in the area and "created" several "obligatory" visiting sites in the region, such as: Linnamägi, the historical stronghold of ancient Estonians, later a Christian bishops castle; the Väike-Munamägi Hill as the starting point of the River Emajõgi; the health resort on the shore of Lake Pühajärv; the Kääriku sports center of the University of Tartu, etc. All of these attractions involve a certain amount of legends created about them and all have a memorial meaning for many older people, not only locally but all over Estonia. Altogether, the image of Otepää is developing and being updated. It is a mixture of recreation, skiing and nature conservation all making use of the scenery once dropped there by the melting glacier edge.

Tourism, sports, and recreation have been the priority of local government in terms of developing the area since 1990 (Prede 1997). Recreation itself can be considered traditional in the area – however, certain aspects bring more and more elements replicating "what is everywhere else" – such as downhill skiing centers, GSM stations providing communication facilities for tourists, and higher demands from visitors leading to new international hotels. The opening up of Estonian society and the development of fiscal ability provides for the reconstruction of houses and farms, bringing in new internationally-unified and untraditional building materials, thus resulting in a new (globalised) look to the houses. However, much of this can be seen as a normal reconstruction of the region, trimming the landscape from its laxity.

4. MATERIAL AND METHODS

The study included revision of the major historical tourism attractions mostly based on participatory experiment, as the authors have been involved in different aspects regarding the development of the area for decades (including working as tour guides). The results of this analysis were used in selecting particular objects important as key features in the image of the region. These features were used in a questionnaire to capture people's attitudes to changes of landscape and selected landscape objects.

To assess the perception of the Otepää landscape by people (both local inhabitants and visitors), a questionnaire was compiled and used. The objectives of the questionnaire were to understand:
– what makes the image of Otepää as a recreation area;
– why people come to Otepää; how they value the landscapes of Otepää;
– what expectations people have and how they find the reality to match these expectations;
– what the challenges are for the development of the resource and the conflicts around consuming it;
– who is responsible for the sustainable use of the landscapes in the area.

The questionnaires were available during 2000 and 2001 at the tourist information center. They were distributed to housing establishments and given to participants of certain events in the area, etc. Altogether, 136 answered questionnaires were returned. The questionnaire included both questions with suggested possible answers and open-ended ones.

Respondents consist of 50.7 percent female and 47.1 percent male (2.2 percent did not answer the question). 28.7 percent of them were aged under 20, 41.9 percent between 20 and 40 years of age, and 27.2 percent between 40 and 60. 64 percent of the respondents were local, and 36 percent were visitors. In terms of the seasonal timing of visits, 19 percent of the respondents visit the region in summer, 16 percent in winter, 35 percent in both summer and winter, and 13 percent only on weekends. The portion of people that did not respond to the seasonal visitation question is a relatively high – 17 percent.

In trying to assess the representativeness of the material, the number of responses was compared with the number of local inhabitants and visitors. There are over 4,300 inhabitants in the Otepää community (47 percent male and 53 percent female), thus the percentage of responses among local inhabitants reaches two percent (Eesti … 2002; Otepää 2003). Although there might be a slight variability between those who considered themselves local while answering the questionnaire and the formal number of inhabitants in the parish, the number of responses allows one to consider the representation satisfactory. The number of visitors is more complicated to assess, but it is much higher than the number of responses received from visitors. Thus the results have to be interpreted more carefully. Definitely, the total number of responses is too small to group the respondents by several parameters simultaneously. So grouping by only one parameter (either by local-visitor, age group, gender or visit timing) at a time was used in analysing the responses.

Only part of the questions used in the questionnaire are represented in this paper aiming at bringing out the most important tourism attractions as compared to personal favourite sites, state of the sites, and share of responsibility in maintaining the sites. For two selected sites their state and the main problems were more thoroughly studied. All the problems in question reflect different aspects of change in landscape.

5. RESULTS AND DISCUSSION

5.1 The Main Tourism Attractions

Based on the previous experience of the authors the main tourism attractions in Otepää can be put forth as follows. *Nature in general* (hills, lakes, views, beauty of nature, silence, etc.), and selected *natural monuments as favourite places* (Lake Pühajärv, the Väike-Munamägi Hill, the Linnamägi Hill, Lake Kääriku, etc.). *Historical and archeological sights* (the Linnamägi Hill, Otepää Church, ski museum, etc.) represent Estonian cultural memory. *Today's legends/myths, connected with a certain person –* Energy Column, Raidal's House (home and tourist sight in Otepää, personally advertised by the former (in the early 1990s) young radical mayor of Otepää. This is an example of a whole layer of imagery being created within a decade, marked in the physical landscape by half a dozen buildings. *Outdoor activity attractions* (cross-country skiing tracks, slalom slopes, hiking traces, riding areas, etc.) represent certain aspects of consuming landscape as scenery. *Architectural attractions* can be found (Sangaste Castle, modern Otepää Secondary School) in the region. Basically, the same categories of objects are mentioned by the Otepää tourist information officer answering the question about the main tourism attractions in the region (Estravel 2002). Similarly, the high importance of the variable landscape has been evinced in a similar study for the whole of Southeast Estonia (Laas 2001). The same aspects in regards to image come pretty much out also from the questionnaire discussed here.

5.2 Perception of the State of Sites

In answering the question regarding the most important tourism attractions/sights in Otepää, clearly the most favoured is Lake Pühajärv, mentioned by 25 percent of respondents. The Linnamägi Hill (historical stronghold) is mentioned by 15 percent, the Väike-Munamägi Hill (a historical natural monument and ski slopes) by ten percent, the Tehvandi Olympic Training Center by seven percent, the Otepää Church by six percent, the "Energy Column" (a place as a source of positive energy) by five percent, and the Sõjatamm (The War Oak) by five percent. 41 other objects share the remaining 27 percent. These answers probably include a certain amount of "learned" knowledge about what is important in Otepää, i.e. being influenced by the image of the region created by the educational system and media coverage. Somewhat different is the share of objects in terms of personal favourite places/sights. Lake Pühajärv dominates even more with 41 percent, followed by the Linnamägi Hill with nine percent, the Kääriku Sports Center of the University of Tartu near Otepää with eight percent, and the Väike-Munamägi Hill with six percent. Another 28 different preferences share the remaining 36 percent. Remarkably, the list of personal favourites is shorter (32 versus 48) and even more dominated by Lake Pühajärv. Also worthy of mentioning is the importance of the Kääriku sports and recreation center, which people "forget" while mentioning the generally important sites and remember again when thinking about personal preferences. The list of personal favourites probably is influenced by a person's particular experiences (personal subjective perception of image), and of "being used to going somewhere", which may be different from the general image.

In answering the questions about the maintenance of sites two lists are brought out. The least-maintained sights according to the respondents are headed by the Väike-Munamägi Hill (eight cases), the Linnamägi Hill (six), the Hobustemägi Hill (five), and the building of the (formerly popular) Pühajärve restaurant (five). The best-maintained sights start with Lake Pühajärv (18), the Linnamägi Hill (ten), the Pühajärve Resort (eight), the Tehvandi Olympic Training Center (five) and the Pühajärve Beach (five). Finding the Linnamägi Hill on both lists probably demonstrates the high variability of people's expectations. Being located almost in the center of town and having clear historical value, it has been a popular tourist site for years. Remarkably, it is also a clear domination of Lake Pühajärv on the positive side. Furthermore, let us look at two examples of major and historically important tourist attractions.

5.2.1   Case One – Lake Pühajärv

It is the largest lake in the region, a favourite recreation area since the 1920s that hosts large tourist events like *the Beach Party* – a music festival for youth held on the beach in early summer – and the *Goldfish*, an ice-fishing event held in winter; as well, Pühajärve Park has been a nature conservation object for years.

As can be seen from Table 1, visitors evaluate the state of Lake Pühajärv more critically than the local people. Also, elderly people are much more critical than youth, and women are more critical than men. To what extent the fact of the visitors being more critical can be related to their "learned knowledge" or image of the site they have read about in the papers remains speculative. Women being more critical to changes in the environment than men could be related to different perception of space. According to Schmitz (1999) gender differences in environmental knowledge may be related to "stylistic preferences" in the use of different environmental strategies rather than to different competencies (Ward et al. 1986). Women prefer a more landmark-based strategy, whereas men prefer a more configurational (Euclidean) strategy (Galea & Kimura, 1993). Changes in landscape mean loss of landmarks. Men see the loss of view a bigger problem (and view is more important for getting the general direction) whereas once existed hiking trails (with a lot of landmarks designed for wayfinding) being demolished is a much bigger problem for women (Table 2). The elderly people being more critical than the youth could be to certain extent explained by the nostalgia driven "everything was better when we were younger" attitude. However, the share of realising for example, the water quality worsening caused by eutrophication (Prede & Oja 2001) or changes in the surrounding landscapes in unsatisfaction with the state remains unclear. Decrease in the viewability being the main problem for elder people and less important for the younger might refer to remembering the more open surrounding with better views.

Table 6-1. Share of answers (%) to the question evaluating the state of Lake Pühajärv and the near surroundings.

	Perfect	Good	Satisfactory	Bad	No idea
Men	7.8	37.5	32.8	7.8	4.7
Women	1.4	37.7	47.8	0	5.8
<20 yr	103	51.3	28.2	2.6	5.1
20-40 yr	1,8	35.1	38.6	7	3.5
40-60 yr	2.7	27	56.8	0	8.1
Local	6.8	43.2	40.9	4.5	2.3
Other	0	26.5	42.9	4.1	10.2

In asking about the main problem of the site, the total range of answers is quite wide (Table 2).

The biggest problem is seen by almost all groups as being the open view growing into bush lands – you cannot see the open landscape over the lake from many places anymore. The same problem is also demonstrated by mentioning the deconstructed viewing towers. Only younger people mention changes in the shoreline, as the main factor that can be related to their higher willingness to access the lake from anywhere. Privatisation has touched parts of the shoreline with some very hotly disputed objects, but the share of private land closing access to the shore is not high, so change in the natural conditions of the shoreline affects access more.

Table 6-2. Share of answers (%) to the question as to the main problem of Lake Pühajärv. v – spoiled (closed) view; s – shoreline growing into bushes; a – limited access due to privatisation of land; i – infrastructure (loss of once-existing hiking trails and camping sites); b – buildings on the shore; t – towers (GSM) and other technical constructions spoiling the landscape; n – numerous new buildings; f – extensive cutting of forest with resulting change in natural landscape; 0 – no problem.

	v	s	a	i	b	t	n	f	0
Men	32.8	18.8	7.8	4.7	6.3	10.9	7.8	1.6	3.1
Women	23.2	15.9	11.6	13.0	2.9	4.3	8.7	2.9	1.4
<20 yr	17.9	25.6	5.1	17.9	12.8	7.7	10.3	5.1	2.6
20-40 yr	31.6	12.3	14.0	7	1.8	8.8	5.3	1.8	0
40-60 yr	32.4	16.2	8.1	2.7	0	5.4	10.8	0	5.4
Local	30.7	21.6	10.2	9.1	5.7	5.7	9.1	2.3	1.1
Other	18.4	14.3	10.2	8.2	2.0	10.2	4.1	4.1	4.1

5.2.2   Case Two – the Väike-Munamägi Hill

As a second example, the Väike-Munamägi hill was studied. It is one of the main tourism sites, and the source of the Väike-Emajõgi River, with relevant legends leading to it becoming a major visiting site by the1970s. A downhill skiing slope has been on the hill since the mid-1990s – for this purpose half of the slope was deforested, which caused a clear change to the (visible) landscape, and along with that – much opposition. It can be considered one of the most contradictory sites. Again, general satisfaction with the state of the site is expressed in the share of answers (Table 3).

Table 6-3. Share of answers (%) to the question evaluating the state of the Väike Munamägi Hill.

	Perfect	Good	Satisfactory	Bad	Value lost	No idea
Men	4.7	14.1	23.4	23.4	21.9	3.1
Women	4.3	11.6	18.8	18.8	24.6	11.6
<20 yr	15.4	30.8	25.6	10.3	17.9	0
20-40 yr	0	5.3	22.8	26.3	21.1	7
40-60 yr	0	5.4	13.5	24.3	32.4	16.2
Local	6.8	17.0	28.4	20.5	19.3	5.7
Other	0	4.1	8.2	26.5	28.6	10.2
Winter	0	4.5	13.6	45.5	27.3	0
Summer	11.5	23.1	26.9	19.2	7.7	3.8

Again, women are more critical, visitors are more critical, and elderly people are more critical. It is interesting that winter visitors are more critical – the possible explanation for this is that there may be fewer answers given from active skiers, although it may also be a stochastic variability arising from the small number of answers. Analysing the state of the site from a natural science point of view, it is not in that bad a condition at all – the spring starting the Emajõgi River is still there, and while in the forested part access to it has eroded a little, this is not due to the ski slope. Potential for the development of the site is very high and depends mostly on the good co-operation of different uses and different owners, which has been lacking so far. The big problem reflected in the answers might be more a problem of fast change, compared to the image of the site created as a legend during the 1960s-1970s.

The answers about the main problem of the site are partly similar to those mentioned about Lake Pühajärv, and are partly specific problems (Table 4).

Table 6-4. Share of answers (%) to the question on naming the main problem of the Väike Munamägi Hill. f – extensive cutting of forest with resulting change in natural landscape; v – changes in general view; e – erosion; b – new buildings; r – changes in the surroundings of the spring starting the river; l – value is lost; 0 – no problem.

	f	v	e	b	r	l	0
Men	34.4	14.1	12.5	9.4	1.6	9.4	10.9
Women	14.5	20.3	13	4.3	5.8	11.6	8.7
<20 yr	56.4	15.4	17.9	5.1	0	2.6	2.6
20-40 yr	8.8	22.8	12.3	5.3	5.3	10.5	14.0
40-60 yr	13.5	10.8	8.1	10.8	5.4	18.9	10.8
Local	29.5	20.5	10.2	6.8	2.3	12.5	10.2
Other	10.2	14.3	18.4	8.2	6.1	8.2	8.2
Winter	18.2	9.1	9.1	9.1	9.1	18.2	4.5
Summer	34.6	23.1	7.7	7.7	3.8	3.8	7.7

The most important problem mentioned here is clear-cutting, which has affected a big part of the hillside – this is unusual, since many people remember what the hill looked like and refuse to accept a different image of the site. This is also found in the second choice of response – changes in the usual view of the site. Erosion is a problem that can occur on steep slopes with clay soils after damaging the plant cover – however, there is less actual erosion than there are speculations about possible erosion. For some elderly people, the change in the site is so large that they refuse to accept it, and consider the value as being totally lost. They simply know that it is not "the" site any more.

5.3 Sharing the Responsibility

Also, of interest was to understand who should be expected to take care of the landscape, the development of the region and improving its image. Traditionally, the role of the Nature Park and its predecessors has been high in the area, which makes some people see the Ministry of Environment as a responsible actor. The role of local municipalities was negligible during the Soviet times, and is something new for many regions of Estonia.

Table 6-5. Share of the answers (%) to the question as to who is responsible. NP – nature park; LM – municipality; ME – Ministry of Environment; CE – county environmental officers; LP – local people; LO – local organisations; O – other.

	NP	LM	ME	CE	LP	LO	O
Men	26.6	28.1	15.6	6.3	6.3	3.1	0
Women	37.7	29.0	4.3	4.3	4.3	13	1.4
<20 yr	28.2	17.9	10.3	15.4	5.1	17.9	2.6
20-40 yr	28.1	33.3	10.5	0	7	5.3	0
40-60 yr	43.2	32.4	8.1	2.7	2.7	2.7	0
Local	35.2	27.3	9.1	8	5.7	10.2	1.1
Other	26.5	34.7	10.2	0	4.1	4.1	0

The Nature Park administration and the local municipality are expected by respondents to share the biggest part of the responsibility. Elderly people and women see the Nature Park administration as being mainly responsible for the overall image, as it has been historically so. This is also true for very young people. Elderly people are probably more conservative and since they remember the situation that according to them used to be better. Middle-aged people and visitors see the role of municipality being higher (as it probably should be in matters of local development).

6.   CONCLUSIONS

The image of the Otepää region depends mostly on the state of landscape basis of which was once formed by nature and later reshaped over and over again by human activities. The main attraction of the Otepää region is its diverse landscape and natural biodiversity. The landscape serves as an arena for outdoor activities, sports and culture events. The main hot spots of the region are Lake Pühajärv with its environs, the Linnamägi Hill and the Väike-Munamägi Hill. Two equally important tourism seasons – summer and winter – characterise the region, and both require further development of the region. The main factors decreasing the value of the landscapes and worsening the image of Otepää as a recreation area are: 1) socio-economic changes in the last decade in rural life (land reform, privatisation, collapse of the agriculture) that cause afforestation of abandoned cultivated lands, over-cutting of up-forests on good sites, and insufficiently maintained villages and farms; 2) human activities exceeding the tolerance limits of either the natural ecosystem or local social environment – big fests, extended construction activity, waste dumping; 3) a weak nature protection system – weak control and supervision by the state institutions, and weak administration by the Otepää Nature Park. Positive impact concerning the image of the region arises from a balanced development of the whole region, and maintaining the best-known sites and tourism objects of the region in good shape. Therefore, it is crucial to find sustainable management plans for the controversial and in some sense the most damaged sites like the Väike-Munamägi Hill, and to decrease the high human pressure on the Lake Pühajärve area. The responsibility for the development has to be taken by the Otepää Nature Park and the local government.

Physical landscape dictates much of the imagery of the landscape in the region; however, certain aspects of the landscape imagery are subjective and fragile depending on how they are presented and how they are affecting the degradability and renewability of physical landscape.

ACKNOWLEDGEMENTS

This paper was supported by the Estonian Ministry of Education, SRT 0181788s01, the support scheme to doctoral students and Estonian Science Foundation, grant No. 5261. Discussions and critical remarks by Helen Sooväli and Hannes Palang while preparing the manuscript are thankfully acknowledged.

REFERENCES

Cosgrove, D. (1998). Cultural Landscapes. In T. Unwin (Ed.), *A European Geography* (pp. 65-81). London: Longman

Eesti Linnad ja Vallad Arvudes (2002). Tallinn: Statistikaamet.

Estravel (2002). Eesti siseturism http://www.estravel.ee/estraveller/jul_sept2002/-kysitlus.html, 20.12.2003.

Farina, A. (2000). *Landscape Ecology in Action*. Dordrecht, Boston, London: Kluwer Academic Publishers.

Galea, L.A.M. & Kimura, D. (1993). Sex Differences in Route-Learning. *Personality and Individual Differences*, 14, 53-65.

Hanssen, B.L. (2003). Ethics and Landscape, Values and Choices. In T. Unwin & T. Spek (Eds.), *European Landscapes: From Mountain to Sea* (pp. 201-209). Tallinn: Huma Publishers.

Karukäpp, R. (1997). *Gotiglatsiaalne morfogenees Skandinaavia mandriliustiku kagusektoris*. Dissertationes Geologicae Universitatis Tartuensis, 6. Tartu: Tartu Ülikooli Kirjastus.

Laas, A. (2001). *Lõuna-Eesti mainekujunduse uuring*. Tartu: SA Lõuna-Eesti Turism, OÜ Laas&Laas.

Mander, Ü. & Oja, T. (1999). Eesti maastike omapära ja sellest tulenevad ökoloogilised iseärasused. In T. Frey (Ed.), *Loodusliku mitmekesisuse kaitse viisid ja vahendid* (pp. 14-30). Tartu: IM Saare.

Olwig, K. (2004). "This is not a Landscape": Circulating Reference and Land Shaping. *This volume.*

Otepää (2003). Otepää valla koduleht, http://vald.otepaa.ee/index_et.phtml. 20.12.2003.

Prede, M. (1997). Landscape Protection in Estonia: The Case of Otepää. In J.G. Nelson & R. Serafin (Eds.), *National Parks and Protected Areas: Keystones to Conservations and Sustainable Development. Proceedings of the NATO Advanced Research Workshop on Contributions of National Parks and Protected Areas to Heritage, Conservation, Tourism and Sustainable Development, 1996* (257-264). Vol.40. Berlin Springer.

Prede, M. & Oja, T. (2001). Sanitation of Lakes in Otepää for landscape Restoration. In Y. Villacampa, C.A. Brebbia & J.L. Usó (Eds.), *Ecosystems and Sustainable Development III. Advances in Ecological Sciences* (pp. 605-614). Southampton, Boston: WIT Press.

Remm, K. & Oja, T. (2003). Stepwise Modelling of Rural Housing Pattern Near Otepää Using Neighbourhood Corrections. *Ecological Modelling.* (submitted).

Shmitz, S. (1999). Gender Differences in Acquisition of Environmental Knowledge Related to Wayfinding Behavior, Spatial Anxiety and Self-Estimated Environmental Competencies. *Sex Roles: A Journal of Research*, July, http://www.findarticles.com/cf_0/m2294/1_41/57590493/p1/article.jhtml. 20.12.2003.

Sooväli, H., Palang, H., Alumäe, H., Külvik, M., Kaur, E., Oja, T., Prede, M. & Pae, T. (2003). (Traditional) Landscape Identity – Globalized, Abandoned, Sustained? In E. Tiezzi, C.A. Brebbia & J.L. Usó (Eds.), *Ecosystems and Sustainable Development*, Vol. 2 (pp. 925-935). Southampton, Boston: WIT Press.

Ward, S.L., Newcombe, N., & Overton, W.F. (1986). Turn Left at the Church, or Three Miles North. A Study of Direction Giving and Sex Differences. *Environment and Behavior*, 18, 192-213.

Chapter 7

COUNTRYSIDE IMAGERY IN FINNISH NATIONAL DISCOURSE

Maunu Häyrynen
School of Cultural Production and Landscape Studies, University of Turku, Finland

1. INTRODUCTION

This article is based on a notion of national landscape imagery understood as a visual signifying system, representing national space and linking physical sites with national ideology. According to Stuart Hall, national identity is produced by a cultural system of representation essentially based on stereotypes. These are conventional by nature, yet have the power of directing individual thought and action (Hall 1992; Moscovici 1984).

Landscape imagery forms a part of the national representational system. The power of representations is derived from their rhetorical evocativeness, wherein verisimilitude plays an important part. This is certainly the case with landscape, which as a convention claims truthlikeness, hiding its social and historical origin (Moscovici 1984; Green 1991; Mitchell 1994). Its meaning is nevertheless defined within a mesh of references to both other landscapes and regions and territories, and to other discourses such as those of history, nature or national ideology.

Imagery arranges the visual representations into a spectacle. This entails their framing, classifying and organising into hierarchical taxonomies and canons (Duncan 1993; Baehr & O'Brien 1994). Imagery is ultimately an apparatus of power, justifying existing spatial orders and suppressing others. It counterposes the centre and the periphery, the country and the city, defining the national identity in terms of the inclusion or exclusion of others (Mitchell 1994; Helsinger 1997).

H. Palang et al. (eds.), European Rural Landscapes:
Persistence and Change in a Globalising Environment, 113-122.

The empirical study behind this article set out to examine the national landscape imagery, understood in the sense discussed above, as it appears in popular publications dealing with Finnish landscape (see Häyrynen 2000). The empiria thus defined consists of fifty publications, dating back from the early 19th century to the 1960s, and comprising about six thousand landscape images. These have been analysed in terms of their spatial references (locality, region, generic landscape types), as well as in terms of themes, such as historical, peripheral, urban and rural imagery. The material has been roughly divided into three periods – the autonomy under Russia, the time between the two World Wars, and the post-war period. This is not to say that national landscape imagery developed in a uniform fashion over each period; rather, that the macro-historical turning points have also changed the context regarding the interpretation of imagery.

Within the space of this article, it will not be possible to present the theoretical stipulations and empirical results of the background study. Suffice it to say that Finnish national landscape imagery shows considerable structural unity, starting from its early formation in the mid-19th century and lasting well into the post-war era. This unity is paralleled by the relative homogeneity of Finnish national culture, ensured by the strong position of the national movement on the one hand and by the central role of the state on the other. Thus, the strength of cultural hegemony has always been considerable, if not uncontested, and has only been broken during major historical points of rupture. However, national landscape imagery has never been monolithic. Neither has it been an independent cultural formation, but rather it has been a framework of representation, readily utilised by the dominant constellations of each period.

2.   THE POSITION OF RURAL IMAGERY IN FINLAND

The article at hand concentrates on the theme of rural landscape imagery. It aims at sketching out how countryside has been constructed as a particular topos of Finnish nationalistic landscape imagery. It will argue that the different aspects of imagery, such as rural landscape, have played different narrative roles within this landscape imagery, and that these aspects may have drastically changed over time. Their ideological dimension has been the colonisation of national space, as well as the naturalisation of the national community. Narrative models such as the representation of countryside as an essentially national *heartland* show similar traits to other nation states, reflecting a common nationalistic vocabulary employed in the demonstration of national uniqueness, as will be illustrated in the context of Finnish countryside imagery.

The topographic identity of Finland has often been described as based on natural landscape features. During the early national movement in the 19th century, panoramic lake scenery was grasped as a geographically unique phenomenon that allegedly contributed to the formation of Finnish folk character. At the turn of the 19th and 20th century, it was accompanied by the Karelian wilderness, at that time presented as the cradle of original Finnish culture. Later on, the area of Lapland acquired an increasingly significant position in the visual definition of Finland. Thus it would seem that Finns as a nation principally identify themselves with nature, and a rugged and peripheral nature at that (Häyrynen 2000).

However, another prevalent idea about the formation of the Finnish national identity has been that it has promoted peasant culture and values as the fundament of national community. In early Finnish-language literature, as well as in the activity of the Fennoman political party, the focus was placed on the Finnish-speaking peasant population in general, and on freehold farmers more specifically. The fatherland of the national movement was equalled with cultivated land. This is reflected even in the present-day culture, as even post-industrial urban Finns may perceive themselves as still rural or even backwoods people under the surface, trying to live up to agrarian ideals such as thrift, perseverance and firmness (Paasi 1986; Apo 1996).

There is yet a third image of Finland, namely that of its urban core. Both natural and agrarian imagery are urban creations, projections of a polite visual order on the rest of the country, as has been demonstrated by Leo Marx and many others. Marx speaks of "pastoral design", a structure of thought and feeling comprising the ideal, counterbalancing the idyllic vision of the countryside with its more gruelling aspects. In the American literary tradition, the countryside has played the part of a middle landscape between the corrupted cities and the untamed wildernesses, offering a setting for a harmonious society in the New World (Marx 1964; see also Cosgrove 1984). By following the writings of Denis Cosgrove, one may distinguish the three layers of mythological landscape in Finnish landscape imagery: the *city* or the urban core; the *garden* or the intermediate cultural landscape; and the peripheral *wilderness*. The middle landscape represents the harmony between the society and nature, enabling its interpretation as national landscape (Cosgrove 1993).

3.   STUDYING THE REPRESENTATION OF THE COUNTRYSIDE

The representation of the Finnish countryside has not been systematically analysed in a social context. Here the critical vein of British landscape studies may offer some tools for analysis, although it goes without saying that both the social and the artistic development of Finland have differed substantially from that of Great Britain. It is not intended here to show historical parallels between the two countries, but to suggest new approaches for the interpretation of Finnish rural imagery.

According to James Turner, the topographical image of a landed estate – the "propriety landscape" – became an allegory of a well-ordered society in 17th-century England. This served as a means of naturalising the social order into an ahistorical pastoral harmony, a return to the lost "Golden Age". In this ideal landscape, the rural community was eclipsed or marginalised or alternatively theatrically idealised (Turner 1979). As Malcolm Andrews notes, the 18th-century fashion of the Picturesque landscape marked a shift from the preceding moralised image of the countryside to an aestheticised one. This change of attitude highlighted the natural features and rejected all signs of work, rural or industrial. The Picturesque representation of rural poverty nevertheless offered visually interesting items. Even more importantly, it set standards for the visual appreciation of "ordinary" countryside, based on artistic pictorial models (Andrews 1989).

Elizabeth Helsinger has shown how rural views became a way of representing the national within the local in 18th-century England. Unlike the previous map-like landscapes of property, these stressed the opposition between the urban and the rural. At the same time, they replaced earlier idealisations with accurate renditions of local characteristics, presenting the nation as scenically diverse. Private ownership was thus turned into a collective national proprietorship (Helsinger 1997). Similar development could already be discerned in the 17th-century Dutch series of landscape prints that represented the rural surroundings of major cities. These lacked the magisterial attitude of the contemporary English landscapes of property and offered a more generalised view of the countryside, which gradually assumed the status of a Dutch national landscape. Even here, however, the approach was not rural – it was that of the bourgeois urban city-dweller (Alpers 1987; Andrews 1999).

Stephen Daniels presents the well-known *Hay Wain* of the painter John Constable (1821) as an epitome of the English rural landscape, raised into the status of a national icon. There the countryside appears as a timeless pre-modern setting, populated by a harmonious *Gemeinschaft* and unaffected by the ongoing industrial and urban revolution – corresponding to the imagined national community of Benedict Anderson (Daniels 1993; compare with Anderson 1983). Finally, Nicholas Green has traced the origin of the modern rural imagery to 19th-century Paris, where it constituted a metropolitan "spectacle of nature" based on the commodification of the countryside. This would produce a consumption-oriented new reading of the rural landscape, converting it to visual images of holiday resorts, scenic routes and *maison de campagne* settings (Green 1991).

Next, I shall sketch the development of Finnish countryside imagery, which occurred in a very different context from the British or Dutch ones and must therefore be examined against the background of both the Finnish national movement and the socio-economical situation of the Finnish countryside.

4. THE FINNISH COUNTRYSIDE AND ITS IMAGE

Miroslav Hroch has – in his path-breaking study of national movements among the small East European nations – noted the importance of the peasantry in relation to what he terms the second phase of national mobilisation, during which a limited group of patriots strived to promote national ideology. Transition to the next phase, i.e. political mass support for the national cause, would have been impossible without first winning over the peasant population. However, in the particular case of Finland, the social background of the second-phase patriots was predominantly urban intelligentsia, concentrated in the larger cities of the south (Hroch 1985).

The political programme of the Fennoman party relied heavily on promoting the Finnish language, establishing a national culture based on it, and civilising the Finnish-speaking peasant population by means of a general primary school system and association activities; which would also serve to integrate them into the state administrative system. The undertones of the Fennoman movement were paternalist and anti-urbanistic, maintaining the tradition of the corporativist society. Their national cultural models were the toiling and God-fearing peasants, as promoted in early Finnish literature and paintings (Liikanen 1995).

The peasant population was the target, rather than the origin, of rural imagery. The aim focussed especially on the freehold farmers, who were entitled to political representation; leaving out the growing numbers of landless population. Even so, it was the Fennoman intellectuals rather than the freeholders who voiced political opinion, the former claiming the position of a mouthpiece for the latter (Liikanen 1995). While the contents of Fennoman politics were drawn from below, from the everyday life and popular culture, its form originated from above, i.e. the conceptual constructions of nationalism (Anttila 1993). It was at this time that the peasant values were redefined as being national, assuming the significance they still hold in today's Finland.

When looking at the evolution of Finnish national landscape imagery, it becomes obvious that its principal stages correspond to the phases of national mobilisation as observed by Hroch. The elitist patriotic imagery of the Fennofile period – viewing the countryside from an idealising distance – became replaced by more openly ideological representations of the Fennoman movement that portrayed the peasant population as the source of national identity. Yet the prevailing image of the countryside remained a polished Sunday view, lacking the representation of its inner experience and contradictions. These became registered only after the shock of the Civil War in 1918, which led to a new image of agrarian nationalism.

5.   THE HÄME REGION AS AN EXAMPLE OF THE IMAGINED HEARTLAND

The Finnish peasant culture was not unisonal, which was acknowledged by the early nation-builders. On the contrary, the diversity of regions and local characteristics was highlighted in the literature and landscape imagery as part of their integration into a national whole. A key role in this portrayal was played by the national writer and historian Zachris Topelius, who presented the historic regions of Finland as a basis for its identity as a nation. This did not prevent some of these regions from appearing more Finnish than others, and it was the people of the inland Häme, or Tavastland region, who Topelius eventually came to describe as the "average Finns" (Paasi 1986).

The stereotypic image of the people of Häme depicted a steadfast and tenacious, if also conservative and a bit slow-witted, independent peasant. As Topelius saw it, they had remained unspoiled by the foreign influences and urban ways of the coast (Topelius 1875). Häme had been an area where certain vestiges of the old peasant society, such as folk traditions or group villages with common fields, had persisted longer than elsewhere (Rasila 1979; Varpio 1979). In the 19th century, however, the region of Häme formed a part of the unified core area of southern Finland, which was predominantly agricultural but rapidly industrialising. Especially after the building of the country's first rail connection from Helsinki to Hämeenlinna, Häme constituted a hinterland and a resource pool for the capital (Alapuro 1988).

In early 19th-century popular landscape imagery, Häme was depicted mostly in terms of its urban centres and landed estates, with some lake views from along the roads connecting them thrown in for good measure. This situation changed as Häme became a regular subject of Finnish paintings and literature. Several of the writers and painters even chose the area as a location for their National Romantic timber villas towards the end of the century (Reitala 1979). The images of estates and historic monuments gave way to panoramic views of tranquil lake scenery, picturesque cottages and villages, as well as folk characters. All in all, Häme was presented as an unchanged heartland, where the peasant community coexisted in harmony with the beauty of nature (accompanied to some extent by the adjacent inland region of Savo). There is a strong resemblance here to the representation of the Dalecarlia region as the national heartland of Sweden, by the artist Carl Larsson and others (Facos 1998).

Against this idyllic background, it may appear surprising that Häme was an area of aggravated conflict between the landowners and the landless population, having had the highest percentage of croft or tenant farmers in the country. During the Civil War of 1918, rural Häme decidedly joined the ranks of the Red Guardists. After the victory of the White government troops, this shattering of the image of rural harmony was not easily overcome. The rebellion was explained as a "volcanic" action, comparable to a natural catastrophe (Alapuro 1988).

The agrarian *Gemeinschaft* ideal was revitalised in post-independence imagery and literature, amplified by the land reform that had been carried out in order to pacify the landless population, and which had turned them into smallholders. Rural imagery of this period was less concerned with ethnographical details, and more focused on the peasant population itself, above all on its healthy-looking and racially-pure exemplars. This, combined with the weakening of didactic historical imagery and the rise of militaristic imagery, constituted what might be termed as pre-war "agrarian nationalism".

After the Second World War, rural imagery underwent another change in meaning. After the 1940s imagery of national recovery and rebuilding, the process of urbanisation gradually turned the Finnish countryside into an object of nostalgia, a place of origin left behind, only to be returned to during summer holidays. This period meant the commodification of the countryside, its imagery turning into a politically-neutralised spectacle for urban consumption – rural landscape as an amenity.

To sum up, throughout the period of study the proportional volume of rural imagery in general, and that representing Häme in particular, remained the same; contrary to the fluctuating imagery of the periphery. At the same time, rural imagery was clearly overshadowed by urban imagery. It appears that rural heartland imagery has been a necessary but not overwhelmingly exciting part of national baggage, a stable backdrop for the changing urban and industrial scene.

6. CONCLUSION

It appears that in Finland, the propriety landscape of the landed gentry was followed by an aestheticisation of the working countryside, masking the social inequalities in a manner resembling a Picturesque painting. What came next was a "realistic" appropriation of rural landscape in all its diversity, maintained from an urban bourgeois point of view as a source of national identity. The most recent phase has been the construction of a consumerist image of the countryside, largely devoid of social tensions or overtly ideological contents.

The image of the countryside in all its cliché-likeness seems to be a crucial element of Finnish national imagery at large. It serves to attach the abstract idea of nation to the concrete ground of rural society and scenery, whatever that may be. At the same time, it frames and highlights particular aspects and features of the countryside, while hiding or marginalising others. The concept of a Georgic *heartland* is one necessitated by the structural logic of national imagery. Together with similar concepts of *borderland*, *core* etc., it serves to establish an illusory hierarchy of places that legitimises the new spatial order imposed by the nation state.

The imagined countryside occupies the middle ground between the urban and the wilderness. Cities are often represented as proof of national achievement, but they are also cosmopolitan links to the world outside. The represented countryside acts as an "authentic" contrast to the city – it constitutes a setting dwelt upon and shaped by the allegedly pure and unspoiled inhabitants, as well as being a repository of tradition. It also contrasts to the wilderness as an inhabited, cultivated and civilised milieu, bearing traces of cultural continuity. The image of the countryside is an intellectual creation, but sufficiently anchored to physical and social reality to lend it credibility even among the rural population.

Neither national nor rural imagery is wholly logical and one-dimensional, but rather they entail various shifts in meaning. An interesting point of conflict is offered by the propriety landscape, which – though evidently rural and a long-time bearer of cultural continuity – was dismissed as a national symbol in Finland. Another may be found from the small towns, which had neither proper urban pretensions nor a truly rural character. The definition of "real" countryside has varied both in terms of extension and location. Even the perceived visual continuity of rural scenery has proven deceptive – a 19th-century crofter community or a 20th-century smallholder village differ drastically from a present-day EU-subsidised large-scale farming landscape, even when occupying the same site. Yet rural imagery lives on, now fuelling landscape conservation as a desperate attempt at retaining fragments of the deep countryside – a fundament held standing by props.

REFERENCES

Alapuro, R. (1988). *State and Revolution in Finland*. Los Angeles, London, Berkeley: University of California Press.

Alpers, S. (1987). The Mapping Impulse in Dutch Art. In D. Wood (Ed.), *Art and Cartography: Six Historical Essays*. Chicago: Chicago University Press.

Anderson, B. (1983). *Imagined Communities. Reflections on the Origin and Spread of Nationalism*. London: Verso.

Andrews, M. (1989). *The Search for the Picturesque. Landscape Aesthetics and Tourism in Britain, 1760-1800*. Aldershot: Scholar Press.

Andrews, M. (1999). *Landscape and Western Art*. Oxford & New York: Oxford University Press.

Anttila, J. (1993). Mitä on suomalaisuus? In T. Korhonen (Ed.), *Käsitykset suomalaisuudesta – traditionaalisuus ja modernisuus* (pp. 108-134). Helsinki: Suomen antropologinen seura.

Apo, S. (1996). Agraarinen suomalaisuus – rasite vai resurssi? In P. Laaksonen & S.-L. Mettomäki (Eds.), *Olkaamme siis suomalaisia*. Kalevalaseuran vuosikirja 75–76 (pp. 176-184). Helsinki: Suomalaisen kirjallisuuden seura.

Baehr, P. & O'Hanlon, M. (1994). Founders, Classics and the Concept of a Canon. *Current Sociology*, 42 (1), 1-49.

Cosgrove, D. (1984). *Social Formation and Symbolic Landscape*. London: Croom Helm.

Cosgrove, D. (1993). Landscapes and Myths, Gods and Humans. In B. Bender (Ed.), *Landscape, Politics and Perspectives* (pp. 281-306). Providence/Oxford: Berg.

Daniels, S. (1993). *Fields of Vision. Landscape Imagery and National Identity in England and the United States*. Cambridge & Oxford: Polity Press.

Duncan, J. (1993). Sites of Representation: Place, Time and the Discourse of the Other. In J. Duncan & D. Ley (Eds.), *Place/Culture/Representation* (pp. 39-56). London & New York: Routledge.

Facos, M. (1998). *Nationalism and the Nordic Imagination. Swedish Art of the 1890s*. Berkeley, Los Angeles, London: University of California Press.

Green, N. (1991). *The Spectacle of Nature. Landscape and Bourgeois Culture in Nineteenth-century France*. Manchester & New York: Manchester University Press.

Hall, S. (1992). The Question of Cultural Identity. In S. Hall, D. Held & T. McGrew (Eds.), *Modernity and its Futures*. London: Sage.

Helsinger, E. (1997). *Rural Scenes and National Representation. Britain, 1815-1850*. Princeton, NJ: Princeton University Press.

Hroch, M. (1985). *Social Preconditions of National Revival in Europe. A Comparative Analysis of the Social Composition of Patriotic Groups Among the Smaller European Nations*. Cambridge University Press, Cambridge.

Häyrynen, M. (2000). The Kaleidoscopic View: The Finnish Nationalistic Landscape Imagery. *National Identities 2* (1), 5-20.

Liikanen, I. (1995). *Fennomania ja kansa. Joukkojärjestäytymisen läpimurto ja Suomalaisen puolueen synty*. Helsinki: Suomen Historiallinen Seura.

Marx, L. (1964). *The Machine in the Garden: Technology and the Pastoral Ideal in America.* New York: Oxford University Press.

Mitchell, W.J.T. (1994). Introduction. In W.J.T. Mitchell (Ed.), *Landscapes of Power* (pp. 1-4). Chicago & London: University of Chicago Press.

Moscovici, S. (1984). The Phenomenon of Social Representations. In R.M. Farr & S. Moscovici (Eds.), *Social Representations* (pp. 3-70). Cambridge & New York: Cambridge University Press.

Paasi, A. (1984). *Kansanluonnekäsitteestä ja sen käytöstä suomalaisissa maantiedon kouluoppikirjoissa: tutkimus alueellisista stereotypioista.* Joensuu: Joensuun yliopisto.

Paasi, A. (1986). *Neljä maakuntaa. Maantieteellinen tutkimus aluetietoisuuden kehittymisestä.* Joensuu: University of Joensuu Publications in Social Sciences 8.

Rasila, V. (1979). II Katsaus Hämeen historiaan. In *Häme taiteilijoiden kuvaamana 1818–1940.* Tampereen taidemuseo 13.6.-16.9.1979 (pp. 49-67). Tampere: Tampereen taideyhdistys.

Reitala, A. (1979). IV Hämeen kuva taiteessa. In *Häme taiteilijoiden kuvaamana 1818–1940.* Tampereen taidemuseo 13.6.-16.9.1979 (pp. 103-129). Tampere: Tampereen taideyhdistys.

Topelius, Z. (1875). *Boken om vårt land. Läsebok för de lägsta läroverken i Finland.* Helsingfors: G.W. Edlund.

Turner, J. (1979). *The Politics of Landscape. Rural Scenery and Society in English Poetry 163-1660.* Oxford: Basil Blackwell.

Varpio, Y. (1979). III Kirjallisuuden hämäläinen maisema. In *Häme taiteilijoiden kuvaamana 1818–1940.* Tampereen taidemuseo 13.6.-16.9.1979 (pp. 68-102). Tampere: Tampereen taideyhdistys.

Chapter 8

RELIGIOUS PLACES – CHANGING MEANINGS.
THE CASE OF SAAREMAA ISLAND, ESTONIA

Taavi Pae & Egle Kaur
Institute of Geography, University of Tartu, Estonia

1. INTRODUCTION

The reverence for places (e.g., religious sites) cherished by native people is increasingly being disturbed and replaced by more pragmatic notions and purposes, mostly connected to tourism. How does this change influence the identity of a place? In order to demonstrate the possible effect of such a turn, we looked at the example of churches on Saaremaa Island – the biggest island of Estonia, situated on the west coast (see Figure 1). We studied the ransformation of religious sites, from their symbolic roles in the past to their contemporary roles.

Before discussing how sacred places and associated meanings have changed on Saaremaa, some related theoretical concepts will be discussed.

For centuries, Christianity has shaped the European landscape. It was first expressed in landscape by the basilica and campanili, such as those of Ravenna in the west and the mushroom domes of Byzantium in the east (Jellicoe & Jellicoe 1995). Illustrating the influence of Christian culture on landscape, Unwin (1998) highlights the formalisation of a congregation[1] and the creation of strong symbolic elements in the landscape. In medieval times, when the majority of buildings were constructed of wood, the firm structure of stone-built churches was one of the dominant elements of the visible and spiritual landscape. Even today, in many rural parts of Europe the church remains the central and most visible landmark in most villages. Kouleshova

---

[1] With *congregation* we mean a group of people who attend the church, without any territorial or administrative functions.

H. Palang et al. (eds.), European Rural Landscapes:
Persistence and Change in a Globalising Environment, 123-135.
© 2004 *Kluwer Academic Publishers. Printed in the Netherlands.*

(2003) describes the impact of monastic culture on the appearance and functions – as well as on the semantic features – of the landscape on the Solovetski Islands. Temples, chapels and small monasteries became focal points in the landscape: as a result, travelers most often moved to an area within sight of a sacral symbol.

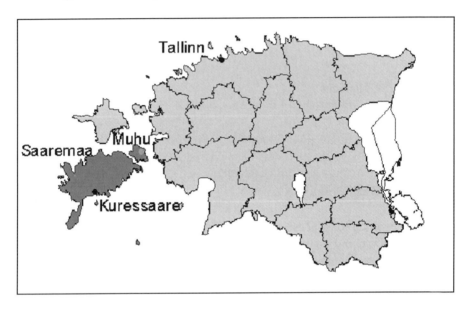

Figure 8-1. The location of Saaremaa Island, case study area.

The way people feel about religious establishments is connected with their understanding of sacred place. Sacredness by no means applies exclusively to religious sites – it may also involve ordinary places that are ritually made extraordinary or valued for some specific quality. Religion is just one among many possible sources of the sacred (Demerath III 2000). However, officially-sacred sites such as churches and temples receive more attention than informally revered ones (e.g., indigenous sacred sites). In many places, a blend of pagan traditions and formal religious practices occurs. Unwin (1998) points at local religious festivals – especially in Catholic countries – that continue to reflect the surviving importance of folk religion, and the way in which Christianity in its earliest forms subsumed previous religious expressions into its culture. Paulson (1997) argues that in Estonia, the Christian affiliation and customs have induced a number of phenomena operating synchronous with rituals of Estonian folk religion. The meaning of a holy place must be understood not in terms of the place itself, but in terms of the social and religious practices of the local communities and the identities generated by those activities (Kong 2001).

1.1   Religious Background of Saaremaa

Although the ethnic composition of the population of Saaremaa is uniform, religious convictions are diverse. The Christian background of the island of Saaremaa is rather peculiar, combining the Lutheran and Byzantine culture. Altogether there are 15 Lutheran congregations and 14 Orthodox congregations on Saaremaa. The location of the churches of Saaremaa is presented on Figure 2.

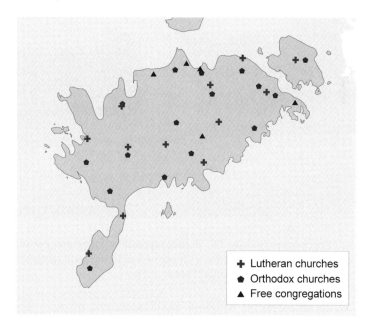

Figure 8-2. The location of churches on Saaremaa.

The native people of Saaremaa were extensively Christianised after they lost their fight for independence in the first half of the 13th century. The island as well as the whole country became submitted to German rule, which lasted – despite frequent rebellions – until 1559. The new authorities erected a stone chapel in Valjala in the central part of Saaremaa, the walls of which form the lower section of the present church choir. It is the oldest known stone building in Estonia. Art historians consider that it marks the beginning of Estonian monumental architecture (Alttoa 1997).

The consequences of the conquest and Christianisation can be evaluated differently; but undeniably as a result of these events Estonia joined the Western European cultural arena. Contacts with the cultural centres of Western Europe are best demonstrated by medieval art and architecture. The churches of Saaremaa are fine examples of that (Alttoa 1997).

Compared to other regions of Estonia, there are relatively many medieval churches on Saaremaa. The churches from the 13th century appear in striking contrast to the rustic surroundings of the island and embellish the landscape (Figure 3).

Figure 8-3. The churches from the 13th century appear in striking contrast to the rustic surroundings of the island. St. Katherine's Church of Karja is the most visited church on Saaremaa. Photos: T. Pae.

In the 1840s an extensive religious conversion took place in the areas of Ancient Livonia. In total, about one-third of the population found its way to the Orthodox Church. Saaremaa was the most active region in terms of accepting the Orthodox Church. Different studies suggest opposing views on the motives of such religious conversion. According to Plaat (2001), the religious reasons receded into the background. In the middle of the 19th century, when Estonia was a part of the Tsarist Russian Empire and the Tsar implemented persistent Russification, rumours spread that those adopting the Tsar's faith would be given land. Still, most likely the conversion stimulus was not linked to practical reasons only. The transition from a "masters' church" to a "peasants' church" became a popular way of expressing dissatisfaction with the German nobility.

In Northern Europe, Saaremaa can be considered as the westernmost region influenced by Byzantine culture through religion. Most of the Orthodox churches of Saaremaa were built around 1870. The Orthodox buildings are more colourful than the modest Lutheran churches. Their onion-shaped domes and spectacular towers form the most important eye-catchers for tourists; foreigners in particular (Figure 4). Nevertheless, the oldest medieval churches are among the most visited tourist destinations on the island.

Figure 8-4. Orthodox churches with their onion-shaped domes and spectacular towers diversify the rural landscape of Saaremaa. Photo: T. Pae.

In 1922, the share of the population in Saaremaa that considered themselves religious was 99 percent, of which Lutherans made up 57.9 percent and Orthodox people 39.6 percent (Marksoo 2002). According to various sources, the number of believers today lies between 31-38 percent of the population, of which the share of Lutheran is 18.0-23.9 percent and that of Orthodox 5.0-8.9 percent (Marksoo 2002).

Even as the modern retreat of religiousness reflects a general pattern in Estonia and other newly-independent countries of the former Soviet Union, the despotic alienation of people from church by authorities during the Soviet period had a decisive role in shaping the current situation of broken traditions and transformed meanings. Due to the large-scale atheist propaganda and hostile attitude towards church during Soviet times, church membership decreased drastically and Christian values became lost (Altnurme 2001).

1.2 Multiple Uses of Churches

Tourism is nowadays understood as a significant and signifying cultural phenomenon and landscape is considered as a mirror of produced meanings reflecting beliefs, values and social ideologies of people (Urry 2002). Tourism transforms landscapes, both physically and mentally. Antonen (2003) argues that along with the mental changes the associations to landscapes are transforming. Landscapes become tourist places through meanings and practices ascribed to them by tourism industry. The transforming situation in regards to religious convictions and the changing utility of sacred places has brought up debates on the economical exploitation of religious symbols. This includes commercial use in relation to Christian times of feasts, such as Christmas and Easter. Likewise, religious places have been promoted as famous tourist sites and highlights in terms of cultural sightseeing (Raivo 1997). Many tourist guidebooks and postcards all over the world outline buildings such as churches and monasteries as major places of interest. The appropriateness of this highlighted interest in religious sites can be questioned, as secular visits to sacred places sometimes result in conflicts with local worshippers.

2. THE SAAREMAA CASE STUDY

2.1 Data and Method

A survey was carried out on Saaremaa in 2002 in order to identify the actual meaning of churches in local communities, and to see how tourism affects places of traditional Christian worshipping. Open in-depth interviews were done with pastors and priests of Saaremaa, as they represent the best available pool of knowledge on the subject in concern. The representatives of multiple congregations were selected, in order to include a variety of locations and as many different views as possible. Altogether, representatives of six Orthodox congregations and 10 Lutheran congregations were interviewed (for a total of 16), including the dean of the Islands' Deanery of the Estonian Evangelical Lutheran Church. Interviews covered the following topics: (1) the significance of church for local community, (2) the opposition between different functions of church buildings, (3) the economic and operational management of churches, (4) and the pressure of increased tourism on local life and identity. The data from interviews were analysed by content analysis, a research technique that is often applied for processing data obtained via open-ended questionnaires or interviews (Holsti 1969). The answers were categorised according to topic, and within each topic the units of analysis (i.e. meaningful parts of the communication content) were identified and summarised according to the frequency of references to the identified categories within each topic (Table 1).

Table 8-1. The identified content categories and frequency of references.

	Frequency of references		
	Lutheran	Orthodox	Total
TOPIC 1. SIGNIFICANCE OF CHURCH			
Historic importance	10	6	16
Clerical ceremonies	10	6	16
Scenery value	4	6	10
Architectural value	6		6
Social centre (education, culture)	4		4
TOPIC 2. OPPOSITION BETWEEN DIFFERENT USES OF CHURCH			
The sacred vs. commerce	6	6	12
Tourism vs. congregation activities	5		5
TOPIC 3. MANAGEMENT OF CHURCH			
Significance of activeness of local pastor	4	6	10
Tourism donations	9		9
Congregation membership fee	2	6	8
Secondary activities (accommodation, selling souvenirs etc.)	6		6
TOPIC 4. PRESSURE OF TOURISM ON LOCAL LIFE			
Significant when not regulated	9		9
Insignificant	1	6	7

2.2 Results

The results show a clear unity of opinions among congregation representatives. While only 50 to 100 years ago the main role of churches was to serve local communities as shrines and centres of cultural, social and welfare activities, at present many churches do not assume this function. Moreover, many have become tourist attractions. The decrease in the number of religious practitioners and the associated decline in reverence for churches is a common development. While elderly people still perceive religion as a natural part of their lives, the majority of young people claim to have no ties to church. The role of church as a shrine, as a place for spiritual experience, and as a landmark shaping the identity of the local community has been lost.

In the summertime, many of the Lutheran churches of Saaremaa receive many more tourist visitors per week than there are congregation members going to the churches on Sundays. Some easily accessible churches receive numerous people daily, while others more distant from main tourism routes – Orthodox churches in particular – remain closed to visitors and deteriorate. The most visited church – St. Katherine's Church of Karja – and regarded as the most beautiful due to its splendid architectural design, subsists completely on tourist resources. Annual tourist donations constitute up to five times more than church fees paid by congregation members.

Pastors admit that the conflict between sacred and commercial aspects requires attention. As long as tourists behave respectfully, they do not consider tourism troublesome. On the contrary, congregations even encourage tourism, as this provides them with extra income. In fact, for some congregations of Saaremaa tourist donations are vital. Some churches do some commerce themselves; e.g., asking entrance fees for visiting a church and for visiting a tower, selling souvenirs, and/or providing accommodation. Although these practices raise objections, they are often the only option a congregation may have in order to be able to carry on. The pastors strictly object to undignified behaviour by tourists, especially during sacred ceremonies.

The general attitude of churches towards visitors is courteous. Furthermore, congregations encourage tourism practices conducted in appreciation of local values.

The project called Wayfarers' Churches (*Estonian Evangelical Lutheran Church* 2002), which also involves other Estonian churches, is one example of guided tourism supporting the preservation of values associated with Christian tradition on Saaremaa. The idea is based upon the medieval pilgrimage, one of the earliest modes of tourism, which sometimes brought masses of people to churches. The churches included in this project are open for everybody, for those who come to pray or just to observe, and not only on Sundays but on a daily basis. The project aims at offering visitors conscious stops in churches – for them to experience and admire the handiwork of previous generations, to become acquainted with the activities of local congregations, to learn about the history of the place and the people, and to meditate upon spiritual values. People, when entering a church, can receive part of the church's blessing in the tradition of the ancient pilgrims.

Pastors of the Lutheran churches of Saaremaa do not consider the tourism-oriented development plans of the island harmful when it meets both the needs of locals and tourists. However, unregulated tourism and ignorant visitors still impose threats to the island's customary life.

Orthodox churches have experienced a remarkable drop in the number of Orthodox believers, a lack of visitors, and the deterioration of church buildings, which are not open to visitors.

Unlike other parts of Estonia, where a number of Orthodox churches have been closed and have fallen into ruin during the last decades, on Saaremaa no Orthodox church has been shut since the end of the Second World War. In total there have been approximately 120 Orthodox churches in Estonia, of which one-third have been closed by now and are left falling apart. In 1939, there were 157 Orthodox congregations – by 1992 this number had dropped to 80 (Sild & Salo 1995). The Orthodox churches on Saaremaa are today in an exceptional but critical situation.

3. DISCUSSION

Christianity has been an important force in the development of Europe's identity, thus greatly influencing the shaping of its cultural landscapes (Unwin 1998; Jordan-Bychkov & Bychkova-Jordan 2002). Even today, the clerical congregation in Estonia plays an important role in the identity of people who define their identity through the congregation they come from. Although the once-predominant role of the clerical establishment has declined, religion continues to play a significant part in local cultural identity through values formed in the past. Paul (2002) attests that for most Estonians, the church is of no use; but he assumes that it is useful for society to preserve the "useless". Churches carry on traditions that often are centuries old. Religious beliefs are central to the construction of identities and the practice of people's lives (Holloway & Valins 2002).

The loss of the sacred, of religious and sacred places and of a sense of local identity and community goes together with modern development. To preserve religious places, special management and maintenance is required. According to Kong (2001), politics are evident here. Sacred space is the object of many competing interpretations that are intimately linked to social, economic, and political conditions and backgrounds. One of the critical issues, as pointed out by Carmichael et al. (1994), is the negotiation over the management and maintenance of sacred places between native people who revere the site and external forces that want them for pragmatic, commercial or even alternative religious purposes. Thus, religious objects are taken into everyday lives, through museumisation and transformation of meaning, but also through other means – such as the mass production of religious artefacts or the promotion of religious buildings as tourist attractions.

The following paragraphs present the division of the functions of churches as recognised in this study.

3.1 Church as a Religious Place

The ties of people to the church as a shrine are weakening. This alienation on Saaremaa as well as elsewhere in Estonia is very much due to the negative attitude of the former Soviet power towards religion and churches. Along with the national awakening in the late 1980s and early 1990s, a religious awakening took place. For a short period of time the number of baptised grew and other religious ceremonies were practiced again; but this trend soon went into decline (Sild & Salo 1995). The notion of church as a religious place, as a space for spiritual experience, and as a shaper of the identity of local community has become more and more lost.

Traditionally, one of the central functions of the church was to serve as a social centre, where adherents gathered not only to pray but to engage in social activities as well. This has changed drastically. The reasons are multiple: there are alternative places for social activities, growing individualism, and new modes of communication. As long as people prayed in the same place and had common interests and activities, they felt a sense of belonging together as a community (Kong 2001). In a number of congregations, especially in those of small rural communities, church still plays an important role in local social life as a single establishment of gathering and cultural events. However, on Saaremaa only in one of the studied congregations active social activity is practiced. Although several concerts and celebrations take place in others as well, these are often random events not capable to fulfill the conventional function of church as a community center. The question remains as to whether the loss of community feeling due to the loss of religious practice may also contribute to rural decline.

3.2 Churches as Landmarks

A study that surveyed the landscape preferences of Estonians revealed that churches were considered as significant features contributing to landscape aesthetics (Palang 1993). In many rural parts of Europe, the church remains the central and most visible element in most rural areas, and often determines the identity of villages. Often situated on the highest ground, with its tower or spire raising the eye "heavenwards", the church serves as an ever-present reminder of the importance of Christianity (Unwin 1998). On Saaremaa, both Lutheran and Orthodox churches are important visual attractions in the island's terrain of flat relief and sparse settlement.

3.3   The Church as an Attraction

Churches in Saaremaa have become increasingly important as tourist attractions during the last ten years, both as monuments and as symbols of local history.

There is increasing concern as to whether tourism can stimulate local development. A study conducted by Kaur et al. (2004) reveals a contrast between the perception of the advantages of tourism by authorities and media sources, and the perception held by local people on Saaremaa. The major issues addressed in the media as positive developments (e.g., expansion of infrastructures and boosting tourism) are perceived as threats by local people. Also, some of the major threats identified by locals (e.g., rural decline, loss of traditionality) are almost completely ignored by the media. Prioritising tourism in regards to regional and local development implies several threats; such as a dependency on external factors and the sacrifice of other local needs and values to those related to tourism. Therefore, regional development and planning should also aim at preserving local culture and environment. Furze et al. (1996) suggest that in terms of local development, tourism may provide a vehicle for translating the values held by others in relation to a particular area into benefits for those who live in or near that area. The ideal situation is, however, problematic, as it assumes that tourism does not impose problems and costs for local resources and people, or it assumes that the benefits of tourism are always greater than the problems it creates.

3.4   The Church as Cultural Heritage

Despite the transformed meaning of churches, their affiliation with the present cultural landscape is clear. Contemporary culture springs from religious traditions. The church was an important carrier of heritage values. On occasion, religious places were recognised officially as heritage centres, and thus became attractions. Even ceremonies could become special tourist events. Thus, as expressed by Raivo (2002), heritage involves using the past as a resource for the needs of the present. This opens the debate regarding the conservation, maintenance and function of heritage sites, and the purpose of related activities. These activities often provide a simplistic interpretation of history – aimed at a better interest and understanding of the past – but are not always scientifically sound or authentic – the idea of a spectacle dominates. In essence, the real-life experience of cultural tradition has been lost, but the external experience has been maintained.

4.   CONCLUSION

Seemingly, the present meaning of church in Estonia is indistinct and unclear. Estonian churches in the countryside, often abandoned, belong more to landscape scenery and history than to people's lives.

It can be concluded that the main shift in the religious landscape of Saaremaa includes the exploitation of sacred structures for secular use. At present, it is not so much about religious convictions and customs as it is about carrying forward cultural and historical values accumulated during centuries of Christian tradition. The alternative function of church could be to present these values to the visitors, as a way of reminding them of the past importance of religious places. We also argue that churches have an immanent value that should be kept. How this can be achieved depends on the protection of landscape and national heritage.

ACKNOWLEDGEMENT

Our sincere gratitude belongs to Anu Printsmann for her kind help.

REFERENCES

Altnurme, R. (2001). *Eesti Evangeeliumi Luteriusu Kirik ja Nõukogude riik 1944-1949*. 2nd Ed. Tartu: Tartu Ülikooli Kirjastus,.

Alttoa, K. (1997). *The Churches on the Island of Saaremaa*. Tallinn: Kunst.

Antonen, M. (2003). *Imaging the Wonderland: Tourism-Related Representations of Finnish Lapland*. Paper presented at Annual Meeting of the Association of American Geographers, 5.3-8.3.2003, New Orleans, USA.

Carmichael, D.L., Hubert, J. & Reeves, B. (1994). *Sacred Sites, Sacred Places*. London and New York: Routledge.

Demerath III, N.J. (2000). The Varieties of Sacred Experience: Finding the Sacred in a Secular Grove. *Journal for the Scientific Study of Religion*, 39, 1, 1-11.

Estonian Evangelical Lutheran Church (2002). *Wayfarers' Churches* http://www.eelk.ee/kirikud_teeliste.html. 23.11.2003.

Furze, B., De Lacy, T. & Birckhead, J. (1996). *Culture, Conservation and Biodiversity. The Social Dimension of Linking Local Level Development and Conservation through Protected Areas*. Chichester: John Wiley & Sons.

Holloway, J. & Valins, O. (2002). Placing Religion and Spirituality in Geography. *Social and Cultural Geography*, 3, 1, 5-9.

Holsti, O.R. (1969). *Content Analysis for the Social Sciences and Humanities*. Reading, MA: Addison-Wesley Publishing Company.

Jellicoe, G. & Jellicoe, S. (1995). *The Landscape of Man. Shaping the Environment from Prehistory to the Present Day*. London: Thames & Hudson.

Jordan-Bychkov, T.G. & Bychkova-Jordan, B., (2002). *The European Culture Area. A Systematic Geography*. New York: Rowman & Littlefield Publishers.

Kaur, E., Palang, H. & Sooväli, H. (2004). Landscapes in Change – Opposing Attitudes in Saaremaa, Estonia. *Landscape and Urban Planning*, in press.

Kong, L. (2001). Mapping 'New' Geographies of Religion: Politics and Poetics in Modernity. *Progress in Human Geography*, 25, 2, 211-233.

Kouleshova, M. (2003). Relict Cultural Landscapes of the Russian North: Evaluation, Protection, Development. In T. Unwin & T. Spek (Eds.), *European Landscapes: From Mountain to Sea. Proceedings of the 19th Session of the Permanent European Conference for the Study of the Rural Landscape (PECSRL) at London and Aberystwyth (UK) 10-17 September 2000* (pp. 71-77). Tallinn: Huma Publishers.

Marksoo, A. (2002). Rahvastik ja asustus. In H. Kään, H. Mardiste, R. Nelis & O. Pesti (Eds.), *Saaremaa. Loodus, Aeg, Inimene* (pp. 377-366). Tallinn: Eesti Entsüklopeediakirjastus.

Paulson, I. (1997). *Vana eesti rahvausk. Usundiloolisi esseid*. Tartu: Ilmamaa.

Palang, H. (1993). *Mentaalsed maastikud, nende kujunemine ja sõltuvus looduslikest teguritest*. Manuscript. BSc thesis at the Institute of Geography, University of Tartu.

Plaat, J. (2001). *Usuliikumised, kirikud ja vabakogudused Lääne- ja Hiiumaal*. Tartu: Eesti Rahva Muuseum.

Paul, T. (2002). Kasutu koorem. *Sirp*, 23.08.2002.

Raivo, P. J (1997). Comparative Religion and Geography: Some Remarks on the Geography of Religion and Religious Geography. *Temenos*, 33, 137-149.

Raivo, P.J. (2002). The Peculiar Touch of the East: Reading the Post-war Landscapes of the Finnish Orthodox Church. *Social and Cultural Geography*, 3, 1, 11-24.

Sild, O. & Salo, V. (1995). *Lühike Eesti kirikulugu*. Tartu.

Unwin, T. (1998). Religious Dimensions of European Culture. In T. Unwin (Ed.), *A European Geography* (pp. 51-56). New York: Longman.

Urry, J. (2002). *The Tourist Gaze*. 2nd Ed. London, Thousand Oaks, New Delhi: Sage.

Chapter 9

OF OAKS, ERRATIC BOULDERS, AND MILKMAIDS

The Poet Imants Ziedonis and Art as Mediator Between Discourses About Rural Landscapes

Edmunds V. Bunkše
University of Delaware/University of Latvia, The United States of America, Latvia

1. INTRODUCTION

In the study and care for rural landscapes and their inhabitants a perpetual dilemma is knowing the different discourses those landscapes embody for a culture group, or "discourse community," as Siri Aasbo (1999: 148) calls it (after Eco 1977). It is a well-known truism that a gap exists in the understanding and evaluation of landscapes between insiders and outsiders, natives and visitors, actors and observers, inhabitants and experts. Since this is known territory, I shall not revisit it, except to restate the obvious – expert opinion, even when well-intended, rarely agrees with the local inhabitants in what is good for them. As Sverker Sörlin expresses it, landscape is a "contested terrain" (1999: 103). The expert will tell the native that his sacred cow is useless, but the native will keep it anyhow. It is a question of what is good for the flesh (i.e. the economy) and what for the soul. This is a very real situation in Europe today, especially in countries such as England, France, Estonia, and Latvia, where the so-called agricultural sector represents economic and political problems, with fundamentally complex issues of the flesh and the soul for farmers.

The theme of this paper is the role of the poet, and art in general, as a mediator between different landscape dialogues, as well as being an active agent in rural and urban, local and national landscapes. We have not really trusted poets outside the realm of poetry ever since Plato "banished the

H. Palang et al. (eds.), European Rural Landscapes:
Persistence and Change in a Globalising Environment, 137-149.

poets" (Murdoch 1977: 2), because by nature poets were untrustworthy in the world of rational and pragmatic affairs. However, especially since the Romantic era, individuals in the arts have been instrumental in shaping how people perceive, relate to, and form landscapes. Arguably, the most famous (and ultimately notorious) was the epic poetry of Ossian brought out of the Scottish Highlands in 1759 by James Macpherson. Although it turned out that Macpherson had written the verses himself (albeit intertwined with Scottish folklore), their appearance as ostensibly the work of the 3rd century AD Gaelic bard Ossian was sensational and had a wide-ranging influence in Europe and beyond, stimulating interest in life in wild landscapes. For example in Denmark the heath landscapes of Jutland were celebrated in poetry and prose by Blicher, as well as by Ingemann, Grundtvig, and Ochlenschlager, so that the heaths "became established as that landscape which preserved, under its peat and moss, the heroic character of a nation" (Olwig 1984: 23-26). Well known examples are Chateaubriand's contribution to the sentiments associated with ruins (and therefore to the preservation of historic landscapes), William and Dorothy Wordsworth's to the appreciation and preservation of the English Lake District, Victor Hugo's to the preservation of historic segments of Paris. In the United States the Hudson River School of landscape painters helped to create appreciation and love of indigenous landscapes. The minor painter, George Catlin, contributed to the "invention" of national parks, with Yellowstone as the first such park.

In this paper the case in point is Imants Ziedonis, a leading poet, writer, and activist in Latvia during the Soviet occupation. In his work he has effectively mediated between the rural and urban, expert opinion and ordinary folk, the totalitarian regime and the rejuvenation of deep-seated, often forgotten cultural and ecological values. He had, and continues to have, wide appeal among all strata of Latvian society. Even to milkmaids. Perhaps most of all to milkmaids.

2.   WHY ZIEDONIS?

I came to Ziedonis because of a long-standing conviction that ultimately landscape and nature values and perceptions of any people can be accessed and understood not by outsiders (scholars, opinion researchers) but by expressions of the people themselves. Initially I found folklore a useful source to study of historic folk attitudes, especially of women (Bunkše 1978). Eventually, in order to study modern attitudes, I realized that all art forms are excellent sources of landscape and environmental values (Bunkše 1990, 1996). In recent studies of Latvian attitudes I used both folklore and literary sources (Bunkše 1998, 1999a, 1999b). In a sense this essay on Ziedonis completes a cycle of research into Latvian attitudes. What is especially valuable in his writings is that often he combines his observations and evocations with Latvian folklore. But more than that, he is a brilliant and original poet (he was nominated for the Nobel prize in the 1970s) who is also, without realizing it for most of his life, a keen and knowledgeable cultural-humanistic geographer.

Figure 9-1. Imants Ziedonis in conversation with a farmer, ca 1969. Ziedonis Collection. Photographer unknown.

Although references cited here range over the entire corpus of Ziedonis' writings, I relied primarily on three works by him: a two volume account (published in a single volume in the collected works: 1995, vol. 2) of his peripatetic exploration of the western region of Latvia known as Kurzeme (Kurland or Courland); and his story (1998, vol. 9) of a group that he organized to "liberate" large oaks, erratic boulders, and similar actions carried out over a span of some twenty years.

3.   TO FIRST SET THE SCENE

The location is of course Latvia; the time period stretches from the early sixties into the present. By the sixties, one way or another, against opposition and through mass deportations to the GULAG, collective farms – *sovkhoz* collectives answering directly to Moscow; and *kolkhozes*, answering to the republic government, were being set up. Experts in Moscow and Rīga determined rural work and life. An enormous gulf existed between the local and the national, the local people and the experts, between the oppressed and the oppressors.

Ziedonis came onto the stage as a student during the1950s. He had been recruited into the Communist Youth Organization as a model of hard work and intelligence. With a talent for writing and as a handsome youth, he occupied a privileged position, becoming secretary of the Writer's Union in 1965. Indeed so privileged, that later on he was allowed to travel to Canada with his wife without the requisite KGB minder. But it was not party work that made Ziedonis one of the most popular and respected individuals in Latvia: it was his poetry, prose, public performances, and twenty one year-long physical labor to save oak trees, large erratic boulders, and other elements of rural landscapes. Without question Ziedonis started Latvians on the path to recovering and healing their national identity and culture, as well as to rekindle interest in their regional and local landscapes.

Born in a small coastal village (Ragaciems) in 1933, Ziedonis was the son of a fisherman. Uncharacteristically, as a boy he looked neither to the sea nor to Rīga, the usual destinations for such adventurous lads, but to the rural countryside. He tells of driving into that countryside with his mother to sell fish:

> …at each turning [of the road] there is a wonder and it tests me: do my eyes see?

> It seems that we swim in each bend of the road as into a creel and there is no wish to return home. We are fish that do not return. We wind ever deeper into the away – past daisies, barns, fruit trees, and fields of dark red clover. My mother by my side and there's no turning back (Berkis et al. 2001: 7).[1]

Ziedonis wound up living and working in Rīga anyway, but he made the country his spiritual and intellectual home. It was, and is, the object of his curiosity and muse.

The road and looking deeply into things are among the themes that preoccupy the poet for the rest of his life. In a reversal of Antigone's

---

[1] All translations are by the author, unless noted otherwise.

mythical unwinding of a ball of twine in a maze "I wind the road into myself in a ball [of twine] ..." (1995, vol. 1: 13). Most recently, in a prose poem (2003: 5), Ziedonis expresses his philosophy of seeing:

> Wonders ... Is there anyone who doesn't wish to see them? But must we always name them so, with such a much promising word? ... we can call them [i.e. wonders] the joys of noticing [here the Latvian *ieraudzit* is untranslatable] ... Surprise comes not from what you see, but from that which you notice. I saw and it disappeared. I noticed and ... it stayed. Glued onto me. Doesn't leave. Is not forgotten.

4. POETRY AND PROSE

Ziedonis was, and is, a prolific writer of poetry and prose (the collected works number twelve volumes, 1995-2002), making his name also with children's prose and adaptations of his works for the theater and television. He read poetry in public to large audiences. The first major book of poems called *Motorcycle* appeared in 1965 and immediately gained him fame (1995, vol. 2). His motorcycle was freedom and escape from the city into the rural countryside and he shared that freedom with his readers. He loved to ride at night: "I speed through waves of the night. / In layers and zones the night air stands: / Through damp and fulsome breath of bogs, / Through the sweet aroma of clover ..." (1995, vol. 1: 153). "I drove Latvia into my lungs and sweaters ... The motorcycle carried me away from an ideological explosion" (1995, vol. 1: 12). Freedom meant escape from the horizon, which to most people seems "an innocent line." No, "The horizon has always / Strangled man's thought ... The horizon is around my neck / Like the iron ring around the neck of a slave ..." (1995, vol. 1: 229). (Strong words in the context of totalitarianism.) He asked simple questions (e.g., how does grass grow, a candle burn?) and was able to see and express ordinary things "in bold and even far fetched metaphors" (Ekmanis 1978: 336); for example: "My hand becomes ever more delicate; soon it will fly like a spider's web on an autumn morning" (1978: 64). Ziedonis possesses the rare gift of synesthesia, the blending of the senses, in which the stimulation of one sense stimulates another (Ackerman 1991: 289). In one prose poem (an epiphany) he tells about the scents of a raindrop, calls his nostrils a violin: "My nostrils play lovely songs...polkas, nocturnes, old popular tunes ...". He elaborates this theme: "Doesn't the *baravika* (the edible mushroom boletus) have the scent of a baritone? Aren't milk-agaric mushrooms altos? Aren't puff-balls sopranos? Doesn't a chanterelle dance like a ballerina?" He explains all this: "Aromas are as quiet as the footsteps of ants in a forest … as fine as the spawn (undersides) of mushrooms. I could feel where a mushroom will rise out of the moss … I could almost hear the ants laughing on the other side of the stream …" (1995, vol. 3: 70-71). He is also able to connect the mundane of daily life with much broader contexts. Thus a 1977 book called *A poem about milk* (1997, vol. 7) includes a history of the world seen in association with milk, as well as bucolic conversations with milkmaids. He was an angry young poet (Ekmanis 1978) who was looking for ethical and aesthetic values that had been obliterated by the regime. Like poets elsewhere within the Soviet sphere, Ziedonis had to cow-tow the party line, but at the same time he could express what was implicitly on people's minds. Reading his work today, in the safety of a free country, it is astonishing how audacious he was at times in his criticisms of communist ineptitudes. According to Ekmanis (1978: 336), Ziedonis was "convinced that he who cannot embrace truth can never be a true poet."

Arguably his most influential work from the standpoint of Latvian rebirth was *Kurzemīte* (Little Kurland or Courland), a two volume factual account, published in 1970-1974 (1995, vol. 2), of his search for historic and contemporary rural values in the countryside of Western Latvia. These books replaced some of the forbidden literature about rural life and work that had existed in the pre-war years. The project started out of dissatisfaction with the dull and oppressive bureaucracy of the Soviet regime. As he told me in July of 2002, one day he smashed all the telephones at his office ("save the red one") and walked literally away, into the landscapes of western Latvia, following the "call of the road" (1995, vol. 2: 15). In *Kurzemīte* he relates how he got the idea to roam Kurzeme. Like Ishmael in Melville's *Moby Dick,* Ziedonis becomes restless and fears that he will do something self-destructive. He worries how it would be if one morning he would confuse "what is allowed with what was not." And then he dares to criticize the regime. In describing his work for the Writers' Union and the Minister of Culture (1995, vol. 2: 20), he writes:

> … as I sat at long conference tables, it seemed that each speaker was driving nails into my skull. Everyone driving nails into each other's skulls … Most tormenting – when a nail has been driven in and it can go no further, and still, it is being hammered.

Ziedonis walks away and starts traveling, with his wife, first to the Black Sea ("where people with big paunches stroll about"), then to the mountains nearby, but in the end he and his wife walk into Kurzeme, without a plan, without a map. Eventually he travels alone, on foot. In the mountains they had realized that "in the world there is a certain place that we must know the best, that we must realize … that there is a land that people from foreign parts will not discover for us (1995, vol. 2: 24)." That land was Kurzemīte.

Even though in *Kurzemīte* he was ostensibly publicizing the good deeds of people in the context of the communist system – workers, specialists, and chairmen of collective farms (large segments of the two books do consist of tedious communist propaganda) – really he was investigating the rural landscape, particular places and individual people, to find moral values that had been lost. He especially sought out individuals for he believed that "you cannot gain anything from human uniformity, but you gain much from human variety" (1995, vol. 2: 491). The people that he sought out, in hindsight (2002: 12-14), were those who inspire: "… good, active, beautiful [people], with whom you want to linger, with whom you wish to deepen the moment of meeting." The values that he looked for were the noble, sublime, honorable (*cildens*) – in people, farms, and landscapes. "And, if I don't see [values] myself [in the landscape], they [the people] show me" (2002: 14).

In hindsight, the publication of *Kurzemīte* was a major turning point for many Latvians, from hopelessness to hope. (It must be remembered that the early 1970's was an especially grim time in the Soviet block). Ziedonis was in fact reconnecting his generation of Latvians with the grandparent generation and generations before (*Old folk, wonder of wonders, are beautiful and wise* 1995, vol. 1: 13). The dignity of hard work, ethics, and aesthetics in the rural landscape were the principal values that he discovered and expressed.

5. CRITICISM OF THE SYSTEM

Ziedonis was able to criticize the Soviet system quite openly and effect the preservation of some historic buildings and possibly bring about some compromises between economics and aesthetics in rural landscapes. His strategy was to accept the fact that the system was in Latvia to stay and to describe in great detail the specific accomplishments and intentions of the rural collectives, their chairmen, and workers. He played on the standing Marxist-Leninist philosophy that everything was moving towards utopian perfection. Then he pointed out undesirable acts in the landscape or in work routines (1995, vol. 2). For example, if the regime had planned to urbanize the rural population and landscape by creating collective farm centers with their cluster of large industrial farm buildings and multi-story apartment houses, how could it allow the chairman of a model *kolkhoz* to contemplate the destruction of a house of a famous turn-of the century painter? He may have earned his keep by painting duchesses and countesses, but had not he also painted ordinary folk "with down-turned corners at the mouth that come from difficult lives?" The city inhabitant needs that house by the river, argues Ziedonis. And soon that rural fellow, "the urbanite-to-be will also need that house." "A strong, understanding party secretary is needed, who would realize the aims of the party in the rural countryside" (1995, vol. 2: 236).

On a different tack, without any overt criticism whatsoever, Ziedonis recounts an interview with a milkmaid who held the milking record for that particular year. Her milking statistics are obviously the model for a system in which getting people to work hard was a major problem. But the moral lesson that Ziedonis draws forth from the milkmaid is not the mechanics of milking; it is that work with animals entails love and care. Milk "must be caressed out of a cow … Just as you pet a cat, stroke affectionately the cow's udder, massage it, have a calm conversation with cows. They have their nation, their power, their glory, and eternal life" (1995, vol. 2: 233-234). In this fashion Ziedonis literally juxtaposes human touch with arid official statistics. The critique of collectivized, industrial farming is out in the open, but he gets away with it.

Knowing the different "party lines" Ziedonis was able to play off one arm of the system against another and argue for humane values in an otherwise crude political and economic milieu.

6.   THE POET AS LANDSCAPE ACTIVIST

Ziedonis did not wish to be merely the poet, the intellectual, the publicist – his eventual official title under the regime. He wanted to become an agent of actual change in rural landscapes. To this end in November of 1976 he called up some friends and asked them if they would like to spend the weekend liberating some noble oak trees from encroaching brush. So began the ad hoc activities of a small, informal group of people that called itself, also informally, *The Liberators of Noble Oaks*. For 21 years, the loosely knit group made between nine and 12 weekend-long excursions all over Latvia each year, liberating not only great oaks, but also large erratic boulders (once sacred) left by the last glaciation, sites associated with historic cultural figures and writers, castle mounds, some abandoned parks, and old sacred sites. They also planted trees in some places, placed green sun symbols at other sites (Ziedonis holds mythical beliefs about certain geographic locations). At one site they built a pyramid eleven meters high from the stones of destroyed traditional family farm buildings. In all, 64 people participated over that time span, although on each excursion there were probably around 15, perhaps 20 participants. A small autobus sufficed. They slept wherever they could find donated space. With one exception, the group was made up of the intelligentsia: artists, teachers, foresters, doctors, professors, journalists, and geography graduates (all women!).

Why did Ziedonis start this group? With collective farms having replaced the dispersed family farm settlement pattern, the forest and brush-land – had increased to over 50 percent of the land area of Latvia. *Homo sovieticus* was in full bloom. Many farm laborers were lazy and unwilling to work, lacking in initiative; drinking was widespread. Except for some well-lead collective farms, the system had degraded cultural traditions that had encompassed pride in hard work, nurture and care for the land, and for living things.

The most obvious inspiration came from the burning and dynamiting of great oaks. This was done by the tractor drivers in the creation of collective farms. As a boy after WW II he had witnessed the burning of oaks. "Those were death years for oak trees," he wrote (Ziedonis 1998, vol. 9: 34). "Where a tractor driver drives, there follows fire … [they] were happy burners. They even poured diesel oil into oaks." (It was not necessarily hatred of oak trees that prompted such behavior. Rather it was disconnectedness from the land and its people.) But the fate of trees was part of a broader problem with landscapes. In a New Year's address (really Christmas) he singled out Jūrmala, a string of seaside resorts outside Rīga that had become an all-Soviet Union vacation paradise, as a particularly horrid example of crippled nature from a lack of ethical and aesthetic responsibility. He called Jūrmala "a lager, a concentration camp of crippled trees" and blamed rootless, culture-lacking chairmen "of communal establishments," and a labor force that was indifferent if not outright nasty (Jūrmala was, for all practical purposes, Russian). Ziedonis then put the fate of trees in the context of alarming world deforestation rates. But, he said, "such abstract figures do not move the heart. One is moved by the presence of cruelty right here. What is happening in front of our eyes, my eyes … Who ordered that cow to urinate under that noble tree? Who ordered him to pour his motor oil under that tree?"

He spoke of two dead linden trees by a department store, two dead oaks by a collective farm raising cattle. "Why aren't they cut down?", he asks. "They are unseen. The whole civilized world knows that corpses must be buried …! Of course!" (Ziedonis 1998, vol. 9: 49-53).

Remarkably, in the same breath, he was able to criticize the whole Soviet-Latvian landscape. The people who create it:

are their own enemies. With their tilting utility poles, which reel chaotically all over Latvian landscapes like drunkards. With formless additions on structures end on end. With small hot houses made of flimsy cellophane. With no sign of caring in construction sites that have debris and discarded materials polluting the environment for great distances. With animal farms where the offal flows out onto the road, beneath the wheels of their Ladas and Volgas. But in the center of the farm there are cement paths, metal rod gates, a cement threshold, a flower-bed with cement borders (Ziedonis 1998, vol. 9: 52).

Liberating trees for over two decades was a demonstration of positive values: of work, of caring for and nurturing nature; of man's love of nature for its own sake. "We had agreed that at least we will nurture the idea of nurture within ourselves," writes Ziedonis (1998, vol. 9: 12). Thus the group liberated some oaks in the middle of forests that few would ever see. They were, in fact returning home, themselves and the Latvian people.

What Ziedonis and his little band did was quixotic. Even though they persisted for 21 years, they could not possibly change the Soviet landscape system in any significant way. However, the symbolic value of their actions was enormous. Although great oaks and erratic boulders had long ceased to have sacred meaning for most Latvians, they were remembered as a vital part of Latvian cultural history (Today's Latvian automobile touring guides show the location of many such oaks and boulders).

7.   ZIEDONIS' INTERPRETATIONS OF THE MEANING OF TREES

Ziedonis' criticisms of the system and acts of liberating trees and boulders were but a small part of his oeuvre, which consisted of expressing positive culture-nature-human values that he found in folklore and in his own observations of nature. As he has said "You cannot act as a destructive force in a destructive world. You cannot despair … You must be a positive witness" (Ziedonis 1995, vol. 2: 14). His interpretation of the meaning of trees is representative. Describing Latvian love of trees, he cites from a well-known folksong "The tops of forests shouted in joy ….". Ziedonis asks "Who else could say it this way, if not a human being who has inherited in his genes special hearing? An ear for trees inherited from countless fathers." Ziedonis then shows his own hearing in a characterization of the nature of human interaction with oaks, linden trees, and birches. Thus the noble oak is aloof. It is not a comforting, warm place to sleep under, for it:

> rustles sharply and a bit harshly. The linden tree is the most loving of trees. In the summer Latvians took midday naps under linden trees in their farmyards. Children, who grew up napping under lindens at noontime are different: they listen far and think beautifully. From … linden trees…a heavy aroma of honey rains down over the sleeper … The birch is the most musically refined – a tree of songs. No other tree speaks the way a birch does – in many tongues, continuously changing, bending to every touch of the wind. "Through a silvery birch grove I walked …"

[first line of a folksong]. That's not only the silver of twisting leaves. It is the whispering, the expression of a glittering pattern (*nirbona*), the murmuring of the sky unique to a birch grove. Everything here is in everything ... One must be silvery inside in order to reverberate in the silvery grove. It happens only so: she, who is silvery inside, receives the gift of silvery ness ... Silver calls to silver. You are on the same wavelength: trees, the whispering (*salkas*) of light, and the breaths of trees. Nothing can be grasped, the light is yet to materialize, has yet to happen. "Not a branch did I break ..." [the next line in the folksong]. Everything ... is flowing, free, and untouchable. "Had I [a girl, a woman] broken off a branch ..." Then, of course, everything would become visible. "Then I [a girl, a woman] would walk cover'd in silver." Then you would indeed believe it. But you don't have to believe it and can think of it as just a song. About trees as inviolate, about nature protection. Of course, also about that. But also about silver, about silver everywhere and the silver that is in everything. Which reveals itself when you are silver" (Ziedonis 1998, vol. 9: 49).

8. REPERCUSSIONS

The seeds of the Latvian independence movement, not in a political but cultural sense, were planted with the publication of *Kurzemīte* in the early 1970s (it would be interesting to know its impact in Estonia and Lithuania, since it was translated into both languages) and the symbolic acts of liberating great oaks and erratic boulders begun in 1976. That movement became a political force and gained strength in the early 1980s, when focus on ecologic issues became a way to gain increasing independence from Moscow, with eventual independence achieved in 1991.

At the time the actions of Ziedonis' group seemed senseless. Sarmīte Ēlerte (Ziedonis 1998, vol. 9: 356-358), one of its members (and today the editor of Rīga's principal daily newspaper), asks: "were we really caring for landscapes? "What sense did it make to care for an oak tree in the middle of the forest? ... I think what united us was daring to do senseless work." She likens what they did to pouring water into a bottomless barrel. Ēlerte explains that they lived in a senseless system and thus gained meaning by doing something that was senseless to the system. She writes that:

At one point I was overcome by the senselessness of it – if there is no future for Latvia, then what's the point? But Imants said "You will see, after so many years ... (I don't remember how many), the Baltic States will become the brightest point, they will become the reason for huge changes, the interest of the world will be focused here." When that had already happened, I asked him: "How did you know that?" He had forgotten that he had ever said anything of the sort.

Ziedonis assessed his role rather modestly "I thought only about one thing, about how to leave a witness, a witnessing of my people during a small, short and destructive time period. But it was forbidden to put it that way then" (Ziedonis 1995, vol. 2: 14).

9. EPILOGUE

Is there a model or a lesson here for other societies? Yes and no. Ziedonis is a unique individual who was active during unique times when poets were the only ones capable of truly expressing forbidden ideas and thus could compete with a more mundane mass media. Today poetry is hardly a source for mass entertainments. But poetry, and all the arts, should have a role in debates over landscapes and places, because *art can mediate between issues that address the heart and soul of humans and their practical, political, and ecologic concerns.* And not only mediate but take part in the processes shaping landscapes. In free contemporary societies there are competing voices about places and landscapes (Sheldrake 2001; Mitchell 2000; Sörlin 1999) and how they should be nurtured and shaped. Economic and political voices (and forces) tend to dominate, but ecologic concerns based on scientific analysis have gained a strong voice as well. How did that happen? Through innumerable symposia, conferences, writings, and teachings. It is time to welcome to the debate poets and all other artists who address issues of landscapes and places and the values associated with them. At the moment their voices are marginal. Indeed, in a world of pragmatism and *Realpolitik*, Plato's banishment of the poets is very real. It is up to those with deep concerns for landscapes and environments not only to include artists and poets in their symposia and planning boards, but to seek them out. Their voices give expression to our humanity.

REFERENCES

Aasbo, S. (1999). History and Ecology in Everyday Landscape. *Norsk Geografisk Tidsskrift-Norwegian Journal of Geography*, 53, 2-3, 145-151.

Ackerman, D. (1991). *A Natural History of the Senses.* New York: Vintage Books.

Berkis, A., Hānbergs, Ē. & Ziedonis, I. (2001). *Lauku sēta ir gudra.* Rīga: Jumava.

Bunkše, E.V. (1978). Commoner Attitudes Toward Landscape and Nature. *Annals of the Association of American Geographers*, 68:4, 551-566.

Bunkše, E.V. (1990). Saint-Exupery's Geography Lesson: Art and Science in the Creation and Cultivation of Landscape Values. *Annals of the Association of American Geographers,* 80:1, 96-108.

Bunkše, E.V. (1996). Humanism: Wisdom of the Heart and Mind. In C. Earle, K. Mathewson & M.S. Kenzer (Eds.), *Concepts in Human Geography* (pp. 355-381). Lanham, MD: Rowman & Littlefield Publishers Inc.

Bunkše, E.V. (1998). The Case of the Missing Sublime in Latvian Landscape Aesthetics and Ethics. *Ethics, Place and Environment,* 1:1, 235-246.

Bunkše, E.V. (1999a). Reality of Rural Landscape Symbolism in the Formation of a Post-Soviet, Postmodern Latvian Identity. *Norsk Geografisk Tidsskrift – Norwegian Journal of Geography,* 53, 121-138.

Bunkše, E.V., (1999b). God, Thine Earth is Burning: Nature Attitudes and the Latvian Drive for Independence. In A. Buttimer & L. Wallin (Eds.), *Nature, Culture, Identity* (pp. 157-187). Dordrecht: Kluwer Academic Publishers.

Eco, U. (1977). *A Theory of Semiotics.* London: MacMillan.

Ekmanis, R. (1978). *Latvian Literature under the Soviets: 1940-1975.* Belmont, MA.: Nordland Publishing Company.

Mitchell, D. (2000). *Cultural Geography.* Malden, MA: Blackwell Publishers, Inc.

Murdoch, I. (1977). *The Fire and the Sun. Why Plato Banished the Artists.* Oxford: Oxford University Press.

Olwig, K. (1984). *Nature's Ideological Landscape. A Literary and Geographic Perspective on its Development and Preservation on Denmark's Jutland Heath.* London: George Allen& Unwin.

Sheldrake, P. (2001). *Spaces for the Sacred. Place, Memory, and Identity.* Baltimore: The John Hopkins University Press.

Sörlin, S. (1999). The Articulation of Territory: Landscape and the Constitution of Regional and National Identity. *Norsk Geografisk Tidsskrift, 53: 2-3, 103-111.*

Ziedonis, I. (1978). *Epifānijas.* Rīga: Liesma.

Ziedonis, I. (1995-2002). *Raksti* [Collected works, 12 vols.]. Rīga: Nordik.

Ziedonis, I. (2002). Bailes pazaudēt laudis. Imants Ziedonis saruna ar Noru Ikstēnu. In *Karogs. Literatūras mēnešraksts,* 5/2002, 10-21.

Ziedonis, I. (2003). *No patikšanas uz patikšanu.* Riga: Daugava.

Chapter 10

THE BORDER AND THE BORDERED

An Interdisciplinary Comparison of the Origin and Function of the Border of the Parish Gryt in Södermanland, Sweden, and the Border of the Corresponding Hundred

Maria Bergström & Margareta Ihse
Department of Physical Geography and Quaternary Geology, Stockholm University, Sweden

1.   INTRODUCTION

The parish of Gryt in central Södermanland, Sweden, constitutes the area of investigation in an ongoing research project *The Parish Gryt – Landscape and Community in an Ecosystem Perspective* (Figure 1). An ecosystem approach is being employed in this study, in order to analyse the parish community's ecological relation with its territory over time. The intention is to study the human community as a part of the ecosystem, and thus to reveal aspects of its sustainability and adaptation in regards to carrying capacity in the parish. As a first step in this project, the border of the conceived system – i.e. the border of the parish – was examined.

A discrepancy between the border of the parish and the border of the corresponding hundred (Figure 2) provided the opportunity for a comparative analysis of whether there were any considerable geographical differences between the two types of territorial borders.

The result of the analysis did in fact indicate such differences. This paper presents the results from the investigation, and suggests an interpretation based on the differing origins and purposes of the two kinds of territory: the parish and the hundred.

H. Palang et al. (eds.), European Rural Landscapes:
Persistence and Change in a Globalising Environment, 151-175.
© 2004 *Kluwer Academic Publishers. Printed in the Netherlands.*

1.1   Background

In Sweden, there are two long-established administrative units subordinate to the county division. The *parish*, which until 1952 was the administrative unit for the local community, was connected to the church, while the *hundred* served for a long time as the regional unit for the court of law. At the end of the 19th century, the province of Södermanland consisted of 13 hundreds, with each hundred comprising a varying number of parishes.

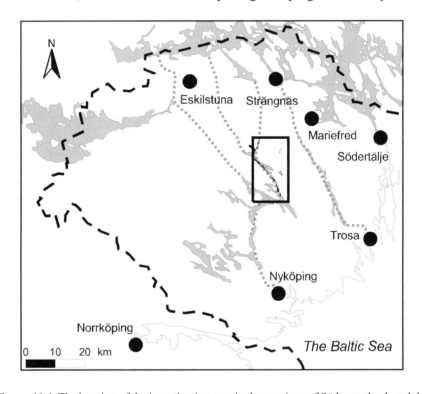

Figure 10-1. The location of the investigation area in the province of Södermanland, and the regional situation according to accessibility between the southern part of Södermanland and Lake Mälaren.

In relation to the study area, the border between the Daga and Villåttinge hundreds divided Gryt parish into two parts until 1889 (Lagerstedt 1973). The larger part of the parish belonged to Daga hundred, and the smaller part (16 percent) belonged to Villåttinge hundred (Figure 2). Such discrepancies involving the two kinds of districts also existed in other parts of Sweden, as for example with the cases of Uppland (Rahmquist 1982) and Öland (Göransson 1982). It is not the aim of this article to analyse the situation particular to Södermanland in its entirety, but – according to a description from 1852 – this is an unusual situation in this province (Tham 1852).

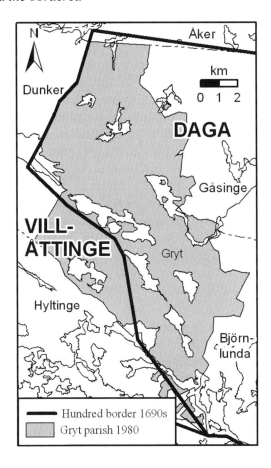

Figure 10-2. Gryt parish, the two hundreds Daga and Villåttinge, and neighbouring parishes.

   The development of hundreds and parishes in Sweden has been the subject of much research and debate over the years among historians, place-names researchers, archaeologists and human geographers (Andersson & Göransson 1982). This has been especially pronounced for aspects pertaining to age, character and origins. Which of these two territorial entities is the oldest and how old are they? Do they correspond to the area or to the people? Are they the outcome of external decree and/or top-down control, or do they represent solutions to an inner (and bottom-up) quest for cooperation and organisation? There are, however, no mutually inclusive or simple answers to these questions. Several authors emphasise that the differing situations in the various parts of the country must be taken into consideration (Smedberg 1982; Brink 1990; Lindquist 2001), but the most widely-held theories seem to prioritise the importance of a nationally-instigated top-down decision of some kind (Wijkander 1983; Brink 1990).

The hundred is the older of the two territorial forms, and is commonly believed to have originated in pre-medieval times. It has been suggested that the hundred served as the district for municipal worship, and it has also been argued that there are connections to the ships and the royal naval organisation during the Viking age (Andersson 1982; Lundberg 1982). In the provincial code from the 14th century (Södermannalagen), the hundred is referred to as the base for the regional court (Holmbäck & Wessén 1979).

In the same provincial code, the parish is referred to as the basis around which to build and maintain a church. Accordingly, the parish may have developed around the 12th and 13th centuries (Brink 1990). There is, however, reason to believe that the parish as a district might be of older origin than the hundred in some parts of the country, and that it originated from a settlement region (Smedberg 1982; Brink 1990).

2.   AIMS

The aim of this paper is to present, analyse and discuss the identified differences between the two corresponding borders. The comparison endeavours to combine ecological, physical geographical and human geographical perspectives in an interdisciplinary way.

The basic assumption is that the combination of methods and materials from these disciplines may facilitate a better understanding of the earlier significance of the landscape, during the times when knowledge pertaining to landscape and nature was a prerequisite for survival, rather than a subject for academic disciplinary study.

The starting questions for the article were:

–   To what extent have the two borders in question been stable over the years?
–   Are there any indications suggesting border connections in regards to the place names in the study area?
–   How are the two borders related to the patterns of settlement?
–   How are the borders related to topography, substrate, vegetation cover, land-use, etc.?
–   If there are differences between the two borders according to the points above, can they be understood to have resulted from the adjustment of cultural "needs" to natural conditions in the landscape, and therefore as representative of the different functions and origins of the parish and the hundred?

3. MATERIAL AND METHODS

The basic method has been to carry out field investigations of the borders and to combine the findings with a geographical analysis based on topographical, geological and cadastral maps. Aerial photographs taken using infrared colour film, at a scale of 1:30,000, have to some extent been used to support the interpretation of the geographical situation. This has mainly been used for instances where generalisations on the maps resulted in disparities between map representation and the findings from the field investigation, or where no visible marks could be found for stretches of the border during the field investigation. The opportunity to gain a three-dimensional view of the landscape from above makes it possible to follow features such as small ridges, elevation, and variations in vegetation that are not shown on the maps, and that are not always possible to distinguish in the field.

Table 10-1. Sources and Material

Maps:
H2: *Geographisch Delineation å Dag Häradz råå och allmännings linier*, Johan Stiernroos. 1698: The regional land survey office in Nyköping;
H14: A title and dateless map catalogued as a map of Commons: the regional land survey office in Nyköping;
U60: Daga härad, Nills Herling: the library of Uppsala University;
Printed map of Södermanland, 1:100,000, from 1804;
Three versions of the Swedish cadastral maps for the area, from 1905, 1958 and 1980;
The Swedish topographical map, 1:50,000;
Geological map 1:50,000 from 1865 and 1980s;
Häradskartan 1905: description of the cadastral map from 1905.
Historical documents, etc.:
RAp and RA C. Pergament letters: the National Archives of Sweden;
Provincial documentation: the National Archives of Sweden;
Register of place names from Gryt parish in old documents, etc., at the Place Names Archive in Uppsala.

In order to establish the situation pertaining to earlier times, medieval documents from the 13th and 14th centuries that document exchanges of land ownership have been examined, as have taxation assessment rolls and cadastral registers from the 17th to 19th centuries, and land survey maps spanning the period from the late-17th century up to recent times. Modern cadastral maps have been used together with registers from The National Archives of Sweden and the Place Names Archive in Uppsala, in order to find out about ancient remains and place names. See Table 1 for a comprehensive list of material that has been used in the study.

The analyses and maps have mainly been developed with the help of the GIS programmes Envi and ArcView, and the diagrams have been developed using Excel. After examining the physical/ecological situation of the border zones, we then propose an induced hypothesis of the *raison d'être* for the development of the zones that would become the borders.

4.   THE STUDY AREA: THE PHYSICAL LANDSCAPE, LAND-USE AND POPULATION

The investigation concerns the border between Gryt parish and the neighbouring parishes of Dunker and Hyltinge, and the ancient border between Daga and Villåttinge hundreds, which runs from Lake Magsjön to Lake Båven (Figures 2 and 3). Gryt parish is approximately 114 square kilometres in an area situated in the province of Södermanland, about 70 kilometres southwest of Stockholm. The highest point of elevation is the peak of Magsjöberget (102 metres above sea level), and the lowest part is Lake Båven, at around 22 metres above sea level.

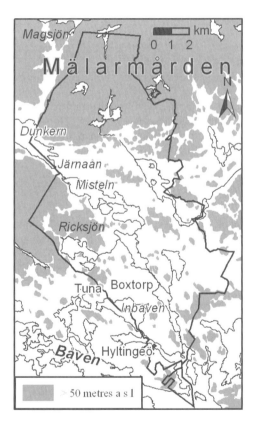

Figure 10-3. Names mentioned in the text.

The bedrock in Södermanland is characterised by fault fissures in a fissure valley landscape, where denudation has created a peneplain with a fairly altitudinal constancy of the commonly gneissic or granite bedrock. The area was glaciated during the Weichselian age, and lay under water as the ice margin retreated (Risberg et al. 1991). During the Holocene period, the continuing isostatic uplift led to a fall in relative sea level – causing a sorting of soils. The resulting landscape comprises both more-elevated areas that are mainly exposed fossil archipelago, and less-elevated areas of fissure valleys that contain a thick layer of clay and elongated lake systems. The predominant soil type is till, existing mostly as a thin soil layer, sometimes with a large amount of boulders – especially on the slopes between the bare bedrock and the valleys. Prominent eskers run through the landscape, mainly in a north-south direction. Mires appear on the higher plateaux of the landscape mainly as raised bogs, and in the lower parts mainly as different types of ferns.

Gryt (Figure 3) is in its northern area a part of Mälarmården, a great horst elongated in an east-west direction. The bedrock is ore-bearing, and a couple of small mines are situated in the northern part of the parish, in the vicinity of the border. In the southern part the parish reaches Lake Båven, and is thus connected to the extensive lake district in the middle of Södermanland.

The land-use in the parish and its adjoining areas has mainly been for cultivation, husbandry, and for the last few centuries, forestry. About three-fifths of the area consists of forest, one-fifth of fields, and one-fifth of lakes.

The vegetation is mainly coniferous forest of pine and spruce on till or bedrock, and agricultural fields and deciduous forests on clay and other fine sediments.

Statistical records show that over the course of the past three hundred years, the population of Gryt parish has fluctuated between 1,000 and 2,000 persons. In 1905 for example, there were 51 farms of at least one-fourth mantal – a Swedish taxation unit based upon assumptions as to the lowest possible carrying capacity of a farm. Of these, almost three-quarters were taxed as *frälse*, i.e. originally exempted from land dues owed to the Crown (Häradskartan 1905).

The nearest town of Mariefred is situated more than 20 kilometres north of the parish church (Figure 1). The earliest-mentioned road (in 1451, according to the Place Names Archives) is *almannawaeghenom*, which passed the church in an east-west direction.

5.   RESULTS

5.1   Evidence of the Borders in Historical Documents and Maps

The oldest records for the parish are found in medieval documents dating from the 13th and 14th centuries, which are kept at The National Archives of Sweden. These can be searched on CD, and places relating to Gryt, Dunker, Hyltinge, Daga or Villåttinge were selected for searches. Among these, places situated in the vicinity of the borders in question were picked out. Four places relating to the western border of Gryt were found (Figure 4). In 1275, the Mora estate is documented as having been affiliated to Villåttinge hundred and Dunker parish (Christiansson 1945; Christiansson 1985). In 1378, two villages of Jälund are mentioned, and their respective parish affiliation is accounted for: Ytterby Jällunda in Gryt, and Överby Jällunda in Hyltinge (RA C4 1378). In 1381, Yttra, Öfra Rösund and Solberga in Gryt are mentioned (RAp 1381). According to the documents, the first two belonged to Daga hundred, while Solberga is recorded as having belonged to Villåttinge hundred. The cadastral register and taxation assessment rolls reveal that Solberga subsequently consisted of three freehold farms, two belonging to Villåttinge and one to Daga hundred.

The two hundreds belonged to different counties during different historical periods (Almquist 1917; Almquist 1934). In the cadastral register and taxation assessment rolls from the 17th, 18th and 19th centuries, homesteads and crofts are registered under their respective parishes. In actuality, they are usually divided into those belonging to Villåttinge hundred and those belonging to Daga hundred; although occasionally the whole of Gryt parish is registered under Daga in the cadastral register (The National Archives of Sweden).

The oldest-known maps of the area are three maps dating from the end of the 17th century; one of Daga hundred (U60), one of Daga common (H2), and one of the part of Gryt parish belonging to Villåttinge hundred (H14). The latter of these is an untitled and undated drawing registered at the regional land survey office under the title of "commons". In the map of Daga hundred, the area of the parish lying outside of the hundred border is excluded. No corresponding maps of Villåttinge hundred from this time have been found as of yet. The untitled map (H14) depicts two dotted lines, one representing the parish border and one representing the border of the hundred. The latter consists merely of one straight line with two turns, while the parish border appears to be more circuitous, with several border poles indicated on the map.

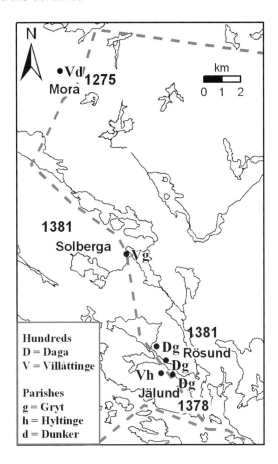

Figure 10-4. Names and localities of medieval homesteads in the vicinity of the borders – the
hundreds are indicated by capital letters and the parish by small letters.

In all of these maps, the western stretch of the parish border, from Lake
Magsjön to Lake Båven, is almost totally congruent with the recent border,
while the hundred border takes a different route, from Lake Dunkern to Lake
Båven (Figure 5).

The most northerly part of the border in these maps is connected to Daga
common on one side, and to Dunker parish in Villåttinge on the other side.
The most southerly part of the border passes through the former island of
Hyltingeö, dividing the island between the parishes (Gryt and Hyltinge) and
between the corresponding hundreds (Daga and Villåttinge).

The oldest printed map showing Gryt parish in its entirety is a map of
Södermanland from 1804. Here the drawings of the parish border and the
hundred border are the same as in the maps described above. In a geological
map from 1865 (scale 1:50,000), the border of the hundred has been altered
so that it follows the parish border from Järnaån to Såsjöstaven (Figure 5),

and, from the end of the 19th century onwards, all maps show a hundred border congruent with the parish border.

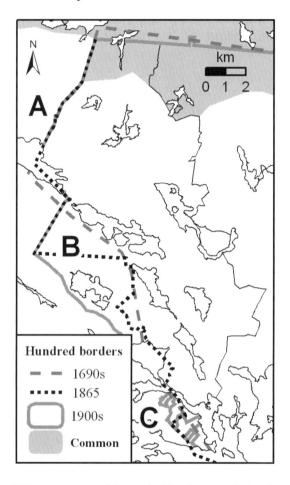

Figure 10-5. The different stretches of the hundred borders over the last three hundred years, and the sections of the investigated stretches.

From at least the end of the 18th century up until the end of the 19th century, taxation assessment rolls incorporate the estate of Rösund within Villåttinge hundred. This represents a discrepancy when compared with older sources and maps.

Officially-decided and executed alterations of the districts were undertaken on three occasions during the 19th century. The first alteration took place around 1838-1840, when the common was divided between the parishes (provincial documentation, The National Archives of Sweden). Then, in 1889, the border between the two hundreds was harmonised with the border of the parishes (Lagerstedt 1973). Finally, the division of the common was reconfigured in 1898 (ibid.) as a result of a request from the people of Gåsinge parish.

Place names are often considered to represent a very old source material. This, however, requires special interpretation, and for the case study at hand the situation can only be tentatively inferred. Along the actual borders, there are several place names which are or might be related to these borders (Table 2).

Table 10-2. Place names related to the border – to be compared with Figure 6.

Type of place name		Swedish place names	English translation
Names of markers	1	Häradsholmen	Hundred-island
	2	Häradsholmen	Hundred-island
	3	Stavåsen	The pole-ridge (a settlement)
	4	Stavdal	The pole-valley (a disappeared settlement)
	5	Såsjöstaven	The pole of Lake Så
Potential names	6	Rösundet	The cairn-sound *or the* reed-sound
Exceptional names	7	Galgholmen	Gallows-island

To summarise, two major changes have been traced concerning the border between Villåttinge and Daga hundreds (Figure 5, Section B), both occurring in the 19th century. The actual part of the parish border (from Magsjön to Båven) seems to have maintained its form over the years, aside from the incorporation of the common. The parts of the borders of both the hundred and the parish around the island of Hyltingeö appear to have had minor, but recurrent, alterations made to them over the years. These alterations seem to incorporate older stretches so that the result is, however, fairly stable over time.

5.2 The Appearances of the Borders in the Field

The current border is normally well demarcated in the field, probably for the reason that it is congruent with boundaries of estates. The border demarcations often consist of mounds of five stones at points where the border alters course, or – where there are more than two estates involved – with smaller markers. It is also common that supporting stone arrangements are situated at intervals along the course of the border. The latter can be mounds of stones, or single raised stones on bare bedrock. Many of the border markers have been renovated and retouched during modern times with bright colour dots, while the marks along some stretches are now difficult to detect, as they seem to have been neglected for many years and are often more or less overgrown.

It has not been possible to discern the old border of the hundred in the field, as there seem to be almost no man-made markings left (at least they have not been found by the investigators). However, by overlaying older maps onto modern maps, it has been possible to gain at least an approximate understanding of earlier locations. The maps from the late 17th century give the names of the places adjacent to the border, and also mention some markers. The geological map from 1865 provides a rather good possibility of reconstructing the border, although the line on this map is somewhat altered in relation to the representation in earlier maps.

In the following description, the investigated stretches of the borders have been divided into three sections: A, B and C (Figure 5). The remaining parts of the parish border – to the north, east and south of the parish – have been studied in the same manner, but will not be described in this paper.

In section A, from Magsjön to Misteln (a distance of approximately six km), the two borders are congruent. On the top of Magsjöberget (the most-elevated part of Gryt parish) is a large mound of five stones with a pole, indicating the meeting point of three hundreds (Daga, Villåttinge and Åker). The old map of the common (H2) identifies the lake as the location of the "corner" of the common(s), while the mountain is the "corner" on the printed map from 1804. The course of the border follows a fault escarpment. In the adjacent parish to the west of the escarpment runs an elongated valley with agricultural fields, at around 30 metres above sea level. In the area of Gryt parish, the altitude is between 70 and 90 metres, and the landscape consists only of lichen-dominated pine forest on thin soils or bedrock with bogs. Along the escarpment follows an esker. Accordingly, the border is easily identified by following these linear physical geographical elements, and they divide the landscape into two different areas: Gryt parish and Dunker parish. These areas are relatively discrete and unconnected, as there is an altitude difference of about 50 metres and disparate nature and land-use types on each side.

In section B, from Lake Dunkern to Lake Inbåven (a distance of approximately 10 km), the two borders follow different courses. Dunkern is a lake without islands, and there are consequently no visible border markers in the lake. The parish border turns southeast in the lake, and subsequently follows a watercourse (Järna River), while the hundred border crosses the lake and follows the esker at the opposite side of the lake, where it changes direction to the southeast.

Järna River is a slightly meandering smaller river, and flows parallel to the adjacent esker. The recent border in the map follows the watercourse except for the northern part, where the border makes a curve, probably indicating the earlier course of the river channel; which is now abandoned, as is common in meandering rivers.

From Järna River the parish border turns eastward, at the point where a bridge crosses the river. It crosses the border of the hundred and continues straightforward for three kilometres, from the esker to Såsjöstaven (Table 2).

Såsjöstaven is a large mound of five stones and a pole. At this place three parishes meet: Gryt, Dunker and Hyltinge. Apart from the existence of modern markers (e.g., coloured plastic bands) and the differences in vegetation (elderly forest and a cleared area), there is no indication that any natural features have been used to demarcate the border, or that any man-made mounds of stones have been raised along this section of the border. The actual line cuts the forested area in a straight line and goes up and down in the rocky terrain, which makes following the border a rather difficult enterprise. There are, however, several extremely large boulders in the terrain that may have acted as points of border reference during earlier times.

From Såsjöstaven down to Lake Båven (a distance of seven km), the recent border still appears as a straight line, but with many more turns and man-made markers (single stones or mounds of stones) along its course. The old map (H14) shows an even more uneven line and, in the vicinity of the actual border, there are also other arranged stones and several natural topographic elements that may well have been used for marking purposes. The most startling of these is a short row of large stones (above one metre in diameter) raised in a line parallel to the actual border, but at some 10 m distance from it. This part of the border also gives the impression of being more adapted to the topography than the aforementioned part, as it follows ridges in the bedrock to a greater extent. As a point of interest, it can be said that the term *badger stone*, given to one of the marks on map H14, probably indicates that a badger's sett has been used as a marker.

The final part of the parish border's section B leaves the coniferous and rocky areas and passes through fields, between the estate of Boxtorp in Gryt and the adjacent estate of Tuna in Hyltinge parish, for about 500 metres.

According to the maps, the direction of the corresponding hundred border continues to follow the esker for most of its course. This esker runs further south along the watercourses to Lake Naten, while the border turns westward on the oldest map. In the region between where the border turns and the point at which it meets Lake Båven, there are two place names deriving from the borders: Stavdal and Stavåsen (the Pole valley and the Pole ridge – Table 2, Figure 5). There are also smaller amounts of glaciofluvial assorted soils in esker form visible in the field. They are not, however, shown on the geological map; probably due to the higher degree of generalisation found on this map.

In section C across Hyltingeö (from Inbåven, a bay of Lake Båven, to Häradsholmen, the southern point of Gryt parish), both of the borders seem to have repeatedly had small alterations made to them over the years. In Inbåven, the borders seem to have been connected to different islands over time. This part of Lake Båven developed into a bay during the 18th and 19th centuries, when the former island of Hyltingeö became connected to the mainland at the wetland close to Tuna. Galgholmen Island, situated in the middle of Rösundet (Figure 6, Table 2), seems to have been used as a border island for both borders on all of the maps that we have come across.

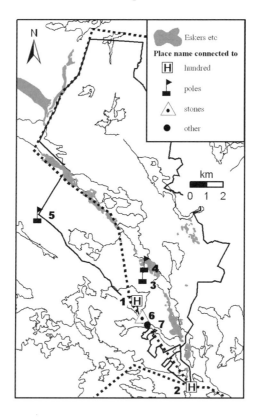

Figure 10-6. Place names connected to the borders (the numbers indicate the markers as in Table 2).

Through Hyltingeö Island, the parish border division cuts between homesteads and villages in a complex manner. There is little or no association with topographic features to explain its course; and the line is exceptionally hard to follow, as there are few markers along its course. The few markers that do exist seem to be "leftovers" from earlier border delineations.

On the map of Daga (U60), the border of the hundred is indicated merely as a straight line extending from Inbåven to Häradsholmen Island, passing the island of Galgholmen and the yard of the Jälund estate.

Both borders end at Häradsholmen. This island has been used as a marking place for both the hundred and the parish in all of the maps we have come across. It is a meeting point of three hundreds (Daga, Villåttinge and Rönö) and of four parishes (Gryt, Björnlunda, Hyltinge and Ludgo).

5.3 Socio-Cultural Aspects of the Borders

The mountainous Mälarmården forest represents a large-scale barrier separating the southern parts of Södermanland and Lake Mälaren in the north. Eskers and valleys extending in a north-south direction have been important for transport and communications towards the towns and marketplaces along Lake Mälaren (Figure 1). An esker forms an excellent foundation for a road, as the sandy soils provide a firm and dry surface; and the ridge form, which rises 10-20 m above the surroundings, affords a good opportunity for surveys to be made. The transit valleys facilitate passage through the otherwise rocky terrain, and sometimes contain navigable watercourses.

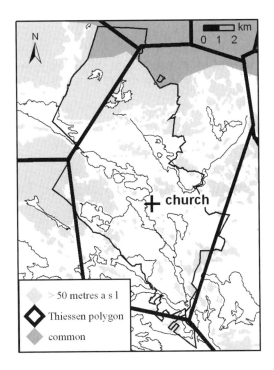

Figure 10-7. A created Thiessen polygon area around the church (based on the distances to neighbouring churches) and the parish.

The best-known route transecting Mälarmården is the fault fissure containing Lake Klämmingen and the lake system of Marvikarna, which offers an opportunity to travel by boat from Trosa on the coast to Mariefred on Lake Mälaren. Along this route is situated one of the few large Bronze Age barrows in Södermanland.

Three eskers connect Lake Mälaren to Lake Båven and further, via Nyköpingsån (the River of Nyköping) to the town of Nyköping. One of them is the esker that runs parallel to sections A and B. This esker also has a barrow, located at Järnaån (Section B), although it is smaller than the aforementioned one.

In order to find out if the form of the parish conforms to the distances from the churches to the border, Thiessen polygons were constructed around the parish church of Gryt and around the neighbouring churches. This method was drawn from a previous investigation carried out in Gotland, which established a significant degree of correspondence regarding distance to churches and parish borders (Lindquist 2001). The result (Figure 7) shows a fairly good degree of correspondence concerning section B of the parish border. However, the topography requires more detailed studies to make a definitive analysis possible. The distance by road from Hyltingeö to the parish church at Hyltinge is about nine kilometres, and to Gryt it is about eight kilometres. Up until the end of the 19th century, when a bridge was constructed, the route to Gryt had necessitated the use of a ferry service at Jälund.

6.   DISCUSSION

6.1   From a Zone to a Line – Analysis and a Tentative Model Concerning Functions

Modern legislation in Sweden presupposes that all land has an owner. Accordingly, boundaries are of central importance when it comes to the division of land between property holders. The situation today, whereby district borders and property boundaries are congruent, eventuates in the risk of taking the need for demarcation for granted. What were eventually to become borders on maps between equivalent territorial units (e.g., between estates, parishes or hundreds) may well have originally developed from other rationale. From such a perspective, a border should be understood more in terms of a process, rather than as the result of a single decision. The configuration of the border itself could thus tell us something about the process involved. Such a process could briefly be described as consisting of three steps: 1) a primary use establishes a zone as something special and discrete; 2) this establishment is underpinned by connecting it to distinctive features or, where these are lacking, by erecting man-made markers; and 3) the zone is delineated and demarcated by a line drawn on a map by a land surveyor using such marks as support.

Evidence pertaining to the two borders examined in this study suggests that they have less to do with dividing desirable land between competing neighbours, as the actual parish border is not in the middle of two adjacent settlement regions. The border in section A leaves very little land to the Mora estate in Dunker (Figure 2). A similar situation applies for section B, where the area of settlement around Ricksjön (Figure 3) has less land than is the case for the neighbouring area of settlement on the other side of the forest.

The hundred border in sections B and C seems to have been drawn with little attention paid to the specific nature of the land that it crosses. Old documents concerning the area indicate that the border separated the three farms of the village of Solberga, and even passed through the courtyard of the Jälund estate.

The printed maps illustrate how the intricate nature of the stretch of the parish border in section C contrasts markedly with the straight line of the hundred border. The parish border follows the margins of the fields, turns across some fields and forms a zigzagged line. It even separated some landowners living in the same village into different parishes. The problematic nature of this line is attested to by statements from local people given to a commission into irregular boundaries that was carried out in the late 19th century. One of these related how: "the parish border between Hyltinge in Willåttinge and Gryt parish in Daga hundred goes through mixed properties of Gälunds village ... such that a proper and obvious parish and hundred border does not exist" (provincial documentation, The National Archives of Sweden).

6.2 To Prevent Getting Lost and to Prevent the Loss of Cattle

Up until the beginning of the 18th century, about three quarters of all land in Sweden was collectively-owned by villages or hundreds as commons or the like (Jansson 2002). In the case of Södermanland, these commons were often outfields, situated either in areas of coniferous forest on lichen-covered bedrock, or in nutrition-poor forests with a thin till layer. In the period before iron working developed into charcoal-needing foundries, the use of such areas had been extensive. The principal uses included: cattle grazing, hunting, timber cutting, the collection of wood for fuel, and the gathering of herbs for food and medication. The areas were also the hiding places of animals of prey and outlaws.

In such areas, there was probably a considerable need for taking measures, for instance, to ensure that the children who were responsible for shepherding did not get lost in the terrain. The possibility to graze cattle in this forestland was also dependent on allowing the cattle to move through the areas. The topography in the fissure valleys district must have limited these possibilities in many areas.

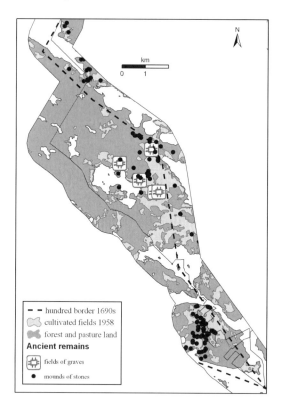

Figure 10-8. The area between the two borders in section B.

One way of understanding more clearly the manner in which the parish border in sections A and B developed is to relate them to settlement regions.

In the case of section A, the border should thus be located at the marginal zone of the settlement region around Mora and the valley at Lake Dunkern. This settlement region is now a part of Dunker parish. The border developed along the fault escarpment, where the possibility of using the land situated above was not considered worthwhile.

Section B should likewise be located at the marginal zone of the settlement region around Ricksjön. In this example, it was not the escarpments that represented the margins, but it was instead the extensive marshy area that fills up a large fissure (Figure 8). In the southern part of section B, the fields in the area of settlement around Tuna in Hyltinge parish reach the border in some instances. This probably explains the more elaborate line of the boundaries. In this part of section B, the border zone is an area requiring negotiation between the neighbours of adjacent villages and settlement regions.

6.3   For Transport and Accessibility

The border of the hundred in section B follows the esker for the most part. The suitability of eskers as roads has been described above.

A border zone, which people could easily travel along, was useful in at least two respects. Firstly, external communication and travel between different parts of the province of Södermanland was necessary; so that men could quickly be called upon, for example, to serve in the defence of the king. Secondly, the provincial code (Södermannalagen) stipulated that the hundred was to be responsible for investigations concerning murder cases where the offender was unknown – *dulgadråp* – either to find the murderer or to pay the damages to the victim's family. An easily-accessible border made it simpler to specify which hundred was responsible in such cases.

For the hundred, a border that divided a region of settlement or a village, as was the case for Solberga and Jälund, did not necessarily constitute a problem. The supposed responsibilities of the hundred – the boats, the royal bodyguard, the court of law – pertained to activities that generally took place elsewhere, and which were independent of the local community.

6.4   Comparison and Inferences

In their northern parts, the two border stretches are similar, following prominent topographical structures in the terrain – such as the fault escarpment and the esker.

The middle parts of the borders differ. The hundred border follows the esker, while the parish border passes through a hilly forested area.

In the southern part, across the island of Hyltingeö, neither the hundred border nor the parish border seems to be related to any topographic or substrate features.

In section B, a comparison of the two borders educes very clear differences in relation to land-use (Figure 9). The parish border is almost entirely drawn through forested areas, lakes or watercourses, while the hundred border – to a considerable degree – passes over fields. The situation on Hyltingeö (Section C) is different, as the parish border crosses fields in a completely distinctive manner compared to the other part of the parish border (Figure 9).

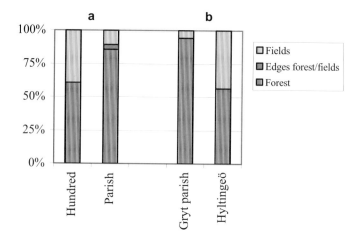

Figure 10-9. Comparisons in section B of the two borders (a); Comparison between the parish border on Hyltingeö and the entire parish border (b) (no differentiation made between edging and crossing fields).

The manifestations of the investigated borders seem to indicate reasons – other than divisions between estates – for the evolutions of what were to become the two borders. For sections A and B of the parish border, we suggest the border be interpreted as having originally been a marginal zone around a settlement region; section A around Mora, and section B around Ricksjön. For the corresponding hundred border, we suggest that the border developed from a zone of accessibility through the landscape. Section C of the examined borders is more difficult to interpret. The appearance of the borders (the parish as well as the hundred) evidences little (or nearly no) connection with natural features. This area has a large amount of ancient remains, such as graves and other remnants (Figure 8). As an island, it seems odd that it should be divided into two parts. One suggestion may be that this partition was the result of an external desire to divide the community located on this island. An alternative suggestion is that it resulted from an internal dispute concerning which church people and/or villages should belong to. The Thiessen polygons demonstrate that for the inhabitants of Hyltingeö, there were three churches situated at approximately the same distance from each other. Before the 19th century, the island could be reached from the mainland either by going across a marsh at Tuna, or travelling by boat from Jälund. The continuous alteration of the border over time could thus be explained as a result of marriages, and new preferences being internalised by families. A more accurate understanding of the situation in regards to Hyltingeö will require further investigation.

We have not found in the relevant literature, any specific discussion concerning the development and the processes of what were to become boundaries in Sweden, but there is many descriptions distinguishing diverse boundary forms as in the example below.

In a study of the development of settlement regions in the landscape around Mälaren, Hyenstrand proposes three types of boundaries: 1) basic boundaries relating to natural linear elements, such as watercourses; 2) boundaries dividing units perpendicular to the basic boundaries; and 3) boundaries dividing outfields from forests (Hyenstrand 1974). The latter type is described as a boundary delineating the outer limits of settlements in regards to the forest area beyond. This seems to correspond to the interpretation suggested above as to the origin of the parish border in sections A and B. The forest boundary is, however, assumed to be fairly indistinct; a presumption which is not supported by our idea concerning the importance of knowing where one is in regards to the landscape.

The fact that boundaries – even in forested areas – appear to have been demarcated as straight lines between prominent markers has been suggested to indicate ancient specific patterns of ownership (Tollin 1999). In the case of our investigation area, we suggest that such straight lines mainly resulted from the mapmaking process itself. It seems unlikely that the boundaries were originally straight in areas such as the northern part of section B, where the borderline crosses ridges and valleys for a distance of three kilometres. The contemporary border ought to have been more adapted to the landscape.

The assumption is, therefore, that the straight stretches of the border are later developments. When people started to make maps, the topographically-adapted stretches of the border in the outfields and commons were also developed as straight lines, as these were more easily managed from a mapmaking point of view. Such praxis could also draw on support from instructions written into the provincial code (*Södermannalagen*), which stipulated that straight lines between markers were to be used in controversial or disputed cases. In the case of our study area, however, it appears that few conflicts arose, as the originally curved line was eventually transformed into a straight line (i.e. the line shown on the maps). If so, this would explain the disparities between the straight lines in the northern parts of section B, and the more elaborate lines in the southern part. When a generally accepted "principal" line was turned into an economically important line, conflicts could arise, resulting in letters being sent to the Court claiming that the altered line failed to take traditional aspects into consideration (Bonow 2002).

In the case of the hundred, the border did not pertain to property rights; and for this reason, it was probably sufficient to indicate a principal line. As conflicts were few (or even non-existent), and because recognition was easy as the zone was a pathway, the visual demarcation of this line seems to have been considered unimportant. The opportunity provided by this zone as a vantage point does, however, appear to have been of some appeal, and there are several ancient remains along the esker – with the barrow at Järnaån representing the most impressive example.

In section A, the two types of borderlines in the study area are congruent. This could suggest the accommodation of a "new" border in regards to an already existing one. Another interpretation, and one which we propose, is that the different landscape elements – the escarpment and the esker – supported the two borders independently.

In prehistoric and historic times, the interest in establishing the recognition and identification of landscapes was probably more crucial than dividing it into individual landholdings, especially in areas that were difficult to access.

7.   CONCLUSION

Boundaries have principally been studied in relation to aspects pertaining either to property rights (e.g., Tollin 1999), or to the establishment of borders of landscapes or countries (e.g., Paasi 1999). In this context, borders or boundaries seem to have developed mainly as a solution to a conflict. There appears to have been less research done in regards to the examination of intermediate-sized territories (Möller 1995). Our findings indicate the existence of motives other than conflict when it comes to the development of such borders.

The findings from the examination of older documents and maps indicate an enduring degree of stability in respect to the manner in which the two (parish and hundred) borders have appeared in the landscape over the centuries. The computer-based comparison between the two borders in section B (Figure 9) evidences differences in land-use in the areas that the two borders penetrate. The findings from the field investigation indicate that the straight lines on the map are a late occurrence. There are often boulders and hillocks – which seem suitable for use in terms of orienting oneself in the landscape – in close proximity to the straight lines.

We suggest interpreting the parish border sections of A and B as being marginal zones connected to the settlement regions of Mora and Ricksjön, respectively. These borders should thus correspond with the suggested forest boundaries (Hyenstrand 1974).

In the case of the hundred border, we suggest that it was originally an access zone, which served as a function of communication between the different regional units (i.e. the hundreds).

The parish border in section C is intricate and difficult to interpret. However, two tentative suggestions may be ventured. Firstly, that the churches were relatively equidistant, and that this caused the villages to divide according to property rights; or, secondly, that the villages were divided as a result of the decree of some external authority, with the intention of splitting up the actual settlement region on Hyltingeö.

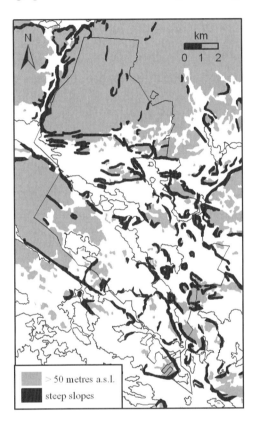

Figure 10-10. The parish border in relation to altitude and steep slopes.

The basic assumption was that the landscape of fissure valleys and escarpments (Figure 10) created a situation where people (and power) were required to subordinate the social structure to the structure of the physical geographical conditions. Our findings regarding sections A and B support this idea, and suggest that the borders of the two types of districts have developed as the result of different needs and circumstances. In section C, we presume it is the social and/or political structures that represent the determinant factors.

A need to map all the areas developed much later, during the 17th and 18th centuries. The desire to connect all areas to ownership, and the enterprise of mapping in hilly and forested areas, may explain the recently-corrected "straight line"-form of borders on today's maps.

Three types of functional zones are thus suggested to have created what were later to become border lines. These are marginal zones, access zones and conflict zones. A fourth variant may also be taken into consideration, i.e. the wish to weaken a community by dividing it.

Two different functions – margins and access – are combined in the watercourses. This might explain why there is such a high degree of congruence between the parishes and the hundreds in Södermanland, a landscape rich in lakes and rivers. At the same time that the water functions as a marginal zone for the parish, it also functions as an access zone surrounding the hundreds.

The findings and inferences presented here must, however, be treated with caution, and their general implications outside of the investigation area have not been examined. More investigation is needed in order to evaluate as to whether the results are more or less generally applicable in regards to other areas.

ACKNOWLEDGEMENTS

Our special thanks goes to Eva Ree, who has been an ever-enthusiastic and supportive companion throughout the fieldwork; Sören Bergström, who has been of considerable support throughout the duration of this project; Lars Gunnar Bråvander and Carl Christiansson for their often challenging discussions; and to Andrew Byerley for improving the English version of this text.

Financial support was made available from Lillemor och Hans W:son Ahlmanns Foundation and Axel Lagrelius Foundation.

REFERENCES

Almquist, J.A. (1917). *Den civila lokalförvaltningen i Sverige 1523-1630.* Meddelanden från svenska riksarkivet. Stockholm: Norstedt.

Almquist, J.A. (1934). *Frälsegodsen i Sverige under storhetstiden.* Skrifter utgifna av svenska Riksarkivet. Stockholm: LiberFörlag/Allmänna förl.

Andersson, T. (1982). Hund, hundare och härad från språklig synpunkt. *Bebyggelsehistorisk tidskrift,* 4, 52-66.

Andersson T. & Göransson S. (1982). Forskning om äldre territoriell indelning i Sverige. En introduktion. *Bebyggelsehistorisk tidskrift,* 4, 3-9.

Bonow, M. (2002). *Contested boundaries of the commons; Landscape, identity and change in Norrra Åsarp parish western Sweden.* Thesis for Phil. Lic at the Department of Human Geography, Stockholm University (mimeographed).

Brink, S. (1990). *Sockenbildning och sockennamn. Studier i äldre territoriell indelning i Norden.* Uppsala. Stockholm: Almqvist & Wiksell International.

Christiansson, H. (1945). Dunker socken. *Södermanlands hembygdsförbunds sockenbeskrivningar för hembygdsundervisning nr 1*. Eskilstuna.

Christiansson, H. (1985). Mora i Dunker – en liten herrgård. In H. Christiansson (Ed.), *Ett stycke Sörmland – Flens kommun i tiden och historien*. Nyköping: Södermanlands museum.

Göransson S. (1982). Härad, socken och by på Öland. *Bebyggelsehistorisk tidskrift*, 4, 97-116.

Holmbäck, Å. & Wessén, E. (1979). *Svenska landskapslagar. Södermannalagen och Hälsingelagen*. Serie 3, Stockholm: AWE/Geber.

Hyenstrand, Å. (1974). *Centralbygd-Randbygd. Strukturella, ekonomiska och administrativa huvudlinjer i mellansvensk yngre järnålder*. Stockholm: Almqvist & Wiksell.

Lagerstedt, T. (1973). *Den civila lokalförvaltningens gränser 1630-1952*. Meddelanden från kulturgeografiska institutionen vid Stockholms universitet. Nr B 24. Uppsala 1973.

Lindquist, S.O. (2001). Kristnandet blir roligare med GIS. In M. Elg (Ed.), *Plats, landskap, karta. En vänatlas till Ulf Sporrong* (pp. 74-75). Kulturgeografiska institutionen, Stockholms universitet.

Lundberg, B. (1982). Äldre indelningssystem i Uppland. *Bebyggelsehistorisk tidskrift*, 4. 24-31.

Möller, A. (1995). *Gränser i upplösning. En studie av by- och sockengränsändringar med exempel från södra Sverige*. Institutionen för kulturgeografi och ekonomisk geografi vid Lunds Universitet. Rapporter och notiser 137.

Paasi, A. (1999). *Territories, boundaries and consciousness. The changing geographies of the Finnish-Russian border*. Chichester: WILEY cop.

Rahmquist, S. (1982). Härad och socken – världslig och kyrklig indelning i Uppland. B*ebyggelsehistorisk tidskrift*, 4, 89-96.

Risberg, J., Miller, U. & Brunnberg, L. (1999). Deglaciation, Holocene Shore Displacement and Coastal Settlement in Eastern Svealand, Sweden. *Quaternary International*, 9, 33-37.

Smedberg, G. (1982). Stift, kontrakt och socken. *Bebyggelsehistorisk tidskrift*, 4. 89-96.

Tham, W. (1852). *Beskrifning öfver Nyköpings län Stockholm 1852*. Facsimileupplaga utgiven av Maj Eriksson. Nyköping: Mellösa, 1976.

Tollin, C. (1999). *Rågångar, gränshallar och ägoområden. Rekonstruktion av fastighetsstruktur och bebyggelseutveckling i mellersta Småland under äldre medeltid*. Meddelande nr 101 Kulturgeografiska institutionen. Stockholms universitet.

Wijkander, K. (1983). *Kungshögar och sockenbildning. Studier i Södermanlands administrativa indelning under vikingatid och tidig medeltid*. Sörmländska handlingar 39. Södermanlands museum. Stockholm: Nordstedt.

Chapter 11

A HIDDEN WORLD? A GENDERED PERSPECTIVE ON SWEDISH HISTORICAL MAPS

Elisabeth Gräslund Berg
Department of Human Geography, Stockholm University, Sweden

1. INTRODUCTION

In learning about historical landscapes and the historical dimensions of today's landscapes, historical maps are a rich source of information. In academic research maps have for long been used by geographers, but they are also becoming more commonly used within non-geographical disciplines such as archaeology, ecology etc. In different types of conservation work, concerning both cultural heritage and nature conservation, their use is today seen as almost compulsory. Historical maps are also often used by non-academics who take an interest in for instance local history and genealogical studies. The attention paid to the information derived from historical maps is often complementing other sources, particularly regarding landscape issues.

Like other sources of information, historical maps need to be analysed critically. This chapter hence provides a critical reading of three large-scale maps from 17th, 18th and 19th century Sweden. Based on these maps, three questions are posed: Whose worlds were mapped? What landscapes do these maps mediate? How do we interpret the maps? These questions will be discussed from a gendered perspective. Within a Swedish context there is almost a total lack of critically assessing the gendered dimensions involved when representing the world through maps and the art of map-making. This chapter thus represents an attempt to provide such a critical assessment. Hence there is also a need to discuss the role of historical maps as academic sources in general more critically. The chapter will end by a brief comment

H. Palang et al. (eds.), European Rural Landscapes:
Persistence and Change in a Globalising Environment, 177-190.
© 2004 *Kluwer Academic Publishers. Printed in the Netherlands.*

on some of the consequences that can be identified of the use of historical maps within landscape and cultural heritage work.[1]

2.   A GENDERED PERSPECTIVE

The concept of *gender* denotes sex to be socially constructed rather than biologically ascribed. The linguistic origin of the term derives from the meaning of *sort* or *class* and it thus pertains to a situation of classifying and sorting *differences* (Scott 1988; Haraway 1991). Gender is part of a socially agreed system that can be called *the gender-order*, which implies how the sexes/genders are kept apart – that the male and the female are clearly defined as different from each other – and that there is an hierarchic order between the sexes where the man or the male constitutes the norm (Hirdman 1988; Scott 1988). Like any other social system, the gender-order is more of a continually ongoing process than a static structure.

In relation to this process, gender research has been linked to a wider discussion concerning the phenomenon of sorting and viewing the world from the perspective of oneself as the norm. As such, the Self can be denoted as "the master subject", i.e. the one who assigns the norm for that which is valuable, interesting, what should be visible and what one should do. Postcolonial and feminist studies question whose norm is considered as being the valid one. With regard to the gender aspect, this concerns how the man or the male comes to represent the norm whilst the woman or the female represents "difference" (Rose 1993).

This gender-order, like other social structures and relations, is embedded in the landscape (Monk 1992; Duncan 1996; Forsberg & Grenholm 2000). As examples can be mentioned how the physical world is structured so that there can be different places where women's and men's activities respectively take place, e.g., daily work or duties. There can also be different ideas connected to how men and women can behave in certain places, like for instance public places, or where they can feel safe and confident. I will therefore argue that also landscapes that are depicted in historical maps can be seen as gendered.

Maps have also been seen as social constructions, an idea that has been developed most convincingly by geographer J.B. Harley (see Harley 1992). Harley has argued that maps are always and thus in general, created within specific contexts and must be read with all the subtexts they contain firmly in mind. He pointed out that maps and cartography are means of power. This means both that a hegemonic power will use maps for its own exercise of

---

[1] This chapter is also published in Swedish: Gräslund Berg, E. (2004).

power, and that an internal power is exercised within the mapping process, as for example in relation to what is mapped and what is not, and how this is then depicted. However, this approach has not been applied in the discussion on Swedish historical maps. Regarding gender-aspects of maps, it has been shown how the ideas of mapping and surveying was closely connected to the ideas of the Enlightenment and the male "Cartesian Self", which Kirby has exemplified with explorers on colonial expeditions in North America (Kirby 1996). There is also, as already mentioned, a lack of a critical gendered reading within a Swedish context. Hence a "double" lack of critical analysis can be identified. In the following I will take the general aspect as starting-point for a more general – yet critical – discussion of historical maps as sources for understanding landscape change.

3. MAPPING PLACES OF WORK

In order to seek out gender aspects in the chosen historical Swedish maps, I will focus on places of work. With this I mean the physical places, where daily work activities take place. Concerning later historical periods (the 19th and early 20th centuries) and the present times, the division of work between men and women has been central to gender-research (Lindgren 1996; Sjöberg 1996). The division of work has been seen as a fundamental expression of how the sexes were kept apart (Hirdman 1988). Gender-research dealing with earlier periods has not in the same way focused on this aspect of the gendering process, but more upon the work-tasks themselves (Lövkrona 2001).

In summary, male and female work-tasks during the 17th to 19th century Sweden were divided between the sexes, but complemented each other to make the household work as a unit (Sjöberg 1996; Andersson Flygare 1997). One has to be aware, however, that what was defined as male and female work-tasks varied both locally and over time. In general, the female tasks in the Swedish agrarian society were to take care of the animals (except the horses) and most tasks related to the animals, they cooked and stored the food, grew vegetables, cared for the children, the elderly and the sick, they managed the household and performed textile work. Men's tasks roughly embraced work in fields and meadows. They performed work that took place outside of the confines of the homestead such as taking part in collective working parties as, for example, in the case of work at the mills. They also made and mended tools and took care of the horses. The general picture of this period shows a societal hierarchy where the male was superior to the female, in public and in terms of societal power. Women could however participate in many of the male work-tasks and one may argue that women thereby had a broader competence than men did. But it was mainly the male work-tasks that dominated and structured the activities of the household – for instance during the harvest season some of the women's daily work had to be deferred so that the women could assist in the harvesting. In addition, children and the elderly had their own duties and work-tasks respectively. Obviously there were also differences depending on social position – if you were a housewife or a maid, a master or a farm-hand (Sjöberg 1996; Andersson Flygare 1997).

These differing work places produced elements in the landscapes that were to be mapped by the Swedish Crown. The process of mapping and what to map is thus what I now will be turning to.

4.   LOOKING FOR GENDER-ASPECTS IN HISTORICAL MAPS

When the Land Survey was established in Sweden in 1628, the aim was to map every farm or hamlet in detailed, large-scale maps to be bound in books and kept by the Crown. Over the following centuries several hundred thousand large- and small-scale maps were consequently produced. They illustrated a range of themes including resources, ownership, boundary disputes and land reallotments in line with agrarian reforms (on Swedish land survey maps, see articles in Sporrong & Wennström 1990).

Large-scale maps of farms and villages were mainly produced within four time periods. In the first period, lasting from ca 1630 to the 1650s, the detailed maps were bound into books. The second period of large-scale mapping of farms lasted from the 1690s to the 1730s and also resulted in maps with detailed information of land-use and landownership. In the third period, from ca 1748 to the beginning of the 19th century, most large-scale maps were concerned with the land reform called the *storskifte*, which aimed at more consolidated landholdings. From 1803 to 1827 new land reforms of the enclosures of *enskifte* and *laga skifte* resulted in a fourth period when very detailed maps were produced. These enclosures aimed at consolidating the landholdings of the different farmsteads into totally separated plots, and included all types of land, i.e. both arable, pasture, forests etc. were included in the redistribution of land. During these four periods most of the large-scale maps of farms or villages were made in the scale 1:4000 or 1:5000. The maps of the different enclosures often showed forests and pastures in smaller scales like 1:8,000, while a few of the *laga skifte* maps were made in 1:2000 (Sporrong 1990; Tollin 1991).

In order to identify how the gendered division of work was represented in the Swedish historical maps, I have chosen three maps from the first, second and fourth of the above mentioned time-periods respectively. This choice is due to the fact that I want to demonstrate any possible changes in what was mapped both over time and in manners of map-making.

Figure 1 shows a map from 1639 of the hamlet of Hammarby in Uppland (Lantmäteristyrelsens arkiv A8:144). The letter A indicates house-symbols for the three farms that comprised the hamlet. The letters B and C indicate the two fields with arable land and the area D indicates the meadow field. Along the stream there are three small buildings, one mill for each of the three farms. In the text attached to the map, the forest and pasture are mentioned as of good quality, but they are not drawn on the map.

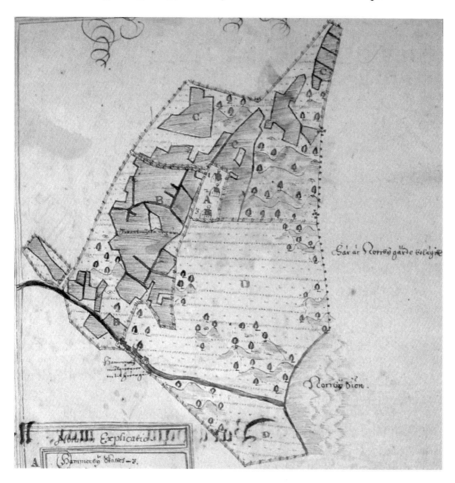

Figure 11-1. Hammarby hamlet, Länna parish, Uppland, in the year 1639 (Lantmäteristyrelsens arkiv A8:144).

What we see in Figure 1 is a map that mainly depicts the work places of men: the fields with arable and meadow land and the mills. The work places of women – places for domestic work, for milking, and the vegetable garden – are not at all represented on this map. Neither are all of the male work places represented, nor indeed are those of many other groups like for instance those of the children.

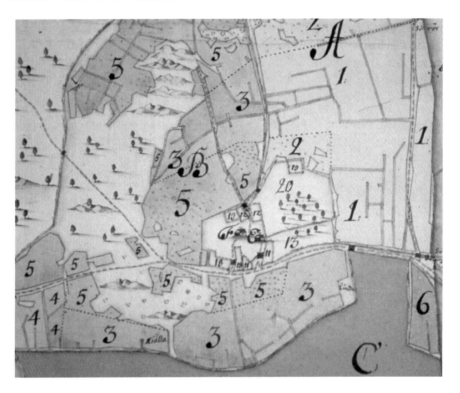

Figure 11-2. Vindsberga hamlet, Tillinge parish, Uppland, in the year 1707 (Lantmäteristyrelsens arkiv B66-48:1).

Figure 2 shows a map from 1707 of Vindsberga hamlet, also in the province of Uppland (Lantmäteristyrelsens arkiv B66-48:1). The letter A indicates the arable field sown in the year of 1707 and the letter B the fallow field. The figures within the fields specify the different qualities of different parts of the arable land. The letter C indicates the meadowland, the southern portion of which is used by the neighbouring hamlet. In the upper left part of the map there is forest and pasture. The accompanying text tells that this was of no importance for firewood collection. The two farms in the hamlet are shown with house-symbols E and F. The blank areas that can be seen surrounding them probably contain the various farmhouses and the farmyard. Some of this area is however marked with figures telling us what is situated there: numbers 11 and 18 are two hop-gardens also belonging to each farm respectively, numbers 12 and 19 are two small herb-gardens belonging to each farm respectively. Two small, enclosed pastures are situated in the area to the right of the buildings. In the southern part of the fallow field there is also a spring or well.

The fundamental features of this map are the same as in the previous one, i.e. focus is placed on the male work places whereby the arable land and meadows are mapped in detail. However, one can also discern the inclusion of more of the work places of women: the garden patches and the enclosed pastures, the latter of which may have been used for calves that needed closer supervision.

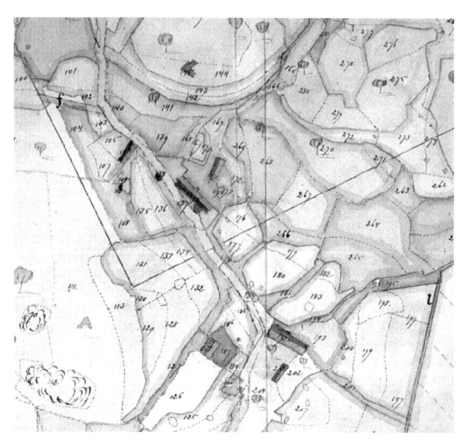

Figure 11-3. Kärnekulla hamlet, Herrljunga parish, Västergötland, in the year 1875 (Lantmäteristyrelsens arkiv O65-6:2).

I now turn to a later map (Figure 3) from 1875 showing the hamlet of Kärnekulla in the province of Västergötland (Lantmäteristyrelsens arkiv O65-6:2). The map was made for the enclosure of *laga skifte*, when the land was to be surveyed in detail in order to be redistributed justly between the landowners. The *laga skifte* maps also focus on arable land and meadows as this was the most valuable land that was to be redistributed. This detailed mapping also included features such as the buildings one can see on this map. The map no longer represents the farm buildings with symbols, but instead depicts them "realistically". The text following the map denotes the dwelling houses (no. 106, 171 and 188), the cowsheds (no. 175 and 192), some woodsheds where firewood was kept for the winter (no. 174 and 203), a smithy (no. 145, situated at a distance from the other buildings because of the risk of fire) and some other buildings about which the map gives no information of their use (no. 110, 138 and 190). The areas for vegetable gardens and orchards are also marked (no. 105, 107, 168, 169, 170, 172, and 173).

This more detailed mapping offer the potential to conduct a more thorough analysis of the places where the larger part of women's work was carried out – in the building(s) and in their immediate vicinity. This is also the case for some male work-tasks, such as the place where the wagons and different tools were kept. If vegetable gardens and similar types of land-use were depicted in these later maps, this seems to have been done in the same mode and with the same detail as with the arable land.

5.   DISCUSSION

When looking for gender aspects in these three Swedish historical maps, it was obvious that in the oldest map from 1639 it was mainly the landscape of male work places that was represented. This focus then continued in the later maps from 1707 and 1875, but there the female work places also became more visible if looked for. In terms of spatial form, maybe one could argue that the male work-tasks more "naturally" belonged to these maps as they particularly denote area-based features such as arable fields and meadows. The female work-tasks, such as the tending of cows and sheep, were also area based and their places (like pastures) are represented in most maps except from the earliest period. But other male work places that were not area based, are in addition to the fields also shown in the maps – as, for example, is the case with the mills in Hammarby (Figure 1). Corresponding female work places such as the washing houses and baking houses are not represented in the maps, except only occasionally in some of the latest maps produced at end of the 19th century or beginning of the 20th century. What was mapped cannot therefore be explained solely by the techniques of measuring and drawing. Rather, one must understand that an active choice was made concerning both what one was interested in showing and what one chose to ignore.

In order to understand how these choices were made, one has to situate the map-making enterprise in its context. The initial purpose of the Swedish cadastral map-making enterprise is not completely evident, but the common argument suggests an economic motive whereby the Crown wished to gain a general over-view of the country's resources (Helmfrid 1959). This is in line with how a combination of economic motives with pure expressions of power and right to land seems to have been the ground for all cadastral mapping in the service of different states throughout the world since the 15th century (Kain & Baigent 1992).

Concerning the economic perspective, one could argue that women were responsible for reproduction while men were responsible for production, and as such, it was only the places where men's work took place that were of interest in this map-making enterprise. Obviously, reality was not that simple. The breeding and raising of livestock, especially in some regions like for example western Sweden, was an important or even dominant part of production on the average farms. As such, the work of women was not delimited to the reproductive sphere. Why then was there such a focus on male work places – "male landscapes" – in the earliest maps?

It is clear that the oldest maps, dating from around 1630 to 1650, were principally concerned with arable land and meadows. Arable land was carefully measured and mapped and done so quite accurately. Meadows were always mapped and represented as an area and not with simple symbols, even if they were not always in a measured manner but often drawn more casually.

If the focus on male work places directly related to an interest in the arable land, one needs to explore the underlying reasons for that interest. In early medieval times Swedish agriculture was intensified, the clearing of land increased and technical advances were made that improved the efficiency of cultivation (Myrdal 1989). The process of State formation at the time meant that the surplus of production was expropriated in order to finance the new State and those in power (Lindkvist 1990). The landowner decided how much was to be paid and in what form, i.e. produce and/or money. It has been suggested that the expansion of grain production was the result of the landlord's interest in levying taxation in grain. The reason for doing so is open to debate, but it has been suggested that the landlord wished to adopt a continental way of living that included the eating of leavened bread. In addition, grains were more easily stored and transported as compared to dairy products (Hagerman 1999). If such aspects contributed to the increase in grain production, the farming of and production from arable land, was to a considerable extent directed "outwards" and "upwards" to the landlord, the Crown or the Church.

I believe that the role of grain production in the taxation system could have influenced the focus of the map-making process initiated during the 17th century. Much of the large-scale mapping carried out on the European continent concerned itself with mapping individual estates, a scale at which the landlord wanted to gain a general view and control (Fletcher 1998). The stately cadastral map-making has often been linked to taxation and to agrarian reforms or development into a more capitalistic agriculture (Kain & Baigent 1992). If the intention with the maps was to provide a better economic overview, it would appear quite rational to concentrate mainly on representing arable land in the maps if the grain production was so central to the taxation. I would thus argue that the Land Survey of Sweden (established in 1628) and the mapping that eventually took place was initiated in a milieu where the arable land was seen as an expression of the socio-economic power-relations between landholders and landlords, and that this was what the Crown wanted to document and visualise.

Maps of regions where grain production was not so dominant – for example in mining districts or coastal areas – follows the same trend with a pronounced focus on areas of arable land even if these were very limited in size. This, in combination with the interest in some specific non-economic features such as ancient monuments (for example Sporrong et al. 1990), suggests that the contents of the maps cannot be explained simply as being rational from a taxation perspective. Instead it reflects the world-view of those in power, of the "Master subject". When the surveyors went out to the different regions of Sweden, having internalised this norm, they did not see – or at least not depict – the reality prevailing there. Instead they held on to their conception or idea of it, and thus mapped what *they* saw and put this into their model of the landscape. So, in the process of map-making, one was not totally guided by different peoples' uses of land per se, but by the power-relation to landowners or rentiers. Whilst maps and surveys are today quite commonly regarded as an expression of power, I would argue that this view is rarely made explicit in the use and interpretation of historical maps.

From a gendered perspective, would I then argue that maps should be seen as an explicit expression of men's power over women? No, I see these maps as evidencing an expression of power within a man's world. The term *homosociality* is an expression of the preference amongst men to socialise and relate to other men (Lindgren 1996). In order to maintain power, it must be expressed or demonstrated in relation to those who were able to challenge it or contest it, i.e. other men. But even if a main purpose could have been to strengthen and articulate power within men's spheres, it often led to a neglect of women's perspectives, which must be seen as a gender-divided hierarchy of power. In line with the argument made throughout the above discussion, the oldest maps were about the economic relation between households and the landlord or the Crown. The initiators of the map-making enterprise were men and it was also usually men who, in their role as heads of households, represented the external relations of households and who were the ones responsible for the payment of taxes. The formal socio-economic relations between different groups in society were thus canalised through men at both ends, and thereby the maps to a large extent came to deal with what could perhaps be called *male aspects of landscape.*

Figure 11-4. Land surveying in northern Sweden around 1910. Photo: Robert Lundgren. Västerbottens Museum, Umeå, Sweden.

In the actual process of making the maps the surveyor did involve the local inhabitants. In line with the general division by gender in society, it was men who represented the local people, as is illustrated in Figure 4. The picture shows the people involved in a survey in the north of Sweden in 1910 in order to reallocate the land in a *laga skifte.* The surveyor and his assistant(s) are men. Around them is also a group of people who are probably landowners or more important landholders in the village, who were called upon to provide information about boundaries, land-use, quality of land and ownership. Also visible are a number of farm-hands and possibly crofters whose information was perhaps also called upon. One can also see younger boys gathering around the adults. There are no women, and as far as is known this picture is representative of the manner in which these map-making processes were carried out. One may wonder how familiar these men were with the women's work places, i.e. how precise was the information they were able to give the surveyor? My suggestions are in line with that shown by, for example, Withers regarding the naming exercise during the mapping of the Scottish Highlands during the 19th century (Withers 2000). Natives as authoritative sources of local naming were exclusively chosen by the officials in a manner that strictly conformed to general ideas of social position in society, and whereby women seem never to have been called upon as informants (ibid.).

6.   CONCLUSIONS AND CONSEQUENCES FOR RESEARCH ON HISTORICAL MAPS

Through my analysis of three selected Swedish historical maps, I have pointed out new ways to discuss the mapping of Swedish historical landscapes, or rather the people of and in those landscapes. This has been done from a gendered perspective. In relation to my argument that the people in historical landscapes are often largely invisible or depicted in a socially unproblematised manner, I see this as important. My aim in this chapter has not only been to expose the landscape of women, even though this is an important task in itself. Rather, I hope to have provided a more nuanced view of the historical landscapes in the historical maps. As I have pointed out above, there were also parts of the landscape of male work that were not mapped. Instead, it was principally the public aspects, which were the focus of attention.

From a gendered perspective, my discussion has also led on to some wider questions concerning the arable land in the maps and also broached the general economic power relations within the Swedish society during historic times. This has opened up for a more principle criticism of the historical maps and their contents as sources, an aspect that I believe is vital for future use and research. My emphasis on the fact that there were many things that were never mapped is not meant to suggest that these maps in general have a lesser value or are less useful. What it *does* suggest, however, is the need to specify and define what the maps pertain to and what information is exactly conveyed (see Harley 1992).

Thus, I have shown that the representation of landscape in the historical maps can be characterised as being gendered. The question then arises as to how this influences our view of (or idea of) the historical landscape today. If historical knowledge is to contribute to identity-formation and roots of people and society, we must try to embrace the different realities in which different people led their respective lives, also in relation to the historical landscape. This also has implications for the way cultural heritage work is carried out, when protecting, maintaining and projecting historical landscapes. The knowledge we acquire through a more traditional analysis of the historical maps, of for example field systems or the development of the settlement structure, does exclude much of the landscape of female work. What cultural heritage could be made known if we use the "gender-glasses" I have used here? And what do we miss if we fail to use them?

Bearing in mind our understanding of the aims of the mapping of the Swedish land survey during the 17th, 18th and 19th centuries, it cannot be considered remarkable per se that not everybody's work places are visible in the historical maps. However, I would argue that it is an important task for historical geography and landscape research to pay attention to such omitted perspectives and hence to try to make them visible.

REFERENCES

Manuscript Sources

Lantmäteristyrelsens Arkiv, Lantmäteriverket, Gävle (The Land Survey Board's Map Archives, Gävle) Sweden:
Map no. A8:144
Map no. B66-48:1
Map no. O65-6:2

Printed and Secondary Sources

Andersson Flygare, I. (1997). Gård, hushåll och familjejordbruk. In B.M.P. Larsson, M. Morell & J. Myrdal (Eds.), *Agrarhistoria* (pp. 183-199). Stockholm: LTs förlag.
Duncan, N. (1996). Introduction. In N. Duncan (Ed.), *BodySpace. Destabilizing Geographies of Gender and Sexuality* (pp. 1-10). London: Routledge.
Fletcher, D. (1998). Map or Terrier? The Example of Christ Church, Oxford, Estate management, 1600-1840. *Transactions of the Institute of British Geographers NS,* 23, 221-237.
Forsberg, G. & Grenholm, C. (2000). *Änglagård(ar) i feministisk analys.* Paper till 20:e Nordiska symposiet för kritisk samhällsgeografi. Hammarö, Karlstad.
Gräslund Berg, E. (2004). Det dolda perspektivet. Genusaspekter på historiska kartor. In U. Jansson (Ed.), *Kartlagt land – kartor som källa till de areella näringarnas geografi och historia.* Skogs- och lantbrukshistoriska meddelanden XX. Stockholm: Kungl. Skogs- och lantbruksakademien, in press.
Hagerman, M. (1996). *Spåren av kungens män. Om när Sverige blev ett kristet rike i skiftet mellan vikingatid och medeltid.* Stockholm: Rabén Prisma.
Haraway, D. (1991). 'Gender' for a Marxist Dictionary: the Sexual Politics of a Word. In D. Haraway (Ed.), *Simians, Cyborgs and Women: the Reinvention of Nature* (pp. 127-148). London: Free Association Books.

Harley, J.B. (1992). Deconstructing the Map. In T.J. Barnes & J.S. Duncan (Eds.), *Writing Worlds. Discourse, Text and Metaphor in the Representation of Landscape* (pp. 231-247). London: Routledge.

Helmfrid, S. (1959). De geometriska jordeböckerna – 'skattläggningskartor'? *Ymer*, 224-231.

Hirdman, Y. (1988). Genussystemet – reflexioner kring kvinnors sociala underordning. *Kvinnovetenskaplig tidskrift, 3,* 49-63.

Kain, R.J.P. & Baigent, E. (1992). *The Cadastral Map in the Service of the State. A History of Property Mapping.* Chicago: University of Chicago Press.

Kirby, K. (1996). Re:mapping Subjectivity. Cartographic Vision and the Limits of Politics. In N. Duncan (Ed.), *BodySpace. Destabilizing geographies of gender and sexuality* (pp. 45-55). London: Routledge.

Lindgren, G. (1996). Broderskapets logik. *Kvinnovetenskaplig tidskrift, 1,* 4-14.

Lindkvist, T. (1990). *Plundring, skatter och den feodala statens framväxt. Organisatoriska tendenser i Sverige under övergången från vikingatid till tidig medeltid.* Opuscula Historica Upsaliensia 1. Historiska institutionen, Uppsala Universitet.

Lövkrona, I. (2001). Hierarki och makt. Genusperspektiv på arbetsdelningen i det tidigmoderna hushållet. In B. Liljewall, K. Niskanen & M. Sjöberg (Eds.), *Kvinnor och jord. Arbete och ägande från medeltid till nutid* (pp. 31-44). Skrifter om skogs- och lantbrukshistoria 15. Stockholm: Nordiska muséets förlag.

Monk, J. (1992). Gender in the Landscape: Expressions of Power and Meaning. In K. Anderson & F. Gale (Eds.), *Inventing Places. Studies in Cultural Geography* (pp. 123-138). Melbourne: Longman Cheshire, Wiley Halsted Press.

Myrdal, J. (1989). Jordbruk och jordägande. In A. Andrén (Ed.), *Medeltidens födelse* (pp. 35-49). Symposier på Krapperups borg 1. Nyhamnsläge: Gyllenstiernska Krapperupsstiftelsen.

Rose, G. (1993). *Feminism and Geography. The Limits of Geographical Knowledge.* Cambridge: Polity Press.

Scott, J. (1988). *Gender and the Politics of History.* New York: Columbia University Press.

Sjöberg, M. (1996). Hade jorden ett kön? Något om genuskonstruktion i det tidigmoderna Sverige. *Historisk Tidskrift, 3,* 362-396.

Sporrong, U. (1990). Land Survey Maps as Historical Sources. In U. Sporrong & H.-F. Wennström (Eds.), *Maps and Mapping. National Atlas of Sweden* (pp. 136-145). Stockholm: SNA.

Sporrong, U., Tollin, C., Roeck-Hansen, B. & Helmfrid, S. (1990). The Home District on the Map. In U. Sporrong & H.-F. Wennström (Eds.), *Maps and Mapping. National Atlas of Sweden* (pp. 146-153). Stockholm: SNA.

Sporrong, U., & Wennström, H.-F. (Eds.) (1990). *Maps and Mapping. National Atlas of Sweden.* Stockholm: SNA.

Withers, C.W.J. (2000). Authorizing Landscape: 'Authority', Naming and the Ordnance Survey's Mapping of the Scottish Highlands in the Nineteenth Century. *Journal of Historical Geography,* 26, 532-554.

Chapter 12

WHEN SWEDEN WAS PUT ON THE MAP

Clas Tollin
Department of Economics, Swedish University of Agricultural Sciences, Uppsala, Sweden

1. INTRODUCTION

The Swedish large-scale surveying of farms, hamlets and villages during the 1630s and 1640s was at that time unique in the world, due to the large scale of its undertaking and homogeneous appearance. Approximately 15,000 maps were created, with most of them put together in so-called *Geometric Land-Atlas Maps* (*geometriska jordeböcker*), consisting of maps from one or more hundreds (*härader*). These maps provide an outstanding source of information in regards to studies on the historical, economical, and spatial details of the old agrarian landscape in Sweden. The material was last addressed in a detailed and complete manner some ten years ago by Elizabeth Baigent and Robert Kain (Kain & Baigent 1992). Recently, a research project was commenced with the aim of making these maps available in a national edition. This project is situated at The National Archives of Sweden, and financed by The Bank of Sweden Tercentenary Foundation and The Royal Academy of Letters, History and Antiquities. The following article provides a brief introduction to this fascinating material.

2. THE MODELS

The oldest preserved large-scale surveyors' maps in existence are from the Netherlands. Among others, there is a map from 1457 drawn up to clarify the duties of dam building in Houtrijk in the northern part of Holland, and another map was made concerning the domains of the manor of des Reiniars in Flandres, signed *Claude mesureur sermenté du Roi* (*Claude land surveyor*

H. Palang et al. (eds.), European Rural Landscapes:
Persistence and Change in a Globalising Environment, 191-208.

of the king). In the year 1600, Maurice of Orange established a course at the University of Leiden consisting of, among other topics, surveying (with a military purpose). In the region of Saxony and town of Brunswick in Germany, systematic large-scale mappings were intended towards the end of the 16th century. The latter were never carried out, however (Helmfrid 1959) Concerning England, there are also in existence occasional maps from the late 16th and early 17th centuries, by Christopher Saxton and others (Tyacke & Hussy 1980).

The Teutonic Order (a German military religious order that formally came into being in the late 12th century) had a long tradition of surveying, and produced numerous maps during the 15th and 16th centuries. Regarding the Livonian Order (a branch of the Teutonic Order), both large-scale and small-scale maps of the Bishoprics of Tartu, Riga, Curland and Rotalia-Ösel (in present-day Estonia and Latvia) were produced (Helmfrid 1959).

Swedes thus had opportunities to learn from Dutch and Livonian cartographers from the late 16th century and onward. The oldest Swedish map that can be dated is from 1593, and shows the island of Dagö (now Hiiumaa, an island of Estonia). It was drawn up by Gerhard Joestinck, a Dutch sheriff then residing on Dagö. In the early 17th century, the Swedish troops in Livonia were commanded by Jacob de la Gardie and Carl Carlsson Gyllenhielm. The latter drew up a summary plan of Tartu (Dorpat) in 1601. There is also a plan from 1609 showing Pernau (Pärnu). In the Livonian War there were numerous maps drawn up by fortification officers of foreign extraction, such as one Georg Ginther Kräill von Bemebergh (who had studied in the Netherlands). Another officer was Georg von Schwengeln from Riga, who among other things mapped the region around the River Daugava/Dvina (Ehrensvärd 2000).

3.   THE SWEDISH LAND SURVEY

In 1628, a central Swedish land survey organization was formally set up under the leadership of the "Mathematician General" Anders Bure (1571-1646), who in 1624 was raised to the level of nobility under the name Andreas Bureus.

Bure and his department had considerable duties to perform. Among others, he was to map all major lakes and streams in Swedish territory, particularly those that could be made navigable and could provide any economic benefits ("… and some profit accrue …"). Knowledge at the time was incomplete regarding ports and natural harbours, where ships and boats could land and anchor securely. Bure therefore was to fathom the approaches and depths of all ports. He was also commissioned with making an inventory of "the nature and properties of all forests, morasses and bogs, naming and detailing them" (Ekstrand 1901: 1). Furthermore, he was to make "secure drawings of all cities and towns in the Realm that now are under the Crown of Sweden or may hereafter be acquired" (Ekstrand 1901: 1).

As for military targets, he was directed to:

> … as far as possible ensure to obtain plans and drawings of foreign, in particular adjacent towns and fortifications, and keep them well and safely (Styffe 1856: 250).

Bure was also made responsible for Sweden's mines and copper and ironworks. This involved both the preparation of mine maps showing breadth, length and depth, and the production of notes concerning the type and quality of the ores. He was also empowered to forbid mine inspectors to break ores at unsuitable places. Finally, he was to peruse the designs and plans for all construction work done for the Crown.

The large-scale mapping of the domains of villages and estates was also mentioned. In the third item listed in the Instruction of 1628, the surveyors are charged with making:

> … special land depictions and drawings, not solely how parishes and districts come together, but also of *the possessions of each and every village with regard to fields and meadows, forest and land* (Samlingar 1, my italics).

This was the point of departure for the unique detailed mapping of Sweden's agrarian landscapes, an activity of which the first phase would last more than twenty years. The resulting maps were collected and bound into volumes, which is why they are called *Geometric Land-Atlas Maps* (in Swedish *Geometriska jordeböcker*).

4. WHAT WAS THE REASON BEHIND THE GEOMETRIC MAPPING OF SWEDEN?

Curiously enough, we do not have definitive knowledge as to the aims behind the great mapping project of the 1630s. Human geographer Staffan Helmfrid writes: "The most remarkable fact about our Geometric Land-Atlas Maps is why at all they were produced" (Helmfrid 1959: 229).

Several earlier scholars assumed that the mapping project was intended as a complete audit of the permanent land levies, which had been accounted for since the 1530s in *the Crown Land Tax Registers* (Lönborg 1903; Williams 1928). The basis for this assumption was summarised by the human geographer Bruno Johnsson in 1965.

Cadastral surveying and valuation of arable land (*revning*) had been done in Sweden since the Middle Ages, by measuring using rods (one rod equals roughly 16.5 feet or 5.5 yards) and lines (a line is roughly one-twelfth of an inch). The land measurement continued into recent times – the cadastre of the Province of Ångermanland was surveyed in 1540 using such means of measurement. The work was carried out by special linemen (*revkarlar*), under the control of the assistant judge and the sheriff. The results were entered into *the Land Tax Registers* for 1542. There were, however, no maps produced associated with these surveys (Westin 1942).

Duke Karl (who eventually became King Karl IX in 1604) in 1585 issued a so-called *Lesson* (an instruction or handbook) regarding the land measurement (*jordrevning*) and cadastral survey of his duchy, Södermanland. According to this instruction, arable land and meadows were to be measured using a rod nine ells long (an *ell* is a Swedish unit of measurement measuring about 0.6 meters long). For fishing waters and pastures, allotments would be granted in proportion to the possession of *öresland* and *örtugland* arable land; that is to say, infields. Regarding new duchy settlements, they were to have six years free from taxation for the construction. After that period, they would be taxed according to the extent of their possessions and utilities, and be recorded in *the Crown Land Tax Registers* (Ekstrand 1902). There was consequently no real need for maps in regards to the measurement and valuation of existing farms or for the taxation of new settlements.

The preserved instructions from the 1630s and 1640s do not give any definitive answer as to why the countrywide mapping process began. In the Surveying Ordinance of 1634, it is stated that all villages (and hamlets) are to be mapped, regardless of land-tenure category. It can be assumed beyond any doubt that there were fiscal aspects involved in the decision to commence mapping. In the authorisation issued by *the Crown Lands Judiciary Board* (*Kammarkollegium*) in March of 1634, the surveyors are exhorted "to examine and to scrutinise the waste condition of all freehold and taxed-tenant farms" (Ekstrand 1901: 7). However, in an authorisation given by Queen Kristina in April of 1635, it is stated that mapping should focus on freehold and taxed-tenant possessions:

> ... appointed surveyors ... [shall] in all parts of the Realm make measure, draw and note each village with regard to freehold and taxed-tenant farm properties and utilities (but not nobility land unless the landowners request so) (Samlingar 15).

The fact that the work was to be countrywide and focussed on crown and freehold land has been interpreted to mean that it was undertaken as an audit of the land taxes (this discussion is summarised by Kain & Baigent 1992:

53-57). It is true that it was the crown and freehold properties that were of primary interest to the treasury, but nobility land was also liable to some taxes, and nobility land was recorded in the regular *Land Tax Registers*. The reasoning behind leaving nobility lands entirely outside of the scope of a new countrywide cadastral survey is therefore difficult to understand. Another weakness of this hypothesis – that the large-scale mapping concerned primarily an overhaul of land taxes – is that it would be an unnecessarily complicated way to accomplish that. If the purpose had been merely to adjust the taxes, it would have been sufficient to measure and note the size and quality of the arable land and meadows; a so-called *jordrevning* (measurement). It is also true that towards the end of the 16th and beginning of the 17th centuries, promises were given of a general land measurement in the whole country in order to adjust the taxes. The countrywide measurement did not come about; however, measurements were done for separate regions – such as for Eastern Dalarna in 1587, Ångermanland in 1604, Östergötland East-of-the-Stångån in 1606, and Öland in 1622 (Lönborg 1903; Williams 1928). To the measurement data accumulated was added a general economic description of the state of arable and meadow lands, and of outfield utilities such as fishery, grazing, forest acquisition etc. (Lönborg 1903).

In the Swedish *Riksdag* (parliament) of 1624, a number of decisions were made aimed at extending the State's administration of the incipient Great Power of Sweden, to make it modern and efficient for its time. Although a State Surveying Authority was imminent, the *Riksdag* decided to initiate a comprehensive measurement, on the basis of which the land tax rates were to be reassessed and a new *Land Tax Register* compiled. In this manner, the system of taxation would be improved (Wetterberg 2002). This demonstrates that, above all, taxation was based on an assessment of the extent of arable and meadow land. That could be accomplished without the need for any maps showing how the different kinds of land were distributed in the terrain.

It can furthermore be argued that what characterised the new taxes in early 17th-century Sweden was that they were *not* connected with land. This is true, for instance, of the mill toll and the cattle taxes from the year 1600 and onwards. The second Älvsborg Castle Ransom (*Älvsborgs Lösen*) in 1613 was dealt with using a six-year levy on individuals and actual fortune in the form of cattle, precious metals and money. Another new tax was the Small Toll, levied from 1622 on goods brought into towns (Wetterberg 2002). Staffan Helmfrid views the maps as something more than simply being the basis for an improved system of taxation, and asserts that the Instruction of 1628 was an effort to create an economic map of the kingdom as a governing tool during the expansionary period of the Great Power of Sweden (Helmfrid 1959).

5.   WHERE DID THE INITIATIVE COME FROM?

The underlying causes of the great mapping project of the 1630s are thus not completely understood. It is also not clear where the impulse for the unique project came from; regarding this, Helmfrid points out two different aspects. In regards to purely technical innovations, including the methodological understanding of how to create large-scale maps, we have seen that there would have been ample opportunities for Swedes to acquire such knowledge from Livonia and the Baltic area, as well as from German states and the Netherlands.

Concerning the conceptual background to the cartographic program, our understanding is much less defined. The question as to why the contents of the landscape should be shown in maps at all has been scantily addressed by research. We have already shown that maps were not necessary for the sake of tax reform. In 1645, surveyor Sven Månsson wrote in a request to Queen Kristina:

> Because such Geometric Jordeböcker (maps) our Ancestors have not known, in which one with haste, *as if one were on that very Spot where the Village is present,* equally well can sit here in Stockholm or elsewhere and from the map or drawing exactly see and know the Property, Condition and location of each and every Farm (Ekstrand 1902: 250, my italics).

The same surveyor writes in 1639, on the title page of *the Land Tax Register*, Volume A8:

> … containing the qualities and utilities of the aforementioned farms … *so that one can with haste get all their condition before one's eyes,* and then of each and every of them grasp and evaluate (*bekoma och dijudiceras*) (my italics).

A similar description is contained on the title page of the oldest *Land Tax Register*, Volume D6, from 1633-1634. Surveyor Johan Larsson Grot writes: "… so that one all their condition at once can put before one's eyes."

Sven Månsson and Johan Larsson stress that maps are a way of realising a direct understanding of a village and its properties without actually having to be present at the location itself. Because a large number of maps were assembled at the Crown Lands Judiciary Board, it became possible for the central administration in Stockholm at any time to quickly get a detailed view of the character, possibilities and limitations of various farms by going through the geometric map volumes.

Another possible reason for mapping is found in the Authorisation of 1635, in which it is pointed out that the surveyors are to measure the villages correctly and accurately, "so that each one shall know his property and thereby several disputes and controversial inspections be avoided" (Samlingar 15). The argument that costly conflicts needed to be prevented is repeated in an instruction from 1649 (Samlingar 37).

Figure 12-1. One of the oldest large-scale maps shows Väversunda village, in the province of Östergötland. The arable land is indicated by the horizontal parallel lines. The hay meadows are marked using small dots. In the south, the farmsteads and the church are seen (D6: 33).

The question is: Who carried to fruition the idea to map and file all of Sweden's villages? It is obvious that many persons may have entertained thoughts of such a plan, but the number of those who could have actually realised it is few. One's first thoughts turn in the directions of King Gustavus Adolphus and Count/Chancellor Axel Oxenstierna. The basis for the idea may have originated earlier, however – perhaps with Duke Karl (who became Karl IX).

6.   STYLE AND SYMBOLS OF THE LAND REGISTER ATLAS MAPS

At first glance, the older *Land Atlas Maps* (Figure 1) are unusually uniform in their appearance and content. The maps were drawn on linen rag paper, and are generally of good quality.

The maps do differ in some ways, however. Older scholars have in many cases passed judgements (at times rather haphazard opinions) on the maps (Lönborg and others). The various surveyors – and perhaps the surveyor henchmen as well – did in fact display personal traits that may have played a role in the making of the maps; but the map differences seem to depend more on when the maps were executed. During the period 1628-1643, a succession of instructions were issued regarding what should be contained in the maps, and what their appearance should be like. In general terms, it is true to say that the maps became more comprehensively uniform over time.

The maps generally were drawn to a scale of 1:5000; but other scales were used – for instance, 1:4000, 1:3333 and 1:2000. The scale was thus mainly based on "round" numbers, which serves to give the maps a modern impression. The maps generally show buildings, the infields with their arable and meadow lands, fences, some hydrography, and certain forms of vegetation. Several other objects may also be included, such as mills, boat sheds, barns, blast furnaces, boundary markers etc.

A systematic description – *Notarum Explicatio* – is generally appended to the map. These descriptions are relatively uniform in their composition (though variations do occur). For one thing, it is usually mentioned how many homesteads there are in the village, including information regarding their respective cameral sizes and land-tenure category. When relevant, there is also an account of *örestal* and *byamål* (village measurements) in ells or rods, after which may be given the size of the arable land measured as barrels of seed (*tunnland*). If there is a fallow system to be measured, the size of each seed year (field) is given. The yield of the meadow land is generally given as *lass* (wagonloads of hay, with one load, or summer-load, being approximately 212 kilograms (in *Måttordbok*)). Then often there follows a farmstead-by-farmstead account, detailing their arable and meadow lands. When exclave fields or meadows exist, they are noted, and reference is made to the pages in *the Land Register Atlas* where they appear. An observation regarding forest and outfields is usually included; and also often receiving mention is the existence of any fisheries, mills, hop gardens, or other items of economic importance. In regards to the forest, three phenomena generally are mentioned: the supply of firewood (for energy), wood for fencing (as protection for growing grain and hay), and timber (for building). Grazing can be evaluated as good, scanty or "sharp" (*meagre*). There are similar evaluation expressions concerning other utilities like fishing-waters and hop gardens.

Sometimes the *Notarum Explicatio* is incomplete, in that one or more of the items mentioned above are lacking.

7. GEOMETRIC MAPPING IN THE 1630s AND 1640s: EXECUTION AND ORGANISATION OF THE WORK

A comprehensive account of the course of geometric mapping in Sweden has not been given since Sven Lönborg's overview in *The Map of Sweden* (*Sveriges karta*), in 1903. In 1945, Olof Arrhenius put together an overview summarizing the coverage of *the Geometric Land-Atlas Maps*, based on information in the card file kept by *the National Land Survey of Sweden*. The map he compiled contains some oddities and shortcomings (Arrhenius 1945). But it is the most recent overview in existence of large-scale mapping in Sweden during the 17th century, and there is now every reason again to inquire into how and when different parts of the country became mapped on a large scale, where the work began, how mapping proceeded, and where information gaps can be found (Figure 2). By doing so, it may be possible to gain a better understanding of the aims and context of the large-scale mapping process.

Anders Bure is often called the founder of Swedish surveying. Even though he was not particularly active in the mapping process during the first years of the survey, it is very probable that he took direct part in the training of the first generation of native Swedish surveyors. In the second and third items of the Instruction of 1628, it is stated that:

> ... (he) may employ young mates, children of honest men, whom he may instruct and train as (surveyors) ... When he has obtained and instructed these young persons, he shall with them map and measure each Province after the other, and of those measurements particular land-maps and drawings be made ... (Samlingar 1: 1).

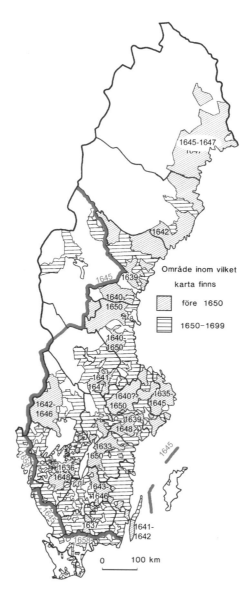

Figure 12-2. The coverage of geometric maps in Sweden, after Arrhenius in 1945, and subsequently updated by Tollin. The years indicate the time of the mapping.

8. THE SURVEYORS: THEIR NUMBERS, SOCIAL BACKGROUNDS, ECONOMIC CONDITIONS, CIVILIAN AND MILITARY TRAINING

Early in 1628, six surveyors were employed by the Crown Lands Judiciary Board in Sweden, with each receiving a salary of 100 *daler*. By June of 1628, the Board had decided that those surveyors were to receive 200 *daler* each, from the budget of the Court of Appeal (Samlingar 2, note).[1]

We know little as to how the training of the surveyors was carried out; but notes in the diary of Johan Bure, a cousin of Anders Bure, may provide a clue. In July of 1629, Anders Bure was travelling with Johan (who would become the first director general of the National Heritage Board of Sweden) and "six geodetics" to Svartsjö, where they were to be taught how to measure land (Lönborg 1903: 20, and sources listed there). Svartsjö Castle and Royal Manor are situated in Sånga Parish on Färingsö Island in Lake Mälaren. Anders Bure was the "Mathematician General", and among his students were engineers, geographers and a *geometricus* (Johan de Rogier).

In 1629 and 1630, there were six surveyors salaried by the Crown: Johan Andersson (2373), Olof Gangius (2619), Jonas Johansson (2371), Jonas (Jost) Månsson, Olof Månsson (804) and Johan Åkesson (1062).[1] Of the earliest known surveyors, Johan Andersson Lenaeus (raised to nobility under the name Vernsköld) went into military service in 1630, and afterwards worked on the building and repairing of fortifications (Ekstrand 1896-1903). In the years 1632 and 1633, respectively, Jonas Johansson and Olof Månsson disappear from the account books. Olof Mårtensson (Gangius) and Jonas Månsson were to perform their duties in Finland[2] (Samlingar 3).

In 1632, there were four salaried surveyors. But soon afterwards, it is clear that new recruits joined their ranks. In March of 1634, broadcloth clerk Samuel Nilsson was ordered to deliver a number of ells worth of black broadcloth to twelve surveyors, as a part of their salary (Samlingar 9).

Around 1633, Bure in practice left his position as director of surveying. At that time, the first generation of surveyors had been trained, and the systematic large-scale mapping of farms and villages had commenced. The surveyors worked during the summer months in the field, measuring and preparing field maps. Then the field maps, or so-called "concept maps", were made into fair-drawn copies. The fair-drawn and coloured maps, the

---

[1] The numbers in brackets after the names of the surveyors refer to the numbers in Ekstrand: *Svenska landtmätare 1628–1900*.

[2] In 1631, engineer Olof Gangius worked in various places for the government. Jonas Månsson (Sträng) (2737) received his Letter of Commission in 1644 as a surveyor in Finland. He was the brother of Anders Månsson (2620), who also worked in Finland.

renovations, were delivered to the Chamber of Accounts in the Royal Palace of Stockholm, where they were filed away (Samlingar 5).

Instruction issued by the Crown Lands Judiciary Board on the second of April, 1634, states:

> ... that the surveyors each year, as soon as the land is bare, will begin their labour and diligently measure the domains of each village ... When thus the work is completed, they will by wintertime come to the Chamber of Accounts and deliver it, and give account therefore (Samlingar 10).

The very-detailed Instruction of April 20, 1643, says:

> Their (the surveyors') duty and obligation therein consist, that they the first day of Spring, when the soil has become bare, in their own persons depart into the countryside ... (Samlingar 34).

As the material increased, it was filed by parishes and hundreds and bound in *Geometric Land-Atlas Map* volumes (*Geometriska jordeböcker*). The administrative division built on cameral land-register parishes, hundreds, and when relevant one type of (coastal) rural court district (*skeppslag*) and a second type of rural court district (*tingslag*), and county. The administrative structure in the 1640s has been reconstructed in detail by Lagerstedt (1974). The prefix *Geometric* was added to distinguish the atlas volumes from the regular *Land Tax Registers*, where the cameral homesteads were noted along with their tax obligations.

We know the names of some tens of surveyors (Figure 3) who were active during the period 1630-1655. It seems unreasonable that they alone managed to map the thousands of homesteads that are contained in *the Land Tax Registers* at the same time as they measured towns, travelled to the national frontier, and reconnoitred suitable shipping routes between the great lakes and the coast. Most likely, the number of persons who played an active part was much higher than the number of those who signed the maps. Some writings indicate that this was so.

In the *Instruction* of 1643, item 16, it is asserted that:

> One can feel great wrongfulness being with the Surveyors, who for this and in the place where they themselves should pursue a good, faithful and useful work, instead use apprentices who travel on and off farms and villages between and measure the domains without much sense (Styffe 1856: 252, 254).

The surveyors were thus not alone. Ever since the surveying service had been set up, they had always had an apprentice or junior surveyor to assist them. The Uppland surveyor Sven Månsson writes in a supplication to Queen Kristina in 1645: "... and have with me a henchman, who has been

with me four years, whom I have also educated, together with two boys"
(Ekstrand 1902: 250). During the fieldwork season, the number of temporary
apprentices and helpers from the villages may have numbered many more.

Figure 12-3. This is probably the only portrait of a surveyor, from around 1640, found on the
backside of a concept map from the province of Östergötland. The writing says: "Emanuel to
God I will you command. Given all tender are you my sweetheart" (D10a).

We also do not know how the fair-drawn copies ("renovations") of the
maps were prepared. In some instances, it is evident that the map sheets were
prepared in advance with frames, north arrows and scale rulers. There was
no need for the surveyor to occupy himself with all of these details; they
may have been done by a helper, perhaps a wife or a daughter. It is true that
the surveying profession at this time was a prerogative of men, but surveyors
did often marry other surveyors' daughters and these women most likely had
a good understanding of surveying and what came with it.

9.   SOCIAL BACKGROUND OF THE SURVEYORS

Anders Bure was directed to recruit as surveyors "young boys of honest
men". The surveyors appear to have been recruited primarily from among
farmer and clergyman households, as well as from lower-rank military
families. They were also required to be intellectually capable. In several
cases, it can be shown that before they were recruited as surveyors, they had

been enrolled at Uppsala University (Andersson 1900). No one had a noble name when he was commissioned; they had only common personal names, with the possible exception of Johan de Rogier. In addition, they also had distinctive surnames, such as Johan Larsson *Grot*, Olof Larsson *Tresk*, Jacob *Stenklyft*, Anders Börjesson *Gadd*, and Peter Johansson *Duker*. Only rarely were they raised to nobility after some time had passed. Johan Botvidsson (2142), for instance, was born in Västergötland in the beginning of the 17th century to the pattern clerk Botvid Larsson and Ingeborg Håkansdotter. He was trained by Anders Bure starting in 1629, and in 1633 he was commissioned as a surveyor in Uppland. In 1647, he left his position as surveyor and became quartermaster at the Nobility Banner (a cavalry regiment), at the same time as he was raised to nobility under the name Gyllensting (Ekstrand 1902).

10. CONTACT WITH THE FARMERS

Most of the surveyors were recruited from the countryside, something that should have been helpful in their relations with farmers. The Crown also placed considerable stress on the surveyors not creating unpleasant feelings among the people. *The Instruction* of 1635 provides definitive directions as to how to treat the peasantry:

> In that connection they shall also interact with the utmost reason, discretion and gentleness with the peasantry, not to render any displeasure or insolvency (Samlingar 14).

In the major *Instruction* of 1643, the paragraph is elaborated upon:

> He shall also know among other things his duty being to interact with the peasants with the greatest reason, discretion and gentleness, thereby preventing anything questionable or causing insolvency, and not pay heed to his own advantage or private benefit (Samlingar 34: 11).

The fieldwork of the surveyor took place during the same season as the growing period. He had to therefore be as careful as possible not to harm growing crops. *The Instruction* of 1635 states, in this regard, that:

> They (the surveyors) must also observe the season so that they, when the corn stands green in the field, work on the corn-field, but when the corn begins to grow higher and form ears, will leave it and work on the fallow field, meadows and other such land, so that the peasantry shall not need to complain of damage due to the trampling of corn (Samlingar 18: 8).

The instruction was developed further in 1643:

... do they have to arrange their labours according to the character of the season, taking great care that, when the corn stands growing or is fully ripe, they do not trample or damage it, but do rather let the corn-fields alone and take on the fallow fields, meadows and other such, until the harvest time is over, so that the peasantry will not be induced to complain and demand the payment of damages by the surveyors (Samlingar 34: 11).

11. THE GEOMETRIC MAPPING

The natural point of departure when looking to study the spatial and chronological progress of the mapping is *the Geometric Land-Atlas Maps* found in the Research Archives of the National Land Survey of Sweden. One might, though, presume that the numbering of the atlases indicates the order in which they were compiled. However, this is not the case. The signatures of the atlases are not in chronological order – thus, a low number on an atlas does not necessarily indicate that it was made earlier than one with a high number. A person also cannot be certain that the volume was delivered to the Crown Lands Judiciary Board in the same year that the mapping was done. Furthermore, some volumes comprise maps from different geographical regions, and occasionally also by different surveyors and from various time periods. Moreover, there are some twenty private land-map atlases, with contents similar to the state-produced ones. Another source of research material is the written instructions and commissions of the Crown Lands Judiciary Board.

12. CONCLUSION OF THE LARGE-SCALE MAPPING PROCESS

By the end of the Thirty Years War (1648), thousands of Swedish geometric maps had been compiled; from Öland in the south to the Torne Valley in the north, from Värmland in the west to the Roslagen archipelago in the east (Figure 4). To this should be added a few sheets and some ten private land-map books of major estates. It can be estimated that between twenty thousand and thirty thousand homesteads had been mapped and annotated, with respect to their sizes and economic conditions.

Despite the huge effort, the project was never completed, and older geometrical maps are lacking for parts of Sweden. At the end of the 1640s, the Crown Lands Judiciary Board regarded the large-scale mapping of Uppland, Västmanland, Småland, Östergötland, Värmland, Dalsland,

Sörmland, Närke, Västergötland and Western Norrland as by and large completed (Samlingar 37). At some time around 1648-49, the project was discontinued. There was some slowness in the reporting, and it was not until early in the 1650s that the Östergötland maps of Johan de Rogier, the Västmanland maps of Johan Åkesson, and the Västerbotten maps of Erik Eriksson – mapped several years earlier – were delivered. Furthermore, for some years into the 1650s some large-scale mapping was continued, covering private estates and their nobility homesteads.

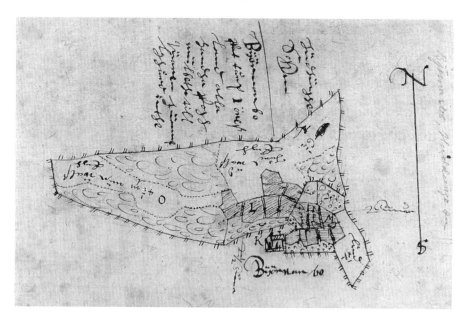

Figure 12-4. Concept map, Huddunge parish. Some hundred concept maps from the province of Västmanland have recently been recovered at The National Archives of Sweden. On the concept map of the Björnebo farm from around 1644, you can see the notes and the outline to the renovation map that was not delivered to the royal castle until 1652.

The remaining surveyors were to transfer their duties to producing geographical maps and maps of the Freedom Mile areas around the Royal Estates:

> … and now hereafter to author … special maps of Hundreds, entirely with properties as a geographical map should be put and formed … (Samlingar 37).

The change from large-scale geometric to small-scale geographical mapping had been partially anticipated by the Crown Lands Judiciary Board, as seen in a subsidiary instruction dated April 20, 1643, concerning geographical maps (Samlingar 35). In 15 points, it was detailed what these so-called *landkort* or *general landkort* were to contain.

The discontinuance of the large-scale mapping by the State meant that some surveyors were discharged; for instance, Håkan Arvidsson in Värmland. Others had already died, such as Olof Larsson Tresk in 1645, Sven Månsson in 1648, and Johan Åkesson in 1652. Some were employed for private mapping assignments. Others were also given new tasks, and a few – such as Johan de Rogier and Per Jonsson Duker – even lived to see renewed large-scale mapping take place around the 1680s.

Finally, it can be said that we have begun to perceive a more-detailed picture of the coverage and course of the mapping; that is to say, what was done and what was not. The most important questions still need to be answered, however: Why was the unique Swedish geometric land-map project initiated, and why was it discontinued?

During the next six years, a systematic survey of these maps will be made, including scanning and statistical compilation. The outcome will hopefully result in a huge increase in the knowledge regarding the historical agrarian landscape in Sweden.

REFERENCES

Manuscript Sources

National Land Survey of Sweden. Lantmäteriverkets forskningsarkiv Gävle.
The National Archives of Sweden. Oxenstierna samlingen, Stockholm.
The Military Archives of Sweden, Stockholm.

Printed and Secondary Sources

Almquist, J.A. (1976). *Frälsegodsen under storhetstiden: med särskild hänsyn till proveniens och säteribildning.* Stockholm: LiberFörlag/Allmänna förl.
Andersson, A.(1900). *Uppsala Universitets matrikel.* Utgiven av A. Andersson. Uppsala.
Arrhenius, O. (1945). Fördelningen av 1600-talets geometriska kartor över Sverige. *Globens,* 29-32.
Ehrensvärd, U. (2000). Svensk kartläggning i Baltikum. In G. Hoppe (Ed.), Öster om Östersjön. Svenska sällskapet för antropologi och geografi. *Ymer,* 120, 225-239.
Ekstrand, V. (1901). *Samlingar i Landtmäteri. Första samlingen. Instruktioner och Bref 1628-1699.* Stockholm.
Ekstrand, V. (1902). *Samlingar i Landtmäteri. Tredje samlingen . Bilder ur landtmätarnes liv.* Stockholm.
Ekstrand, V. (1896-1903). *Svenska lantmätare 1628-1900. Biografisk förteckning.* Umeå och Uppsala.
Hedenstierna, B. (1949). Stockholms skärgård: Kulturgeografiska undersökningar i Värmdö gamla skeppslag. Meddelande från Geografiska institutet vid Stockholms högskola 75. *Geografiska annaler,* 1948, 1-2.

Hedenstierna, D. (1951). Näringslivet i Sotholms härad under 1600-talet. *Geografiska annaler*, 3-4.

Helmfrid, S. (1959). De geometriska jordeböckerna – 'skattläggningskartor'? *Ymer* 3, 224-231.

Helmfrid, S. (1962). Östergötland, 'Västanstång'. *Geografiska Annaler* XLIV.

Jansson, S. O. (1995). *Måttordbok*. Stockholm: Nordiska Museets förlag.

Johnsson, B. (1965). Synpunkter på 1600-talets tidiga geometriska kartering med särskild hänsyn till Västmanlands län. *Ymer* 85, 9-81.

Johnsson, B. (1965). Åkerns omfattning vid 1600-talets mitt enligt de geometriska jordeböckerna: kulturgeografiska metodstudier tillämpade på Västmanlands län. *Ymer*, 85.

Kain R. & Baigent E. (1992). *The Cadastral Map in the Service of the State. A History of Property Mapping*. London: The University of Chicago Press.

Lönborg, S. (1903). *Sveriges karta. Tiden fram till omkring 1850*. Uppsala.

Samlingar se Ekstrand, V. (1901). *Samlingar i Landtmäteri. Första samlingen. Instruktioner och Bref 1628-1699*. Stockholm.

Styffe, C.G. (1856). *Samling af instruktioner rörande den civila förvaltningen i Sverige och Finland*. Stockholm.

Tyacke, S. &. Huddy, J. (1980). *Christopher Saxton and Tudor Map-making*. London: The British Library.

Westin J. (1942). Jordläggning och kartläggning i Ångermanland under 1600-talet. *Svensk geografisk årsbok 1942*. Årgång 18. Utgiven av sydsvenska geografiska sällskapet. Lund

Wetterberg, G. (2002). *Kanslern Axel Oxenstierna*. Stockholm: Atlantis.

Williams E. (1928). Skattläggningsväsendet och lantmätarna. *Svenska lantmäteriet 1628-1928*. Stockholm.

Chapter 13

TYCHO BRAHE, CARTOGRAPHY AND LANDSCAPE IN 16TH CENTURY SCANDINAVIA

Michael Jones
Centre for Advanced Study at the Norwegian Academy of Science and Letters, Norway

1. INTRODUCTION

A little more than 400 years ago, on 24 October, 1601, the great Danish astronomer, Tycho Brahe (Tyge Ottesen Brahe) died in Prague, where he is buried in the Teyn church. Tycho's last assistant, and his successor as Imperial Mathematician to Emperor Rudolph II, was the German astronomer Johannes Kepler, who wrote at the end of Tycho Brahe's log of astronomical observations (Dreyer 1926: 283, translated by Rosen 1986: 313, cited in Thoren 1990: 469):

> At this time … his series of celestial observations was interrupted, and the observations of 38 years came to an end.

With the help of new and improved instruments (but without the telescope), Tycho Brahe achieved astronomical observations of unsurpassed accuracy for his time, and contributed to debates on the nature of the universe. Between 1576 and 1597, Brahe worked from his observatory on the island of Hven (Ven), in the Sound (Øresund) between Denmark and what is now Sweden. Less well-known than his astronomical work, however, is the pioneering contribution he made to the development of cartography. Cartography at the end of the 16th century came to present new ways of "seeing" landscape.

Tycho Brahe's papers in the former Imperial Library in Vienna, now the Austrian National Library, contain just three maps. One is a hand-drawn map of Hven from around 1584. This is the earliest of four maps of Hven from this period, the other three being printed maps. These maps of Hven have the

H. Palang et al. (eds.), European Rural Landscapes:
Persistence and Change in a Globalising Environment, 209-226.

form of estate maps. They are the first detailed local maps in Scandinavia based on systematic survey and triangulation. In the same portfolio as the earliest map of Hven are two maps from Norway. One of these is a draft sketch map of the coast of West Norway from about 1590. It surpasses by far in accuracy contemporary Dutch sea charts of the west coast of Norway, generally considered to be the best of their time. The other is a hand-drawn map of Nordfjord, within West Norway, dated 1594. It shows even greater accuracy, quite remarkable for its period. The two Norwegian maps appear to have been drawn by the Danish bishop of Bergen, Anders Foss, and given to Tycho Brahe as a gift.

The present paper examines the four maps of Hven and the two Norwegian maps, and assesses the contribution to Scandinavian cartography made by these maps, as well as the view of landscape represented by them. Tycho Brahe's interest in mapping Hven is understandable. However, what was his interest in West Norway and Nordfjord? He never visited Nordfjord, nor even Norway. Why would he be interested in "seeing" in particular Nordfjord in the form of a map? These questions will be addressed in this paper. First, however, a few words are necessary on Renaissance cartography and the societal function of maps.

2.   RENAISSANCE CARTOGRAPHY

Until the 16th century, maps were relatively few and far between in Europe. Many of them were intended to depict itineraries. Such were the Italian and Catalan portolan charts, sea charts derived from sailing instructions (in Italian *portolano*), which were based on knowledge of coasts, winds and currents that seafarers had acquired through first-hand, bodily experience (Campbell 1987).

At the end of the Middle Ages in Europe, four systems of cartographical knowledge existed alongside one another. First were the portolan charts. Second were the *mappaemundi*, such as that found in Hereford cathedral, presenting the theological worldview of Christianity. Third were the Ptolemaic world maps, based on co-ordinates of latitude and longitude in the tradition of the Greek astronomer and geographer Ptolemy (Klaudios Ptolemaios) from the 2nd century AD. Fourth, few in number, were local maps of cities and towns, limited regions, and land itineraries (Woodward 1987; Harvey 1987). Important developments in cartography occurred during the 16th century in the Netherlands, Germany and Austria, where triangulation began to be applied to the measurement of land distances (Bagrow 1964; Tooley & Bricker 1969; Crone 1978 [1953]).

During the Renaissance, increasing attention was paid to Ptolemy's cartography (Dilke 1987). Local knowledge, assembled from seafarers and travellers, was located according to the geographical co-ordinates of latitude and longitude derived from heavenly bodies. Astronomical observations of increasing accuracy, of the type made at Tycho Brahe's observatory on Hven, gave cartography an improved scientific basis. Maps developed as increasingly abstract, totalising representations of geographical knowledge, while the seafarers and other travellers who were the original sources of the information on local topography and landscapes became invisible and forgotten (de Certeau 1984).

3. THE POWER OF MAPS

The history of cartography is not simply a matter of giving an account of increasingly accurate representations of the Earth's surface and landscapes. Technical innovations cannot alone explain advances made in cartography. The history of maps cannot be seen independently of the social, political and ideological circumstances in which they were produced. Maps were drawn on the initiative of the learned, many of them churchmen, who had close links to and were supported by the ruling classes. Hence maps have to be understood with reference to the social organization and power structures of their times (Harley & Woodward 1987).

In his book *The Power of Maps*, Donald Wood (1992) showed how maps are a means of presenting a reality that exceeds vision. They present a reality that people have knowledge about rather than what any individual actually sees and feels. Maps tell what others have seen or found out. As a societal product, a map will reflect the interests and biases of the patron who assigned the task of mapping. However, both the cartographer and the patron are generally invisible and often not identified; hence maps are seen as providing objective information about the world. Maps are therefore useful instruments for giving legitimacy to the control of land, exploitation of resources and collection of taxes in complex, hierarchical, non-face-to-face societies. Boundaries that may be invisible on the ground become visible on the map, and thus reinforce claims of ownership to a place.

Historically maps were used by rulers to "see" and hence control territories that they could not always visit. Philip II of Spain, for example, supported cartography in the second half of the 16th century as a means of visualizing and thus controlling his extensive domains in Europe and America when he was unable to visit his widespread empire in person (Short 1998). Political power became experienced, communicated and reproduced through maps. Maps named and located places of political significance. By demonstrating control over territory, maps became increasingly important with the growth of centralized power. They promoted individual property claims, and allowed owners to see widespread and far-flung estates as a whole, thus enhancing control. Maps came to be used as legal documents, while simultaneously having the advantage of presenting a precise, scientifically defined reality. Maps of landed property, drawn with instrumental accuracy, cemented a social structure based on the ownership of land. Maps were simultaneously both a form of apparently neutral knowledge and a means of visual appropriation of nature and landscape. Map-making, in fulfilling the needs of powerful patrons, provided a means by which power was gained, administered and given legitimacy (Harley 1988).

4.   TYCHO BRAHE AND THE FOUR MAPS OF HVEN

Tycho Brahe was granted Hven as a fief by Frederick II of Denmark in 1576 in order to build his observatory. The king was interested in science and wanted to keep Brahe from emigrating to Basel in Switzerland. Between 1576 and 1581, Brahe had built his Palladian-inspired mansion and observatory, which he called Uraniborg after Urania, the Muse of Astronomy. In the following years, Brahe developed and constructed the astronomical measuring instruments on which his fame rests, and, between 1584 and 1590, built a second observatory, Stjerneborg. The 40 farming families inhabiting the island – previously freeholders, paying crown taxes only – were obliged to supply labour to Brahe when the island became his fief. Each farm was obliged to provide two workdays per week. The new burdens led some of the villagers to leave the island. In 1578 they were forbidden to leave without Tycho Brahe's permission. When the villagers complained to the king in 1580, a commission of investigation drafted a village charter, issued over the royal seal in 1581, providing rules and regulations for labour dues, the upkeep of fences and dykes, the use of trees and thickets, and foraging on the commons. It was determined that the village assembly was to meet twice monthly. Previously the villagers had held their land by customary tenure, without documentation. They had administered their affairs and decided on the use of the landscape – the commons and open fields – through their village assembly according to time-honoured custom. Tycho Brahe transformed them into feudal tenants, subject to by-laws that were now formalized by a written village charter (Dreyer 1963 [1890]; Thoren 1990; Wittendorf 1994; Christianson 2000).

At about the same time, Tycho Brahe drafted the first map of Hven (Figure 1), which had not previously been mapped. On the back is a list of geodetic measurements to surrounding towns in Tycho Brahe's handwriting (Christensen & Beckett 1921; Dreyer 1923; Richter 1939; Christianson 2000). Oriented with west at the top, the map provides a plan of fields and paths, while buildings are shown in elevation. Dominating the map is Tycho Brahe's Renaissance mansion, Uraniborg, built on the common grazings in the middle of the island. Also shown prominently are the buildings of the demesne farm. Between the farm buildings and Uraniborg is depicted the new observatory of Stjerneborg. The map shows St Ibb's church on the western shore, and the village of Tuna to the north, surrounded by trees and with a windmill. In the open fields with their strips are finely drawn figures of people and animals carrying out agricultural operations such as ploughing. People are also shown walking along the paths. Animals are grazing on the commons, which cover the central, eastern and southeastern parts of the island. On the common is an enclosure surrounded by trees. Between the village and Uraniborg is a square of stones. The map is drawn in fine pen strokes and beautifully coloured. Drawn in much more crudely is a water mill and a system of millponds, evidently a later addition to the map.

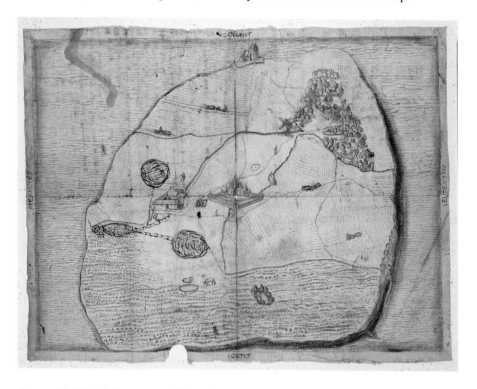

Figure 13-1. Hand-drawn map of Hven from ca 1584, archived among Tycho Brahe's papers in the Austrian National Library (archive ref. 10 688[2]). The map is oriented with west at the top. Photo by permission of the Austrian National Library, Vienna.

This map can be provisionally dated to between 1584 and 1589. It depicts Stjerneborg, the construction of which began in 1584, while it predates the start of the construction of the watermill and ponds in 1589, as these are clearly roughly sketched in later. The mill, dams and ponds, which also served as fishponds, were completed in 1592. The water mill was used to supply power for paper-making and leather preparation necessary for the books in which Tycho Brahe published his astronomical observations. Shortage of paper was a constantly recurring problem. Brahe learnt paper-making from his uncle, Steen Bille, who built Denmark's first paper mill at Herrevad in Skåne (now part of Sweden) some years before (Møller Nicolaisen 1946; Dreyer 1963 [1890]; Richter 1939; Christianson 2000).

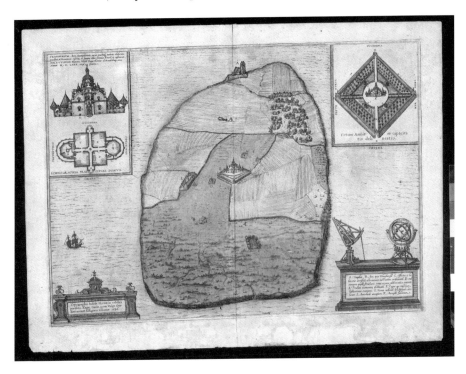

Figure 13-2. Map of Hven, 1586, published in 1588 in Braun and Hogenberg's atlas *Civitates orbis terrarum*, Vol. 4 Map 27. The map is oriented with west at the top. Photo by permission of the Royal Library, Copenhagen.

The manuscript map in Vienna can be more closely dated to between 1584 and 1586, as it was probably a draft for the first printed map of Hven, which has on it the date 1586. This was published in Cologne in 1588, in volume 4 of *Civitates orbis terrarum*, the six-volume atlas of 530 mainly European towns and cities by the German cartographers Georg Braun and Frans Hogenberg. This atlas was described by the British cartographer John

Goss (1991: 5) as "a celebration of the European city." Tycho Brahe was agreeably surprised and satisfied (Richter 1939) to see Hven depicted and described in "one of the most prominent publications of the day" (Thoren 1990: 208).

This second map of Hven (Figure 2) is again oriented with west at the top. The map shows Tycho Brahe's estate and buildings. As on the Vienna draft, the buildings are shown in elevation and accurately placed in relation to one another. The coastline, however, is not based on measurements but on freehand drawing (Richter 1939). The fields and commons, human figures, animals, village houses, windmill and church are depicted largely as on the Vienna draft, although the demesne farmstead is now shown with two gable walls facing the viewer instead of a building with its long side. The main features of the map are indicated by the letters A to K, with explanations at the bottom right. Thus it can be ascertained that the enclosure on the common, shown here with nets, is use for keeping fowl. On another part of the common is a bird-catching device. The square of stones on the common between Uraniborg and the village, indicated by G, is the *forum iudiciale*, that is the outdoor meeting-place of the village assembly. Two of Tycho Brahe's instruments surmount the legend. On the upper left side of the map is an inset showing the façade and plan of Uraniborg, designed and built in Gothic Renaissance style according to Palladian principles of axial symmetry, a major architectural innovation brought to Hven – and Scandinavia – by Tycho Brahe. On the upper right is a drawing of the house surrounded by a square Renaissance garden, divided into four symmetrical parts and oriented with its corners, and entrances, in the four main directions of the compass. The symmetry of the house and garden symbolized the natural harmony of the cosmic order (Christensen & Beckett 1921; Thoren 1990). The map emphasizes the contrast between Tycho Brahe's designed landscape, reflecting the natural order, and the spontaneous landscape of the villagers, the result of centuries of cultivation and grazing. Brahe's new landscape had appropriated both in fact and symbolically the village common, his manor overshadowing the meeting-place of the village assembly, and thus subordinating the villagers' customary rights to the needs of the feudal lord. At the same time, it became a part of the natural order of things that Brahe could appropriate the labour of the villagers in pursuit of his noble objective of scientific endeavour. The map allowed Brahe to see at a glance his estate on Hven, over which the newly enacted village charter confirmed his legal control.

The preceding map in Braun and Hogenberg's atlas depicts the northern part of Øresund, showing prominently King Frederick's newly rebuilt castle at Kronborg, as well as the island of Hven with Uraniborg. This map is oriented with east at the top. Curiously, Hven was placed with St. Ibb's church, on the west side of the island, towards the top, hence appearing to be on the eastern side of the island. The island was here clearly copied on to the Øresund map from the map of Hven, but wrongly oriented.

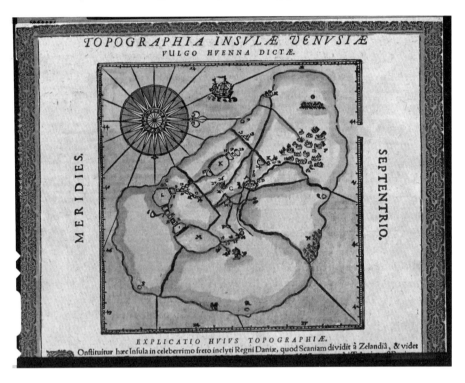

Figure 13-3. Map of Hven first published in 1596 in Tycho Brahe's *Epistolarum astronomicarum libri.* The map is oriented with west at the top. Photo by permission of the Royal Library, Copenhagen.

The third map of Hven (Figure 3) is an improved map printed in 1596 in Tycho Brahe's book *Epistolarum astronomicarum libri* and again in 1598 in his *Astronomæ instauratæ mechanica,* presenting drawings of his instruments; this map was later also reproduced by Kepler in 1627 on the title-page of his *Tabulæ rudolphinæ* (Dreyer 1919, 1923; Richter 1939; Bramsen 1965 [1952]: Figure 38; Thoren 1990). The map must have been drawn at the earliest in 1592, as it shows the paper mill and system of ponds completed in that year (Richter 1939). Once again the map is oriented with west at the top. New features compared with the previous maps include a compass rose and rhumb lines, as on contemporary sea charts, and degrees of latitude and longitude in the margins, but there is no scale bar. This map also has a much more accurate coastline than the previous maps.

Tycho Brahe was familiar with the principles of trigonometry and triangulation (Dreyer 1963 [1890]; Christianson 2000), and had visited cartographers in Germany who were among the first to apply such principles to making maps (Thoren 1990; cf. Tooley & Bricker 1969: 35, Crone 1978 [1953]: 60-61). In 1579 he had undertaken triangulations between the main landmarks of Hven and the surrounding mainland on both sides of Øresund, determining the position of Uraniborg and St. Ibb's in relation to the spires and towers of Copenhagen, Helsingør (including Kronborg), Hälsingborg, Landskrona, Lund and Malmø (Richter 1939, Thoren 1990). The Dutch geodetician N.D. Haasbroek reconstructed Brahe's triangulation in a diagram in 1988 (reproduced in Christianson 2000: 136). Tycho Brahe's maps of Hven are the first in Norden to be based on systematic triangulation (Richter 1939; Bamsen 1965 [1952]: Figure 38). The 1596 map, according to the German cartographic historian Herman Richter (1939), "makes a wholly modern impression", and, in the use of triangulation, "Brahe stands out also as one of the pioneers in the technique of cartographic measurement" (Richter 1939: 58).

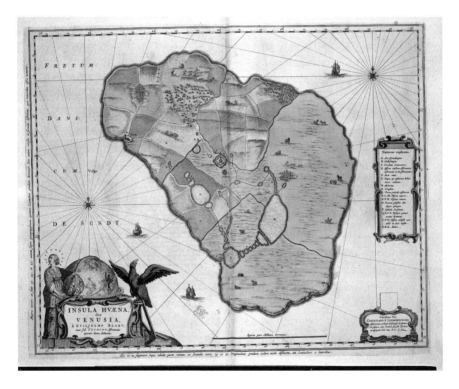

Figure 13-4. Map of Hven drawn by Willem Janszoon Blaeu ca. 1595-1596, published in 1662 in Joan Blaeu's *Atlas maior*, Vol. 1, Map 53, with a tribute to Tycho Brahe. The map is oriented with north at the top. Photo by permission of the Royal Library, Copenhagen.

The fourth map of Hven (Figure 4) was not published until nearly 70 years later. The map is signed by the famous Dutch cartographer Willem

Janszoon Blaeu, but published 24 years after his death by his son, and equally famous cartographer, Joan Blaeu, in his *Atlas maior* of 1662, along with copies of Tycho Brahe's woodcuts of his mansion and observatories, gardens and instruments (Goss 1990). On the map it is stated that it was drawn by Willem Blaeu when he worked with astronomy under Tycho Brahe. Blaeu spent the winter of 1595-1596 on Hven as a student under Brahe, and probably drew it then, perhaps copying Tycho Brahe's map published in 1596, which it strongly resembles. Some authorities (e.g., Christensen & Beckett 1921: 10, Christianson 2000: 136) have attributed the map of Hven published by Tycho Brahe in 1596 to Willem Blaeu. However, Richter (1939) maintains that Blaeu did not survey and map the island himself, but copied an older map, made by Brahe or his assistants before Blaeu came to Hven (cf. van Mingroot & van Ermen 1988 [1987]; Thoren 1990). Blaeu was then a young man at the beginning of his career. During the 20 years that Tycho Brahe was on Hven, Uraniborg functioned as an international research centre, attracting a large number of students. Blaeu came to study Brahe's methods of astronomical and geodetical observation, land surveying and instrument building (Richter 1925; 1939; Keuning 1973; Christianson 2000). The newest methods of triangulation and modern mapping were being applied by Brahe and his disciples for the first time outside Germany.

Unlike the previous maps of Hven, Blaeu's map is oriented with north at the top. It incorporates a compass rose, rhumb lines, scale bar, and latitudes and longitudes in the margin. People are shown working in the fields, as on the Vienna map and the Braun and Hogenburg map of the 1580s, some almost in the same positions. The church, village, village meeting-place (*forum judiciale rusticarum*), Uraniborg and Tycho Brahe's other buildings, as well as the mill and ponds, are shown. Yet Uraniborg and Stjerneborg had been destroyed soon after Tycho Brahe left Hven, and Denmark, in 1597. With the king's permission, they were plundered for building bricks under Tycho Brahe's successors on the island, and 30 years later only ruins remained (Norlind 1970). The map thus showed a past situation. This was probably Willem Blaeu's very first map. The significance of this fact cannot be underestimated. The cartographic publishing-house of Willem Blaeu and his sons was to become one of the most important purveyors of the new way of depicting landscape in the form of maps during the 17th century.

5.   THE MAPS OF WEST NORWAY AND NORDFJORD

The first, manuscript map of Hven is the second of the three maps among the Tycho Brahe papers in Vienna. The other two, the maps of the coast of

West Norway and of Nordfjord, were reproduced in 1908 by the Danish cartographic historians Axel Anthon Bjørnbo and Carl S. Petersen in a collection of previously unpublished maps of Norden. The maps had been recorded by the Danish historian and Brahe-specialist F.R. Friis in 1868. They were described in 1960 by the Norwegian cartographic historian Kristian Nissen (1961a; 1961b), who identified their author as the Dane Anders Foss, Bishop of Bergen from 1583 to 1607, on the basis of a comparative analysis of his handwriting.

Bjørnbo & Petersen (1908) described the map of West Norway as an anonymous draft. It shows the 600-km long stretch of coast from Lindesnes in the south to the mouth of Trondheimsfjord in the north. On the back, in Tycho Brahe's handwriting, are the words *Descriptiones littorum Noruagiae & quedem alia*. The map is 121 cm long and 32 cm wide; glued on the back, perhaps to strengthen it, is a draft receipt made out in 1586 by Anders Foss to Hans Lindenow. The receipt appears not to have anything directly to do with the map, but it does give an approximate date for when the map could have been drawn. Corrections and changes made to the map are visible, especially in the inner fjords, and in some place the fjords and names are shown double, as if not properly rubbed out. The draft is clearly the product of much work, and islands and fjords are particularly well detailed for the parts of the coast within the diocese of Bergen. Mountains are depicted on part of the map. A large number of place-names, and all churches, are marked, indicating that the author had intimate knowledge of the coast. Place-names are written in varying directions, and the map does not have a clear top or bottom, but has to be read by turning it in different directions. The names are aligned in opposite directions on opposite sides of the fjords, suggesting that they were written in during systematic voyages following first one side, then the other side of each fjord. This alignment of names resembles that found on the medieval portolan charts as well as on Dutch sea charts from the second half the 16th century. Unlike these, however, there are no compass roses nor rhumb lines; there is no legend, nor is there a scale bar, and the map shows no boundaries. On the other hand, degrees of latitude are marked along its western margin.

The Nordfjord map (Figure 5), also anonymous, has on it the date 1594. It measures 50.7 by 36 cm. On the back, in Tycho Brahe's handwriting, is the name *Nordttfiordtt*. This map is not a draft like the map of West Norway, but a finished work of remarkable accuracy (Bjørnbo & Petersen 1908). Again there is no legend nor scale, but the four compass directions are indicated in Latin along the edges of the map. The land area is covered by symbols for mountains, shown more faintly for the parts of the map outside Nordfjord. Boundaries of parishes are marked by lines, and the numerous place-names are located by circles. Important buildings are shown with house symbols, and churches have their own symbols. The map is oriented with north at the top, and the main place-names are aligned accordingly.

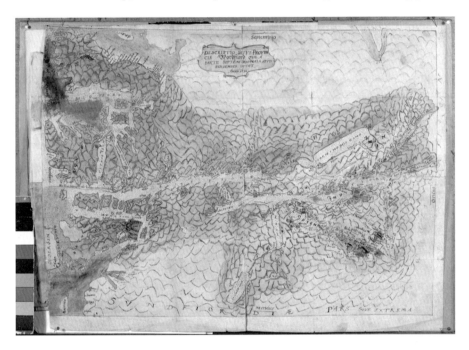

Figure 13-5. Hand-drawn map of Nordfjord, 1594, archived among Tycho Brahe's papers in the Austrian National Library (archive ref. 10 688³). The map is anonymous but handwriting analysis suggests that it was drawn by Anders Foss, Bishop of Bergen. The map is oriented with north at the top. Photo by permission of the Austrian National Library, Vienna.

Bjørnbo & Petersen (1908) wrote that these two maps bear witness to the new epoch in the empirical sciences established by Tycho Brahe, when cartography became dependent on instrumental observations. The maps were drawn at the time that the first serious attempts were being made to solve "the most difficult cartographic problem of Scandinavia" (ibid.), the mapping of the west coast of Norway. During the 1580s and 1590s, the Dutch cartographer Lucas Janszoon Waghenaer published the first printed atlases of sea charts. These contained sailing directions and detailed soundings, compass roses, rhumb lines, bar scales, coastlines in profile, and place-names – but not latitudes and longitudes (Crone 1978 [1953]). Waghenaer's atlas of 1592 had three maps of the Norwegian coast, including two charts of the west coast (Hoem 1986), but not depicted with anything like the accuracy of Anders Foss' map of West Norway.

Nissen (1961a) considered the Nordfjord map to be a revised final version of a part of the map of West Norway, and that it was made so that Anders Foss could surprise his friend Tycho Brahe when he visited him on

Hven in 1595. Nissen dated the map of West Norway to c. 1590. That year Anders Foss was visited in Bergen by Tycho Brahe's former assistant Peder Jacobsen Flemløse, who in 1589, 1590 and 1592 travelled in the south of Norway to determine the latitudes of the main towns for a map of the Danish realm planned by Tycho Brahe (but never completed). He calculated Bergen's latitude at 60° 27′ (compared with today's official determination of 60° 24′). Flemløse spent several months in Bergen, and, according to Nissen, could have taught Anders Foss to make accurate calculations of latitude. Flemløse visited Tycho Brahe together with Foss in 1595 (Friis 1904; Thoren 1990; Christianson 2000). The bishop had evidently collected detailed topographical information about his diocese during visitations, and the map of West Norway provided him an overview. This would explain why the coastal areas outside his diocese to the north and south were less accurately depicted (Nissen 1961a; 1961b). A fair copy of the map of West Norway might have existed but been lost later and, according to Nissen (1961a; 1961b), could have been the source for 18th century hand-drawn maps of Bergen diocese found in archives in Norway and Denmark.

6. TYCHO BRAHE AND NORDFJORD

Anders Foss had an obvious interest in a map of his diocese – but why would Tycho Brahe be interested in receiving a fair copy of a map of Nordfjord? Although he never visited Norway, Tycho Brahe was feudal lord of Nordfjord, having been granted the region as a fief by Frederick II in 1578 (Bjørnbo & Petersen 1908; Richter 1925; Nissen 1952; 1961a). This meant he received the taxes otherwise due to the crown as payment for services rendered, in this case his scientific services. The farmers of the region paid taxes and feu duties in the form of dried fish, hides and skins, tar, butter, cheese, oxen and some money. The income was collected by successive bailiffs, whose efficiency can be inferred from the frequent complaints made against them by the farmers. The income from Nordfjord amounted to 1.000 *daler* a year, accounting for up to one-third of Tycho Brahe's income, thus making a substantial contribution to financing the building of Uraniborg, Stjerneborg and his astronomical instruments. The remainder of his income came from his inherited estates and other fiefs in Denmark granted by the king. However, Brahe did not hold his fiefs without controversy. Nordfjord was, for example, taken from him for short periods in 1579 and 1580, and again in 1586, but restored to him in 1589. He lost the fief again after Christian IV's coronation in 1596, thus losing a third of his income. Because of various conflicts over his manner of administering his fiefs, and his neglect of the duties involved, Tycho Brahe lost royal patronage. Christian

IV appointed a government that supported a strong monarchy and wished to reduce the power of self-willed feudal lords such as Tycho Brahe. The latter left Denmark for good in 1597, moving first to Wandsbeck in Holstein, and then to Prague, where he died in 1601 (Aaland 1898; Dreyer 1963 [1890]; Thoren 1990; Wittendorf 1994; Christianson 2000).

7.   CONCLUSION

Before these two maps, the Norwegian coast had only been mapped by foreign seafarers. The Dutch sea charts of West Norway published at the end of the 16th century were considerably less accurate in representation than Anders Foss' maps of the coast of West Norway and of Nordfjord. Nissen stated that "Norway's national cartography" began with these maps (1961a: 98, 113; 1961b: 79). Bishop Anders Foss had before 1600 tackled the challenge of mapping Norway's west coast, yet these maps did not receive the attention they deserved. Detailed triangulation of the Norwegian west coast was not undertaken for another 200 years (Engelstad 1952). The Nordfjord map and the draft map of West Norway, along with the first draft map of Hven, became forgotten among Tycho Brahe's papers in the Imperial Library in Vienna, and were not noticed until after the mid-19th century (Friis 1868).

The contribution made by Tycho Brahe and his disciples to cartography and new ways of viewing landscape has been largely neglected by historians of cartography. Yet his estate maps of Hven introduced the technique of triangulation to Scandinavia. The most enduring legacy was through the work of his disciple Willem Janszoon Blaeu, not least the latter's map of Hven published in his son's *Atlas maior* in 1662 (Richter 1925; Christianson 2000). Examination of other 17th century maps indicates that Tycho Brahe's cartographic legacy can be traced further.

In 1618, Laurids Clausen Scavenius, the Danish bishop of Stavanger from 1605 to 1626, had a map made of his diocese. He was the son of Claus Lauridsen Scavenius, professor of mathematics and astronomy at the University of Copenhagen, who taught when Tycho Brahe was a student there from 1559 to 1561 (Nissen 1954; Thoren 1990). Although the original map of Stavanger diocese from 1618 has been subsequently lost, it is known through a reproduction in Blaeu's atlas of 1640 (Nissen 1954; 1961b; van Mingroot & van Ermen 1988 [1987]) and again in the *Atlas maior* in 1662. The adjoining parts of Bergen diocese that appear on this map are clearly derived from Anders Foss' map of the coast of West Norway (perhaps from a fair copy of this that was later lost) (Nissen 1954; 1961b). The Scavenius map can be compared with a map of Bergen diocese in the Blaeu atlas of

1662. This is clearly based on another, less accurate source, particularly noticeable in the case of Nordfjord, which seems to be entirely conjectural. Scavenius' map was used in 1626 as a source for the Stavanger area and the adjoining southern part of Bergen diocese by the Swedish cartographer Andreas Bureus on his map of the Nordic countries, which remained the standard map of Norden until the end of the 18th century. North of Sognefjorden, Bureus' map is far less accurate, indicating that he did not have knowledge of Anders Foss' maps of West Norway and Nordfjord (Nissen 1952). In 1668 the French cartographer G. Sanson published a map of Bergenhus county, in which he included the Stavanger area; while the area belonging to Bergen diocese repeats the inaccurate coastline of the Blaeu map, the Stavanger area has the more accurate coastline of the Scavenius map (van Mingroot & van Ermen 1988 [1987]).

Further, in 1630, the German cartographer Philipp Eckebrecht published his *New Representation of the World as Adapted to the … Rudolphine Astronomical Tables.* The Rudolphine Tables were the calculations made by Tycho Brahe and continued after his death in 1601 by Johannes Kepler, and dedicated to the Holy Roman Emperor Rudolph II. Issued in 1627, the tables were an important aid in determining longitude. The prime meridian on Philipp Eckebrecht's world map was a meridian intersecting Tycho Brahe's observatory on Hven (Tooley & Bricker 1969).

Tycho Brahe had contacts with the leading astronomers and mathematicians of his day. He raised the level of observation of the heavenly bodies to a higher level than before. This accuracy was made possible through the instruments he constructed on Hven, especially quadrants, sextants and armillaries, allowing more accurate calculations of the latitudes and longitudes of places on the surface of the Earth. Determining the precise latitude of Hven was a project Tycho Brahe worked on over many years (Richter 1939; Dreyer 1963 [1890]; Thoren 1990). This in turn allowed the construction of maps that provided a new way of looking at landscape from above – a view from the heavens – which has become a standard "objective" scientific way of viewing landscape, as opposed to the view of those working in or travelling in the landscape.

The grant to Tycho Brahe of the fief of Nordfjord was an expression of royal patronage in the partially feudalised state of Denmark-Norway in the 16th century (cf. Imsen & Winge 1999; Christianson 2000). Nordfjord financed a substantial proportion of Tycho Brahe's astronomical activities. These activities made a significant contribution to a new cosmology in which nature and landscapes became mapped in relation to celestial bodies. Tycho Brahe received Nordfjord at the height of his power. Although the map's content – like that of the map of West Norway – reflected Bishop Anders Foss' ecclesiastical interests, the Nordfjord map at the same time gave visual form to and thus confirmed Tycho Brahe's fiefdom, which was at times not always held without conflict. In this way Tycho Brahe could "see" his fief that he never visited. Nordfjord was represented on this map with a cartographical accuracy that was unmatched in its time – while the cartographer and source of information on which the map was based disappeared from sight.

REFERENCES

Aaland, J. (1898). Nordfjords lensherrer og fogder. *Bergens Historiske Forening: Skrifter* 4. Bergen: Griegs Bogtrykkeri.

Bagrow, L. (1964). *History of Cartography*, revised and enlarged by R.A. Skelton. Cambridge, Mass.: Harvard University Press.

Bjørnbo, A.A. & Petersen, C.S. (1908). *Anecdota Cartographica Septentrionalia*. København: Andr. Fred. Høst & Son.

Bramsen, B. (1965 [1952]). *Gamle Danmarkskort. En historisk oversigt med bibliografiske noter for perioden 1570−1770*. København: Grønholt Pedersens Forlag.

Campbell, T. (1987). Portolan Charts from the Late Thirteenth Century. In J.B. Harley & D. Woodward (Eds.), *The History of Cartography*, Vol. 1: *Cartography in Prehistoric, Ancient, and Medieval Europe and the Mediterranean* (pp. 371-463). Chicago: University of Chicago Press.

Christensen, C. & Beckett, F. (1921). *Uraniborg og Stjærneborg – Tycho Brahe's Uraniborg and Stjerneborg on the Island of Hveen*. København: Aage Marcus & London: Oxford Universiety Press.

Christianson, J.R. (2000). *On Tycho's Island: Tycho Brahe and His Assistants 1570–1601*. Cambridge: Cambridge University Press.

Crone, G.R. (1978 [1953]). *Maps and Their Makers: An Introduction to the History of Cartography*, 5th Ed. Dawson, Hamden, Conn: Folkestone, & Archon Books.

de Certeau, M. (1984). *The Practice of Everyday Life*. Berkeley: University of California Press.

Dilke, O.A.W. (1987). The Culmination of Greek Cartography in Ptolemy. In J.B. Harley & D. Woodward (Eds.), *The History of Cartography*, Vol. 1: *Cartography in Prehistoric, Ancient, and Medieval Europe and the Mediterranean* (pp. 177-200). Chicago: University of Chicago Press.

Dreyer, I.L.E. (Ed.) (1919). *Tychonis Brahe Dani Opera Omnia*, VI. Hauniæ: Libraria Gyldendaliana.

Dreyer, I.L.E. (Ed.) (1923). *Tychonis Brahe Dani Opera Omnia*, V. Libraria Gyldendaliana, Hauniæ.

Dreyer, I.L.E. (Ed.) (1926). *Tychonis Brahe Dani Opera Omnia*, XIII. Libraria Gyldendaliana, Hauniæ.

Dreyer, J.L.E. (1963 [1890]). *Tycho Brahe: A Picture of Scientific Life and Work in the Sixteenth Century*. New York: Dover Publications.

Engelstad, S. (1952). *Norge i kart gjennom 400 år med opplysninger om dem som utformet kartbildet*. J.W. Oslo: Cappelens Antikvariat.

Friis, F.R. (1868-1869). Tyge Brahes Haandskrifter i Wien og Prag. *Danske Samlinger for Historie, Topografi, Personal- og Literaturhistorie* (pp. 250-268). IV. Kjøbenhavn: Gyldendal.

Friis, F.R. (1904). *Peder Jakobsen Flemløs: Tyge Brahes første Medhjælper, og hans Observationer i Norge.* Kjøbenhavn: G.E.C. Gads Universitetsboghandel.

Goss, J. (1990). *Blaeu's The Grand Atlas of the 17th Century World.* London: Studio Editions.

Goss, J. (1991). *Braun & Hogenburg's The City Maps of Europe: A Selection of 16th Century Town Plans & Views.* London: Studio Editions.

Haasbroek, N.D. (1968). *Gemma Frisius, Tycho Brahe and Snellius and their Triangulations.* Delft: Rijkscommissie voor Geodesie.

Harley, J.B. (1988). Maps, Knowledge, and Power. In D. Cosgrove & S. Daniels (Eds.), *The Iconography of Landscape: Essays on the Symbolic Representation, Design and Use of Past Enviornments* (pp. 277-312). Cambridge: Cambridge University Press.

Harley, J.B. & Woodward, D. (1987). Concluding Remarks. In J.B. Harley & D. Woodward (Eds.), *The History of Cartography*, Vol. 1: *Cartography in Prehistoric, Ancient, and Medieval Europe and the Mediterranean* (pp. 502-509). Chicago: University of Chicago Press.

Harvey, P.D.A. (1987). Local and Regional Cartography in Medieval Europe. In J.B. Harley & D. Woodward (Eds.), *The History of Cartography*, Vol. 1: *Cartography in Prehistoric, Ancient, and Medieval Europe and the Mediterranean* (pp. 464-501). Chicago: University of Chicago Press.

Hoem, A.I. (1986). *Norge på gamle kart.* Oslo: J.W. Cappelens Forlag.

Imsen, S. & Winge, H. (1999). *Norsk historisk leksikon: Kultur og samfunn ca. 1500 – ca. 1800*, 2. utgave. Oslo: Cappelen Akademisk Forlag.

Keuning, J. (1959). Blaeu's *Atlas*. *Imago Mundi*, 14, 74-89.

Keuning, J. (1973). *Willem Jansz. Blaeu: A Biography and History of his Work as a Cartographer and Publisher*, revised and edited by M. Donkersloot-De Vrij. Amsterdam: Theatrum Orbis Terrarum.

Møller Nicolaisen, N.A. (1946). *Tycho Brahes papirmølle paa Hven.* København: Gyldendal.

Nissen, K. (1952). Gamle kart. In S. Engelstad (Ed.), *Norge i kart gjennom 400 år med opplysninger om dem som utformet kartbildet* (pp. 7-13). J.W. Oslo: Cappelens Antikvariat.

Nissen, K. (19549. Scavenius. *Norsk biografisk leksikon* (pp. 277-279). XII. Oslo: Aschehoug.

Nissen, K. (1961a). Det eldste Vestlandskart. Foredrag i Selskapet til vitenskapenes fremme 14. oktober 1960. *Bergens Historiske Forening: Skrifter*, 63: 1960 (pp. 91-113). J.D. Bergen: Beyer A.S.

Nissen, K. (1961b). Det eldste kart over det gamle Stavanger stift. Foredrag i Rogaland Akademi 19. oktober 1960. *Stavanger Museum Årbok 1960* (pp. 79-96). Stavanger: Dreyer.

Norlind, W. (1970). *Tycho Brahe: En levnadsteckning med nya bidrag belysande hans liv och verk.* Skånsk senmedeltid och renässans, Skriftserie utgiven av Vetenskaps-Societeten i Lund, 8. Lund: C W K Gleerup.

Richter, H. (1925). Willem Jansz. Blaeu – en Tycho Brahes lärjunge. Ett blad ur kartografiens historia omkring år 1600. *Årsbok 1925: Sydsvenska Geografiska Sällskapet* (pp. 49-66). Lund: Gleerupska Universitetsbokhandelm.

Richter, H. (1939). Willem Jansz. Blaeu with Tycho Blaeu on Hven and his Map of the Island: Some New Facts. *Imago Mundi*, 3, 53-60.

Rosen, E. (1986). *Three Imperial Mathematicians: Kepler Trapped Between Tycho Brahe and Ursus.* New York: Abaris Books.

Short, J.R. (1998). Maps and the Renaissance. *Journal of Historical Geography*, 24, 360-363.

Thoren, V.E. (1990). *The Lord of Uraniborg: A Biography of Tycho Brahe.* Cambridge: Cambridge University Press.

Tooley, R.V. & Bricker, C. (1969). *A History of Cartography: 2500 Years of Maps and Mapmakers.* London: Thames and Hudson.

van Mingroot, E. & van Ermen, E. (1988 [1987]). *Norge og Norden i gamle kart og trykk,* oversatt av B. Engen & I. Pommerel. Oslo: Aschehoug.

Wittendorf, A. (1994). *Tyge Brahe.* København: G.E.C. Gad.

Wood, D. (1992). *The Power of Maps.* London: Routledge.

Woodward, D. (1987). Medieval *mappaemundi.* In J.B. Harley & D. Woodward (Eds.), *The History of Cartography,* Vol. 1: *Cartography in Prehistoric, Ancient, and Medieval Europe and the Mediterranean* (pp. 286-370). Chicago: University of Chicago Press.

Chapter 14

NEW MONEY AND THE LAND MARKET
Landownership in 19th-Century Twente, the Netherlands

Elyze Smeets
School of Geography, University of Leeds, United Kingdom

1.   INTRODUCTION

During the nineteenth century the Netherlands witnessed the rise of a newly wealthy class; an urban *nouveaux riches* born of banking, trade and industry. Despite the origins of their prosperity, many of these individuals chose to invest part of their wealth in landownership. This was a phenomenon that was clearly evident in the Twente region of the eastern Netherlands (Figure 1). Here, the chief source of new money lay in the growing textile trade. But it was the coincidence of this industrial development with important changes in the institutional structures of landownership – specifically the radical transformation of long-established systems of communal ownership –, which allowed the newly wealthy the opportunity to establish themselves as an important presence in the land market. Together, the privatisation of communal land and the entry of a new class of investors into the market point to changes with a potentially important effect on the society and landscape of Twente. This paper attempts to increase our knowledge of this development through a brief exploration of the scale and location of land purchases, the use made of newly-acquired land and the motives that prompted entry into the land market.

Until well into the 19th century, Twente was a predominantly agricultural area. Its infertile sandy soils gave rise to a system of mixed arable and pastoral farming. Important in this was the grazing provided by the extensive areas of wasteland, which remained under the control of communal organisations known as *marken* (a title deriving from the Dutch word *marke*, meaning border). This traditional system was, however, placed under

H. Palang et al. (eds.), European Rural Landscapes:
Persistence and Change in a Globalising Environment, 227-243.
© 2004 *Kluwer Academic Publishers. Printed in the Netherlands.*

increasing strain as a result of population growth; between 1675 and 1800 the population of Twente increased from 17,000 to 51,000 (De Vries 1974). In an area with limited natural potential for agricultural development and where the continuing presence of the *marken* initially prevented the transfer of additional land to private farms, many were forced to look beyond farming for their livelihoods. As a result, domestic industry became an important facet of both the rural and urban economy, in turn speeding the growth of towns such as Almelo, Enschede and Hengelo.

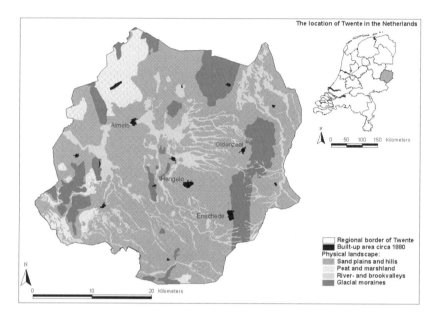

Figure 14-1. The physical landscape and 19th-century urban geography of Twente (Alterra, Wageningen; Het Oversticht/Smeets, Zwolle).

The success of Twente's textile industry during the course of the nineteenth century was reflected in the rise of an industrial élite. At its heart were around a dozen local families, including the Blijdensteins, Van Heeks, Geldermans, Ter Kuiles and Ten Cates. These were the people whose new-found wealth was directed towards the growing ownership of land. From the mid-19th century onwards the *nouveaux riches* of Twente established retreats for themselves in the countryside around the industrial cities of the east of the region, often located on north-south orientated *stuwwallen* or ice-pushed ridges. Typically these properties consisted of a country house with its accompanying garden and park (in Dutch termed a *buitenplaats*); the whole serving chiefly a recreational function. It was common, however, for industrialists also to purchase land with an economic value for agriculture and forestry.

2. RESEARCHING *NOUVEAUX RICHES* ESTATES

The institutional changes which allowed Twente's *nouveaux riches* to acquire much of their land have parallels elsewhere in Europe, including the Parliamentary Enclosure movement in Britain (Turner 1984) and the abolition of the German *Flurzwang* (Haushofer 1972). To date, however, research on the specific causes and implications of the entry of industrialists into the land market in Twente has been limited. Brief mention of the growing presence of the *nouveaux riches* is made in Keiser's (1967) survey of landed estates in the region. Similarly, the first detailed study of the history and architecture of Twente's estates (Van der Wyck & Enklaar-Lagendijk 1983) notes the commercial background of many of the new landowners of the 19th and early-20th centuries. However, the focus of this work is on the history of individual families and the aesthetics of specific estates, rather than any broader analysis of the changing pattern of land ownership.

Of more immediate relevance in the current context is the research of Olde Meierink (1984, 1985) and Olde Meierink & van Dockum (1988), whose studies of the historical development of house and landscape design, and of the commercial aspects of land use have paid attention to the spatial distribution of estates in Twente. Olde Meierink does not, however, link the observed locational pattern to a sustained analysis of change in the land market. Moreover, subsequent work inspired by Olde Meierink has largely concentrated on adding to our knowledge of the aesthetics of garden and landscape design (e.g., Van Beusekom 1991; Jordaan-Jannink 1993) or the family history of industrialists (Jansen 1996). The present study is therefore distinctive in seeking to enlarge upon the geographical dimension of Olde Meierink's work. It aims to show how a changing spatial pattern of landownership resulted from significant alterations in both urban and rural economies. It will also acknowledge the distinctive character of the interest in estate design, agriculture and forestry displayed by industrialists, which tended to set them apart from an older category of titled landowners.

3. DISTRIBUTION OF THE LANDED ESTATES OF THE TWENTE *NOUVEAUX RICHES*

As Figure 2 reveals, the landed estates established by Twente's industrialists were clustered around the eastern cities in which many of them made their fortunes. However, this concentration also reflected the limited influence of the established noble families in local land markets. By contrast, the continuing presence of the nobility as major landowners in western Twente allowed little opportunity for land purchase. Few new estates were established around Almelo, for example, where Count Van Rechteren van Limpurg held over 3,000 ha. Those amongst the city's industrialists who sought to create a country retreat were frequently forced to look further afield, particularly in the vicinity of Oldenzaal. Similar circumstances prevailed in Ambt and Stad Delden, Diepenheim, and other western municipalities.

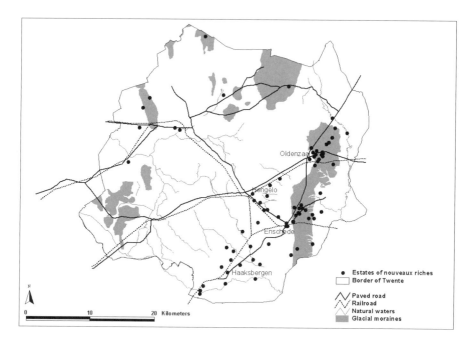

Figure 14-2. The *nouveaux riches* estates in Twente.

Country retreats as an escape from urban life were often located along major roads linking the cities of eastern Twente. Initially these retreats consisted of modest houses for weekend visits, or teahouses and garden pavilions for daytime recreational use. As visits were necessarily short, so was the distance that could be traveled from the city. Towards the end of the nineteenth century, however, it became increasingly common for families to invest in the creation of more substantial country houses, often surrounded by gardens and parkland laid out to a fashionable design, in which they would spend their summers away from the hectic world of the cities. Eventually, many adopted their country seat as a principal residence, retaining their town houses for business purposes.

Even at this later stage (late 19th century) it is evident that industrialists both wished and needed to stay relatively close to their existing urban social and business networks. This is reflected, for example, in the creation of a cluster of late-19th-century and early-20th-century country houses to the north-east of Oldenzaal (Figure 3). The majority of these, for instance De Haer, De Hulst, Kalheupink and Scholtenhaer, belonged to members of the Gelderman family. Other families from the new textile élite with estates near Oldenzaal included the Blijdensteins (Bekspring, Hakenberg and Nijehuss), the Van Wulfften Palthes (De Borg and Kruisselt) and the Storks (Paaschberg and Koppelboer). Over time, however, new estates (for example Poort Bulten, Meuleman and Duivendal) were gradually created further away from the city. In part, this reflected improvements in communications, ensuring quick and easy access along the major roads into Oldenzaal. But it was also a simple reflection of the growing shortage of land close to the city, which forced families to look further afield in developing their retreats.

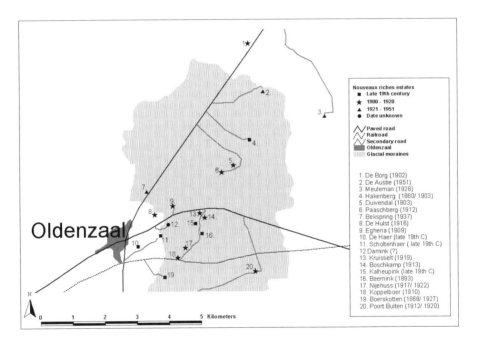

Figure 14-3. The *nouveaux riches* estates to the east of Oldenzaal.

The pattern shown on Figure 3 is complicated by the tendency of some industrialists to create more substantial houses on land that had formerly been occupied by simpler structures. Not all new estate development was thus on the margins of the urban hinterland. The owners of Hakenberg, Beernink and Boerskotten, for example, built new country houses in the early-20th century to replace existing garden pavilions. Some estates also changed hands. Close to Oldenzaal, the Gelderman family bought the eighteenth-century estate of De Hulst in 1916. The site was redeveloped with the construction of a country house designed by the Amsterdam architect Karel Muller, while his friend Leonard Springer added a garden in the fashionable mixed style, incorporating elements of the older formal garden (Moes 2002).

It is also interesting to note that the Blijdenstein family possessed estates near Oldenzaal, given that their business interests were concentrated some distance away in Enschede. The late date of establishment of all their Oldenzaal estates (1903, 1917 and 1937) may offer some explanation. By the early-20th century the land market must have been very tight. After the initial release of land onto the open market as a result of the breakup of systems of communal ownership, it had again become more difficult to buy land. It was also the case that the network of infrastructure was considerably better than it had been in the latter half of the nineteenth century when earlier generations of industrialists had created their estates. Hence, distance probably played a diminishing role in determining the location of land purchases. Further research may clarify the particular logic of the Blijdenstein family's pattern of land ownership.

Figures 2 and 3 suggest that the location of the estates developed by the *nouveaux riches* was related not only to distance from the city and transport infrastructure, but also to the physical landscape. Most estates were situated on *stuwwallen* that lay to the east of Twente's industrial cities. Olde Meierink (1985), and others following his lead (e.g., Jansen 1996) argue that this geographical pattern reflects the undulating character of the landscape, which offered particular potential for the creation of aesthetically-pleasing landscape gardens. Some support for this argument is, indeed, evident in the way that the existing landscape was exploited to create fashionable naturalistic gardens. However, previously mentioned studies may have accorded too great an importance to this motivation.

Any account of the development of new estates in Twente must also take account of changes in the land market caused by the partition and privatisation of communal land, formerly held by the *marken*, during the course of the 19th century. It was this sudden release of land onto the market that enabled so many of the *nouveaux riches* to become substantial landowners. Gerrit Jan van Heek, who bought some 450 ha of heathland in the municipality of Haaksbergen through an auction of former communal lands in 1895, was typical of this new class of landowner (*Natuurlijk Twente* 2003). The overall availability of land thus reflected the distribution of this former common land, much of which was located on the dry, sandy soils of the *stuwwallen*.

4. MEANS AND OPPORTUNITY

At the beginning of the 19th century Twente's textile industry was still domestic (Boot 1935; Hendrickx 2003). Its scale and mode of operation were, however, transformed over the ensuing decades, reflecting technological innovation, in particular the introduction of steam power, a growing input of professional knowledge, and the Dutch government's decision to concentrate cotton production for East Indian colonial markets in Twente (Lambert 1985; Schmal 1995). This commercial expansion provided the Blijdensteins, Van Heeks, Geldermans and other former linen traders with the means to invest in land. Initially, the extension of their landed property was achieved through the piecemeal purchase of individual small farms. Increasingly, however, they were able to acquire much larger tracts of waste as a result of the divisions of communal land executed chiefly between 1840 and 1900.

Around 60,000 hectares of communal wasteland – some 50 percent of the total land area of the region – remained under the administration of the previously mentioned *marken* at the start of the 19th century. The term *marke* denoted both the political unit of the local community and the area it incorporated. The political organisation of the *marke* was formed by the *gerechtigden* or *gewaarde boeren*, the yeomen and large landowners, who had the largest share in the landed properties of the *marke* (Demoed 1987). Together the members of the *marke* regulated the use of the communal lands, which were chiefly heathlands, woods and peatlands, yielding grazing, fuel, honey and peat as fertiliser.

Governmental decisions in the 19th century to divide all the communal lands in the Netherlands sought to extend the privatisation of land, ensuring an increase in the cultivation of the wastelands, and therefore leading to economic growth. Such decisions were supported at local level by existing large landowners. When a *marke* was divided the land entered the market in several ways. Occasionally, the newly-defined individual plots were sold off directly, usually at auction. It was more common, however, for land initially to be distributed amongst the members of the *marke*. The resultant plots were frequently too small, too scattered and located too far away from the home farm to constitute a viable agricultural unit. Moreover, as the lands were generally of poor quality, making profitable cultivation impossible without considerable investments of effort, time and money, most small owners quickly sold off their newly acquired land. In the west of Twente the established nobility profited from this surge of land on to the market; by comparison, in the east, the buyers were predominantly drawn from the industrial *nouveaux riches*.

5.   MOTIVATIONS TO INVEST IN LAND

Thus the combination of industrial and agricultural developments in Twente created opportunities for newly-wealthy industrialists to enter the landowning élite. This observed phenomenon raises a series of questions about their motivation. Why choose land as an investment, rather than devoting capital to other uses? Land was undoubtedly seen as a safe and durable investment. Its purchase was thus consistent with the business instincts that characterised the emergent industrial élite (Keiser 1967). However, their interest in landownership may also have reflected a strong emotional attachment to the region and their own rural origins. Van Heek (1945) suggests that industrialists – many of whom were born in the small villages that clustered around the cities of eastern Twente and who maintained a network of rural employees – retained a traditional agricultural mentality, i.e. their way of thinking resembled that of their predecessors who had combined linen trading with farming. Their resultant interest in farming, forestry and the visual dimension of the landscape found expression in the attention they paid to their new estates.

Typically, estates served both commercial and recreational purposes. At an early stage it was common for estates to be the location of bleach- and dye-works associated with the textile trade, as at Het Schuttersveld and Het Amelink, near Enschede. Initial construction thus focused on the creation of small premises for the supervision of these industrial operations, generally accompanied by a room for the owner. The waterways initially used for bleaching and dyeing often later became ponds within an ornamental garden, while the early buildings formed the nucleus of a more stylish and substantial country retreat.

Later developments were not entirely recreational, however. Land was acquired for cultivation and afforestation. In particular, large-scale pine plantations were established to meet the demand for timber from the coal mining industry in the south of the Netherlands and Belgium. Both individual landowners and companies, such as the *Nederlandse Heidemaatschappij* (Dutch Heathland Company, established 1886) and the *Grontmij* (Ground/Land Company, established 1915), cultivated large tracts of former heathland (Vervloet 1995).

At the same time, more and more families chose to own a house outside the city to escape the increasing industrial pollution and urban congestion which, ironically, they had helped to create (cf Daniels 1981). As a member of the Blijdenstein family reflected in a short poem: "Is it not sweet pleasure, to step into the green? Where there is no smoke, no fumes from the cities" (Translated by the author from HCO Family Archive Blijdenstein, no. 58).

An interest in nature and the countryside was sometimes expressed not only in the creation of gardens, but also arboretums – as at Poort Bulten, near De Lutte – and hunting grounds. Indeed, surviving hunting diaries reveal considerable detail about the scale of activity and the passion clearly felt by some industrialists for hunting as a pursuit (e.g., HCO, Family Archive Blijdenstein, no. 149). The hunting diary of Helmich Blijdenstein shows the application of businesslike precision to recording the type and weight of animal killed, the gun used and the location of the hunting ground.

Figure 14-4. The Hooge Boekel estate, depicted by Anco Wigboldus (1977) (Van der Wyck & Enklaar-Lagendijk 1983: 120).

Woods, both deciduous and evergreen, thus served a variety of recreational and commercial purposes for their new owners. Woodland also played an important part in the aesthetic layout of many estates. As most estates created by the *nouveaux riches* were located close both to the city and to other country houses, privacy was often at a premium. Thus, as at Hooge Boekel, an estate close to the city of Enschede, designed for the industrialist Van Heek in the early-20th century, woodland was planted around the perimeter of many estates to create seclusion (Figure 4). The case of Hooge Boekel also reveals the use made of woodland within the park, creating vistas to give shape and a sense of scale to grounds that were relatively modest in extent.

Few amongst the earliest estates established by the *nouveaux riches* were grand in either scale or design. Often the owners themselves were responsible for the layout of house and grounds. However, as the wealth and landholdings of individual industrialists increased it became more common for the aesthetics of estate development to reflect the work of professionals.

In particular, increasing attention was devoted to the design of parks and gardens, which were often laid out in accordance with the newest and most fashionable tastes. By the end of the nineteenth century the gardens of the *nouveaux riches* were beginning to compete in style with those of the local nobility.

6. COMPARISON WITH NOBILITY

Any comparison of the *nouveaux riches* and nobility of Twente must begin by acknowledging the relative absence of social connections between the two groups of landowners. This suggests a contrast with Britain, where one of the main motivations for the investment of "new money" in landownership appears to have been a desire to gain entry into the social world of the landed nobility (Thompson 1963; Wiener 1981). That this was not a concern for the *nouveaux riches* of Twente perhaps reflects the relative lack of social and political influence of the nobility in the 19th-century Netherlands, when compared with their British counterparts (Van Heek 1945).

As might be expected Twente's *nouveaux riches* were closely bound together by their business dealings. Indeed, their growing landownership provided new opportunities for commercial collaboration. The cultivation company, *Nederlandsche Heidemaatschappij*, was one such instance. These links were mirrored in the social and religious networks developed by industrialists and their families. Many, for example, married within their own circle. Helmich Blijdenstein's hunting diary also reveals that Enschede's industrial élite often hunted together on each other's land. Blijdenstein's hunting companions included fellow industrialists G.J. van Heek, E. ter Kuile, G.J. Jannink and A. Ledeboer jr (HCO, Family Archive Blijdenstein, no. 149). The names of the local nobility are conspicuous by their absence.

The available evidence therefore suggests considerable social differentiation between the nobility and *nouveaux riches* of Twente. Yet the desire for social advancement does seem to have contributed to industrialists' purchase of land. Within their own social circle status was undoubtedly gained through the development of increasingly grand estates, created by professional architects and designers, and maintained by land agents, gamekeepers and lesser servants. The recreational use of land, particularly the establishment of gardens, parks and hunting grounds, provided an important opportunity for the display of both wealth and taste. In this basic desire, at least, there was common territory between "old" and "new" money.

It seems, nevertheless, that industrialists deliberately chose not to imitate the local nobility in the details of their conspicuous consumption. Instead, they developed their own styles in garden design. This was reflected not only in distinct differences in the appearance of the new estates, but also in the choice of designers employed to create them.

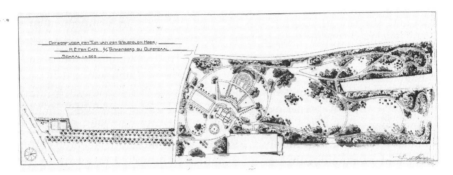

Figure 14-5. Leonard Springer's design in 1908 for the estate Egheria, near Oldenzaal, shows his distinctive mixed style (Special Collections, Wageningen University, No. 01.415.03).

The nobility often retained faith in an established élite of designers, whose ideas reflected their traditional training. At Twickel, for example, the main seat of Countess Wassenaer van Obdam and her husband Van Heeckeren van Kell, a large landscape park was created in 1885-1889 by Eduard Petzold (1815-1891), a designer who also worked for the Dutch Royal Family. Similarly, the Parisian architect Edouard André (1840-1911) and his pupil Hugo Poortman (1858-1953), who worked chiefly for the nobility, were chosen to create a geometric garden for Count Bentick at Weldam in 1885. By contrast, the *nouveaux riches* often employed new, less conventional designers, such as Leonard Springer (1885-1940) and Piet Wattez (1872-1953). Springer, in particular, championed the so-called "mixed style" of estate design, integrating geometrical and naturalistic elements. While some of the features of this combination must have reflected a purely practical desire to create variety and a sense of space within a relatively small area, the aesthetic appeal of the mixed style meant that it was widely adopted by owners of new estates. Springer's work at De Hulst, Kalheupink and Egheria (Figure 5) is typical, with designs consisting of a geometrical garden with roses close to the house, extending into a landscape park with curving paths, clusters of deciduous and exotic trees and naturalistic water features (Moes 2002).

Issues concerning the location, evolution and design of estates will now be explored in more detail through a case study of the industrialist Benjamin Willem Blijdenstein.

7.   CASE STUDY – BENJAMIN WILLEM BLIJDENSTEIN

The Blijdensteins were one of the first industrial families to emerge as substantial landowners, creating Het Amelink estate in the municipality of Lonneker, to the northeast of Enschede and near to the main road to Oldenzaal (Figure 6). During the 17th and 18th centuries members of the family had moved into Twente from the German kingdom of Westphalia. The direct roots of the family's main seat at Het Amelink can be traced to the acquisition by marriage of land in the *marke* of Lonneker in 1741 (Van der Wyck & Enklaar-Lagendijk 1983). However, farming on such infertile soil offered only a precarious livelihood and other activities, including politics, were precluded by their inferior status as followers of the Mennonite religion. The family's fortunes were, therefore, secured through trading and production, first in linen and later in cotton (Van Heek 1977; Wevers 1993). Ultimately, this growing wealth enabled the Blijdensteins to become much more substantial landowners, with considerable commercial interests in the development of both agriculture and forestry.

The Amelink estate was at its largest during the lifetime of Benjamin Willem Blijdenstein (1780-1857), when it incorporated De Welle (purchased in 1830), Het Bouwhuis (purchased in 1844) and part of Het Welna (Van der Wyck & Enklaar-Lagendijk 1983). The estate remained in family hands until 1971. In addition to Het Amelink the family owned land elsewhere in Twente, particularly in the adjacent municipality of Losser, but the present study will focus on their holdings in Lonneker.

Figures 6 and 7 show the growth of Benjamin Willem's land in Lonneker between 1832 and 1854. The former date reflects the introduction of a uniform tax register – *Het Kadaster* – into the Netherlands. This is an excellent source for study of the land market, as it provides a detailed record of landownership within each individual cadastral municipality. Every plot of land – allocated with its own specific number – was described in terms of ownership, function, value and size. Cadastral material has been used to reconstruct the exact location, scale, use and land tax value of Benjamin Willem Blijdenstein's landed property in 1832 (Figure 6). The map reflects the economic and social position of the family in the early-19th century. At this date they were small industrialists who also operated a farm and both economic enterprises are clearly visible on the map. About half of the land area was devoted to arable cultivation, with small portions remaining as woods, coppice and heathland. The property also included a bleach-works – a network of small rectangular canals was used for bleaching the cloth, which was then dried on the bankside.

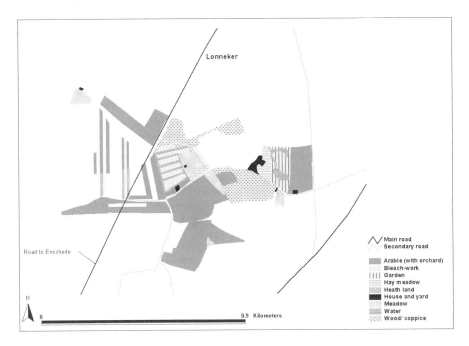

Figure 14-6. The landed property of B.W. Blijdenstein in Lonneker, 1832 (Cadastre of Lonneker, 1832, Kadaster Overijssel, Zwolle).

The main house was situated to the east of the bleach-works. In its immediate environs Blijdenstein created a naturalistic garden with ornamental ponds and a pavilion in the early-19th century. Later alterations have been attributed to the professional garden designer Dirk Wattez (Olde Meierink 1984). The family's personal papers include a copy of a short poem which is suggestive of this interest in landscape design: "Go, planters, make a mountain, a wood, or merry valley. Your diligence is for a generation that is still to come" (HCO; Family Archive Blijdenstein, no. 58). Such sentiments indicate not only a sense of the immediate pleasure to be gained from creating an aesthetically appealing landscape, but also a dynastic interest in the development of the estate for the future.

The mix of recreational, agricultural and industrial land uses found at Het Amelink was characteristic of the initial phase of estate development by the *nouveaux riches*. It also established another basis of differentiation from the local nobility, whose estates were generally larger and lacked an industrial dimension. In 1832 Blijdenstein's Lonneker property extended over 27.5 ha and was taxed at 610.57 guilders per year. In the years up to 1850 Benjamin Willem added to his holding by purchasing a series of farms, each including a house, with yard, arable and pastoral land, orchards and hay meadows.

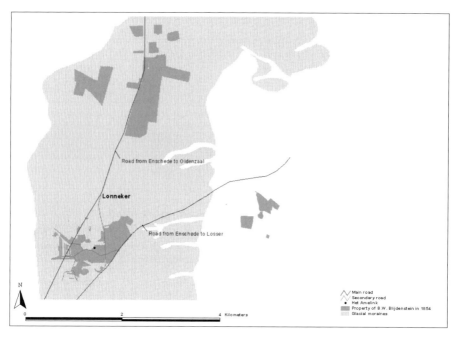

Figure 14-7. The landed property of Benjamin Willem Blijdenstein in 1854 (Cadastre of Lonneker, 1832, Kadaster Overijssel, Zwolle).

These piecemeal extensions, all within a five-mile radius of the family's seat, added gradually to the estate's agricultural activity, but they were completely eclipsed by the major addition to Blijdenstein's landholding between 1850 and 1852, which created an estate of around 206 ha (Figure 8). This sudden expansion reflected the opportunities opened up by the privatisation of the communal *marke* lands in Lonneker. Some of the new land came to Blijdenstein directly as his share of the *marke* division; the remainder was purchased from other recipients. In total Benjamin Willem acquired around 160 hectares from the Lonneker *Marke*. In the main this was relatively poor quality heathland, with an average value of 50 cents per hectare.

In common with *marke* land elsewhere, much of Blijdenstein's new property was located on the *stuwwallen* that were a feature of eastern Twente. This specific case therefore offers support to the general thesis that the geographical pattern of estate development owes much to the process of privatisation of *marke* land. The concentration of activity on these glacial features may in part have reflected their potential landscape value, but we must also take account of the distribution of the available land released by the dissolution of the *marken*. Without this fundamental change in the basis of landownership most new entrants into the land market would have had to content themselves with sort of the modest acquisitions that Blijdenstein secured before 1850.

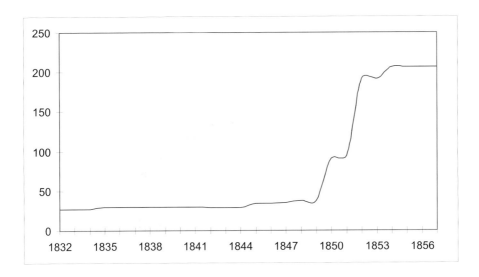

Figure 14-8. The total landed property of Benjamin Willem Blijdenstein in the municipality of Lonneker, 1832-1857 (Cadastre of Lonneker, 1832, Kadaster Overijssel, Zwolle).

8. CONCLUSIONS

The *nouveaux riches* of Twente derived their wealth from the local textile industry, the success of which enabled them to invest their profits in land. Such land was chiefly available from the mid-19th century onwards because of the release on to the market of a significant proportion of the 60,000 ha of formerly communally-owned *marke* land. The creation of new estates by the *nouveaux riches* was thus the result of a combination of advances in both the industrial and agricultural structure of the region.

The availability of former communal wasteland on the *stuwwallen* exerted an important locational influence over the development of new estates. Clusters of country retreats were established in this undulating landscape, chiefly within a five-mile radius of the major industrial cities. The estates were the creation of a class of industrialists who retained strong links with their rural origins. Motivation for investment in land thus in part reflected an emotional bond with the region. The new owners created a landscape that served both recreational and commercial purposes. Gardens and parkland sat alongside agricultural land, and woodland, depending on its type, served to add visual interest to the landscape, to provide cover for hunting, and to provide a crop of timber.

Although they had little in common, either in terms of origins or social connections, with the old nobility, the new estate owners evolved into a parallel élite. Despite the absence of enthusiasm for greater contact on either part, both old and new money had some tastes in common. Both valued their land for hunting and took pleasure in the art of landscape design. But the end result was often rather different, with the two groups employing different designers to create contrasting styles of garden and parkland. The presence of the *nouveaux riches* is, therefore, important both in establishing new economic and social links between town and country, and in setting a distinctive stamp upon the development of estate landscapes during the late-19th and early-20th centuries.

ACKNOWLEDGEMENTS

The author would like to thank Professor Robin Butlin and Dr. Martin Purvis for their inspiring advice on the paper, and is grateful to the following organisations for the kind permission to use data and images:
– Canaletto/Repro Holland, Alphen aan den Rijn, the Netherlands;
– Historisch Centrum Overijssel, Zwolle, the Netherlands;
– Kadaster Overijssel, Zwolle, the Netherlands;
– SC-DLO Staring Centrum/Alterra, Wageningen, the Netherlands;
– Special Collections, University of Wageningen, the Netherlands.

REFERENCES

Manuscript Sources

Cadastre of Lonneker, 1832, Kadaster Overijssel, Zwolle.
Family Archive Blijdenstein, admission 233.1: Historisch Centrum Overijssel, Zwolle.
Family Archive Van Heek, admission 166: Historisch Centrum Overijssel, Zwolle.

Printed and Secondary Sources

Boot, J.A.P.G. (1935). *De Twentsche katoennijverheid, 1830-1873* Amsterdam: H.J. Paris.
Daniels, S. (1981). Landscaping for a Manufacturer: Humphry Repton's Commission for Benjamin Gott at Armley in 1809-10. *Journal of Historical Geography*, 7 (4), 379-396.
Demoed, H.B. (1987). *Mandegoed, schandegoed: een historisch-geografische beschouwing van de markeverdelingen in Oost Nederland in de 19e eeuw.* Zutphen: Walburg Pres.
Haushofer, H. (1972). *Die deutsche Landwirtschaft im technischen Zeitalter.* Stuttgart: Ulmer.

Hendrickx, F.M.M. (2003). Family, Farm, and Factory: Labor and the Family in the Transition from Protoindustry to Factory Industry in Nineteenth-Century Twente, the Netherlands. *The History of the Family,* 8, 45-69.

Jansen, R.W. (1996). *Een heide(ns) karwei: Albert Jan Blijdenstein, president van de Nederlandsche Heidemaatschappij 1889-1896.* Doctoral dissertation, University of Utrecht.

Jordaan-Jannink, E.B. (1993). De Twentse buitenplaatsen. *Tuin Special,* 42-45.

Keiser, J.W. (1967). Landgoederen en andere natuurterreinen in Twente. *Jaarboek Twente,* 119-137.

Lambert, A. (1985). *The Making of the Dutch Landscape: an Historical Geography of the Netherlands.* 2nd ed. London: Academic.

Moes, C.D.H. (2002). *L.A. Springer 1855-1940. Tuinarchitect, dendroloog.* Breda: Uitgeverij de Hef.

Natuurlijk Twente. Retrieved July 16 2003, from the Natuurlijk Twente Web site: www.natuurlijk.nl/twente/.

Olde Meierink, B. (1984). De Twentse fabrikantenbuitenplaats. Een verkenning naar een onbekend verschijnsel van de industriële revolutie. *De woonstede door de eeuwen heen/ Maisons d'hier et d'aujoud'hui,* 60 (4) (pp. 42-68). Luxemburg: Orgaan van de Koninklijke Vereniging der Historische Woonsteden van België en van de Nederlandse Kastelenstichting en met de medewerking van Icomos.

Olde Meierink, B. (1985). De Twentse fabrikantenbuitenplaats. In H. Meindersma & K. de Jong (Eds.), *Jongere bouwkunst in Overijssel: 1840-1940* (pp. 59-76). Utrecht: Matrijs.

Olde Meierink, B. & van Dockum, S. (1988). Buitenplaatsen van de vroegere textielfabrikanten in Twente. *Heemschut 65* (11/ 12), 22-24.

Schmal, H. (1995). Een landschap vol steden. In S. Barends, J. Renes, T. Stol, J.C. van Triest, R.J. de Vries & F.J. van Woudenberg (Eds.), *Het Nederlandse landschap: een historisch-geografische benadering* (pp. 99-115). Utrecht: Matrijs.

Thompson, F.M.L. (1963). *English Landed Society in the Nineteenth Century.* London: Routledge & Kegan Paul.

Turner, M. (1984). *Enclosures in Britain, 1750-1830.* London: Macmillan.

Van Beusekom, J.W. (1991). Beknopt overzicht van de ontwikkeling in de tuin- en landschapsarchitectuur in Twente in de periode 1850-1940. *Tuinjournaal,* 8 (3), 24-33.

Van der Wyck, H.W.M. & Enklaar-Lagendijk, J. (1983). *Overijsselse buitenplaatsen.* Alphen aan de Rijn: Canaletto.

Van Heek, F. (1945). *Op de maatschappelijke ladder. Een onderzoek naar de verticale sociale mobiliteit.* Leiden: E.J. Brill.

Van Heek, N.H. (1977). De Blijdensteins als pioniers van de textielindustrie. *Jaarboek Twente,* 118-123.

Vervloet, J.A.J. (1995). Het zandlandschap. In S. Barends, J. Renes, T. Stol, J.C. van Triest, R.J. de Vries & F.J. van Woudenberg (Eds.), *Het Nederlandse landschap: een historisch-geografische benadering* (pp. 9-26). Utrecht: Matrijs.

Vries, J. de (1974). *The Dutch Rural Economy in the Golden Age 1500-1700.* New Haven.

Wevers, A.L.A. (1993). Benjamin Willem Blijdenstein. In J. Folkert, J.M.M. Haverkate & F. Pereboom (Eds.), *Overijsselse Biografieën,* 3 (pp. 9-12). Amsterdam: Boom.

Wiener, M.J. (1981). *English Culture and the Decline of the Industrial Spirit, 1850-1980.* Cambridge: Cambridge University Press.

Chapter 15

THE LANDSCAPE OF VITTSKÖVLE ESTATE – AT THE CROSSROADS OF FEUDALISM AND MODERNITY

Tomas Germundsson
Department of Social and Economic Geography, Lund University, Sweden

1.   INTRODUCTION

An essential question regarding the understanding of landscape concerns the role of history. While every landscape is considered contemporary, they also possess historical dimensions. Sometimes the historical traits in a landscape are very palpable; but as long as no time machine exists, even the most ancient landscapes must still be regarded as present-day landscapes. "The past is [thus] a foreign country," as David Lowenthal (1985) has so famously argued. Concerning linear time, there therefore exists a duality in every landscape, simply expressed as a "now" and a "then". As the line between now and then is constantly moving, it means that the historical dimension of a landscape is continually reshaped. Furthermore, as time moves on, the past takes on different meanings depending on the position of the observer in the present. This is of course valid for history at large – all history bears the mark of the time in which it is written, but landscape history is special in that it is materialised history forming the totality of the present world.

The impossibility of writing – or representing in other ways – landscape history in a definite or unbiased way calls for constant revaluation and revision of historical landscapes. Otherwise, we run the risk of taking historical landscapes for granted, their history considered written once and for all. These circumstances become crucial especially concerning landscape

H. Palang et al. (eds.), European Rural Landscapes:
Persistence and Change in a Globalising Environment, 245-267.
© 2004 *Kluwer Academic Publishers. Printed in the Netherlands.*

preservation, as preservation means letting history guide present-day landscape management.

Within a Swedish context (from which my case study is drawn), there is a theme within landscape history that is not attended to in accordance with its significance – namely, modernity.[1] Consequently, we find that the era of modernity is underrepresented within the Swedish landscape preservation practice. Interest in the history of modernity has experienced a general upswing in Sweden during the last few decades; this has been connected especially to the emergence of the welfare state. This can be noticed both in academic writings and in popular historical publications.[2] However, the legacy of modernity as found within a preservation perspective has mainly been attended to in an urban context.[3] Modernity is hardly a focus of today's preservation practices concerning the countryside and agrarian landscape[4] of Sweden. I find this to be a serious shortcoming, because I believe that an appropriate knowledge regarding the history of modernity is essential to the understanding of the landscape of present-day Sweden. Changing relations with nature, the ideology of the welfare state, and centralised physical planning are but a few examples of societal processes that have influenced the shape of the Swedish landscape – and people's relations to it – during the modern era.

In this chapter, I will use the example of a south Swedish estate landscape in order to demonstrate how the inclusion of aspects of modernity can provide an alternative view of the historical legacy of such an area. The estate is situated in the province of Skåne, which is widely known as a productive agricultural district, but also as a "non-Swedish" region, characterised by its openness and light beech woods. These features are readily associated with the estates found in the province, and in differing

---

[1] There exists no explicit demarcation defining what would be considered the "era of modernity" within a Swedish context. I use the concept here as a label for the period starting (with an obvious influence from the European Enlightenment) at the end of the 18th century and continuing up to the present, but with an emphasis on the changes in connection to the breakthrough of industrialism and urbanisation during the latter half of the 19th century and the first half of the 20th century.

[2] The literature on modernity is extensive, but some recent examples with a geographical connection are: Johansson 1994; Larsson 1994; Pred 1995; Wahlberg 1996; Hagman 2002; Lisberg Jensen (2002. The more-widespread popular interest for Swedish modernity is reflected in well-illustrated publications such as Bengtsson (1994) and Rune (2003).

[3] This can be noted in preservation and policy documents from regional authorities (an example is Tykesson 2001), but also in other literature, e.g., Lilja 1994; Dunér 2002; Eriksson 1990 and 2001.

[4] This generalised statement could be confirmed by a study of the regional preservation programs in Sweden, but I do not refer to them individually here. References to some of them will be made later in the chapter, when the situation in the province of Skåne is dealt with.

media sources – from school books to preservation plans – the landscapes of the estates are emphasised as typical for Skåne. Not surprisingly, the estate milieus have been identified as a rich cultural heritage. The point is, however, that the important and formative changes that occurred during the breakthrough of modernity in the 19th and early 20th centuries are more or less forgotten themes when it comes to looking at this landscape's history from a preservation perspective (cf. Germundsson 2001). Thus, a vital facet of the historical heritage has been left out. This has, I would argue, resulted in a missing link when it comes to understanding the present landscape. My aim is, therefore, to contribute to the creation of a fuller picture of the Swedish estate landscape by demonstrating how expressions of modernity should be included within the framework of a cultural heritage.

In the following, I first give a general presentation of the province of Skåne and the manorial landscape found in the region. This presentation is followed by a short discussion of the concept of landscape, demonstrating some of its different meanings and its potential to mediate different views of the past. From these starting points, I will then use the example of the Vittskövle estate to illustrate how processes of modernity were expressed in the landscape. The paper ends with a discussion of how the heterogeneous processes of modernity can supply alternative understandings concerning the present-day estate landscapes.

2. SETTING THE SCENE: A PICTURE OF THE PROVINCE AND SOME NOTES ON THE CONCEPT OF LANDSCAPE

2.1 Skåne and the Landed Estates

The province of Skåne (sometimes Latinised as *Scania*) in southernmost Sweden is one of the regions in the country that most palpably bears the stamp of big landed estates. Travel through the countryside gives evidence of this not only by virtue of the rich presence of castles and manor houses, but also via a broader landscape perspective, through which the abundance of the huge fields of modern big-scale farming is perceived. The history of the estate landscape goes back far in time, and can be connected to several formative processes occurring in the region's past. An early sign of a manorial system manifesting itself in the landscape was the establishment of private magnate castles in the politically unsettling period from the middle of the 13th century to the end of the 14th century. Of these buildings only remnants remain, while several manorial castles from the late 15th century and the 16th century are still standing. They thus make up important landmarks in the present Scanian landscape. Since this time, castles and manor houses have been built and rebuilt in great numbers in Skåne, resulting in a "manorial-saturated" landscape characterising considerable parts of the province.

The mid-17th century was a revolutionary period in Skåne (Johanesson 1977; Skansjö 1997). Widespread warfare occurred in connection to political conflicts between Denmark and Sweden. Up until this time, Skåne was a part of the Danish kingdom; but the province was subsequently conquered by Sweden. The shift in nationhood was ratified in a peace treaty in 1658. Among other things, it meant that noble land was exchanged between Denmark and Sweden. Some noble families moved from Skåne to Denmark, while some Swedish noble families were established in the newly conquered province. Even though the big estates afterwards constituted a very important element of the landscape history of Skåne, the manors were commonly not discernable as large coherent land units before the 17th century. Besides the nobleman's house and farm, a manor generally consisted of tenant farms, where the peasants paid their tenancy (land rent) in the form of work or products from the farm. These farms could be quite spread out and localised on vast areas (Skansjö 1997). In this respect, there was of course a territorial dimension regarding the manor – but the basic principle of the system was not built on control over land, but rather on the power, duties and privileges that regulated the relations between lord and peasant.

In early modern times, a tendency towards increasing the production of the demesne land on many manors has been noted (Olsson 2002). In this process, the workforce on the demesne land was supplied according to the corvee system, i.e. peasants (or servants) from farms owned by the lord were burdened with the duty to work on the manor. This process consisted of a change towards what analytically has been labelled a *Gutsherrschaft*, a manor system resembling the early medieval system with serfdom, estate jurisdiction, and corvee duties; although aiming at commercial big-scale farming in later times. The tendency of an increased *Gutsherrschaft* in Skåne (and Denmark) during this period is not clear-cut. In many cases, the *Grundherrschaft* (i.e. the system relying on rents paid in kind or in money by tenant farmers) prevailed. In Skåne, however, a general trend evident from the 17th century onwards was that the big estates gathered their domains in well-assembled property units, and that they increased the cultivation of the demesne land (Olsson 2002).

The aim of collecting and concentrating the estate domains resulted in the estates forming "states within the state". The efforts to make these (e)states fit into a mercantile, and later physiocratic, nation-state had important implications for the landscape. The power and the influence of the manors were increasingly territorialised, and the manor houses and their close milieus were turned into pronounced centres in a surrounding landscape of demesne land and villages under the direct control of this centre. Magnificent buildings, grand parks, and alleys were established during the 17th and 18th centuries, giving rise to visually-rich estate landscapes.

2.2 Modern Changes

These developments laid the ground for further changes throughout the 19th century, when the agricultural revolution was explicitly staged in southern Sweden. Management was then radically rationalised on many of the estates, expressed for instance through enclosures, drainage of wetland and meadows, a tremendous extension of arable fields, the construction of huge cow houses, and the establishment of new settlements for farmhands and agricultural workers. The estates were, in other words, modernised and turned into efficient agricultural enterprises (Germundsson 1999).

In a broader perspective, these changes took place within a nation that on the one hand consciously and wilfully headed towards modernity and urbanism, but on the other hand was still an agrarian land of peasants and crofters (in Sweden, a majority of the population lived on the countryside up until 1930). In such a process, several contradictory traits appeared. On the estates, what was previously a more or less feudal mode of production was gradually influenced by a capitalistic economy. The estates adapted to this development, and in many cases they were forerunners; not only in the area of farming techniques, but also in adjusting to modern ways of securing a workforce, regulating land owning, and so on (Möller 1989). Simultaneously, the authoritarian regime on the estates was sustained in different ways – often with reference to the past – and the estates during this period frequently became the scene for conflicts between an old regime and a growing democratic social system. On the estates, traditional hierarchical social structures often were held up through personal bonds and with reference to tradition. At the same time, worker's unions and other social movements gained ground; first in the cities and industries, but later also in the countryside (Morell 2001). The manorial world could not escape these extensive changes – even though it continued to carry on as a manorial world in many respects.

From a landscape perspective, these processes are challenging. Physically, the landscape underwent substantial changes, corresponding to the economic management of the estates. It is also clear that the landscape was used as a medium to express influence, power, and exclusivity in new ways, relating to both history and current progress. My argument, however, is that the latter aspects are underrepresented in the contemporary narration and presentation of the historical estate landscape. When it comes to issues regarding the representation and handling of these landscapes – for instance, within the tourist industry or preservation documents – a pre-industrial perspective is highly dominant (see for example *Kulturminnesvårdsprogram för Skåne* 1984). There is thus a tendency towards emphasising aspects related to the golden age of the nobility and the high culture of that era. As well, when later periods are dealt with it is often the history of those older elements that is in focus: emphasis is placed on the *re*-construction of castles and manor houses, the *re*-shaping of parks and gardens, while at the same time novel modern imprints are neglected or even seen as bothersome problems (Ambrius 2001; for further discussion see Bjurklint 2001).

In a departure from these observations concerning the modern estate landscape, I will in the next section provide some concrete examples of its changes. Firstly, however, I will present some theoretical aspects in regards to the concept of landscape and its connection to modernity.

2.3   Heterogeneous Landscapes of Modernity

If the Scanian estate landscapes are seen using a historical perspective from the late 18th century onwards, it is quite clear that ambiguity and contradiction are defining characteristics. This has led me to attempt to theoretically understand the landscapes as *heterogeneous landscapes of modernity*, a term used by the sociologist K. Hetherington in his analyses of selected places and landscapes in Europe during the 18th and 19th centuries (Hetherington 1997). A main point in his reasoning is that there was an interplay between freedom and control, and between resistance and ordering, that was fundamental to modernisation during this period. It emanated from the transformations that occurred when utopic ideas formulated in relation to modernity met the practice of everyday life. This had social and spatial consequences: "A process that is utopian in its ideal develops in spaces that are ambivalent and uncertain either because they are new and as-yet unknown or because they are impossible archaic representations of former modes of social order that have become obsolete" (Hetherington 1997: 68). This provides a key to understanding the duality of old and new in a changing modern landscape, because "rather than see a modern social order develop out of the eradication of ambivalence, it is better to see a process of modern social ordering develop into spatial practice through a utopic expression and its effects" (Hetherington 1997: 70). I find this interpretation of modernity's role in a spatial perspective fitting to the situation of a late 19th, early 20th-century estate landscape, as it is exactly such ambivalent processes and experiences that are interwoven in an estate's reorganisation and reshaping during this period.

In order to further develop this line of reasoning, I think it is important to stress that an appropriate understanding of landscape considers it as a many-facetted concept, comprising both concrete, physical phenomenon *and* social constructions (Olwig 1996). The latter indicates that landscapes are subjective and negotiable: the same physical landscape means different things to different people. Landscapes are therefore also representations, which I interpret to mean that they are subject to how individuals or groups within different social contexts describe, understand, symbolise, and communicate their lived-in surroundings. The link between these two meanings is that landscapes are produced. They are the result of human action; of work, but also of the social conditions under which this work is performed (Mitchell 2000). Therefore, landscapes more or less clearly reflect socio-economic relationships, even though a very important aspect of this is that these expressions could be ideological; i.e. more or less created as a *false consciousness* (to borrow Marx's expression). Landscapes thus can possess a tendency to conceal the socio-economic conditions under which they are produced, and instead accentuate other aspects, such as harmony, balance, or order. Landscapes might then appear to be "natural" and easily taken for granted, when instead they very well could be produced under conflicting social circumstances (Cosgrove 1998), or, as Don Mitchell put it:

> If landscapes are produced through specifiable social relations, and hence both the relations of production and the resulting product can be studied, then it is also important to remember … that landscape is additionally a form of ideology. It is a way of carefully selecting and representing the world so as to give it a particular meaning. Landscape is thus an important ingredient in constructing consent and identity – in organizing a receptive audience – for the projects and desires of powerful social interests (Mitchell 2000: 99-100).

The production of landscape and its role as representation thus indicates a reciprocal dimension; landscapes are not just there as an inert scene for human action – they work.

In the following paragraphs, I will demonstrate how different changes throughout the 19th and early 20th centuries affected the estate of Vittskövle in Skåne, and how these changes led to a heterogeneous landscape bearing the marks of both a feudal past and modern influences.

3. REPRESENTATIONS AND VISIONS OF VITTSKÖVLE

3.1 Enclosure

Figure 1 shows the centre of the Vittskövle estate domain as it appeared in the beginning of the 19th century, with the castle to the left and the village of Vittskövle with its church and farmsteads to the right. The church was built in the 13th century, while the earliest known traces of the noble estate date back to the 14th century. The present castle was erected in the 16th century, and surrounded by a moat. The farms in the village belonged to the estate, and the latter was run as a feudal corvee system. This system in principle prevailed up until the 19th century, when significant changes set in.

Figure 15-1. A representation of Vittskövle ca 1806. The map shows the village of Vittskövle with its church and farmsteads (to the right) and Vittskövle estate with its castle and park (to the left). The picture is a detail from a cadastral map produced in order to record and evaluate the village's lands in connection to an enclosure ordered by the estate owner. The distance between the castle and the church is ca 1.2 km (Lantmäteriarkivet, Kristianstad).

An enclosure carried out in Vittskövle village in the beginning of the 19th century can be seen as the starting point for the development of the modern landscape history of the estate. In the enclosure process, almost all farms that for centuries had been situated in the village core were deconstructed and rearranged as new individual farmsteads in the surrounding landscape. The old, collectively organised open-field system was abandoned, and replaced by one that was modern and individual.[5] This was a process quite common in Skåne at the time. The policy of concentrating the estate domains – carried out by many of the estate owners during earlier periods (as referred to above) – resulted in a situation where several villages in the vicinity of an estate were also formally owned by the estate. The farms were then tenant farms, and the land rent was to be paid in both work duty (a certain number of days per year) and in farm produce. During the 18th century, several estate owners made efforts to render the operation on the tenant farms more effective, with the aim of increasing the output to meet the demands of a growing national and international market. These efforts included both rationalisations on the land (such as drainage and the implementation of new and more efficient tools), the introduction of new crops and crop rotations, and land reform programs (Möller 1985). Concerning the latter, a number of statutory enclosure programs proved to be very influential. For the open plain regions in Sweden – like large parts of Skåne – the so-called "one shift" enclosure (in Swedish *enskifte*) confirmed by law in 1803 was the most prominent (Germundsson & Lewan 2003). The first full-scale "one shift" enclosure was carried out on Count Rutger Macklean's estate, Svaneholm, in southern Skåne. The basic idea behind this enclosure was that all land belonging to a farm should be collected into one unit ("one shift"). On Macklean's domains, there were four villages belonging to the estate; and in a radical plan drawn up in 1785, both the estate's demesne land and the land belonging to the village farms were given a totally new layout. The traditional open-field system – with its hundreds of long strips of arable land and each farm's land spread out on fields over the whole village domain – was replaced by a rational checkerboard pattern giving each farm one collected piece of land. This program was carried out under Rutger Macklean's direct initiative and orders. As the land was owned by Macklean, the peasants had no say in the process. After a number of difficult years, the progressive program started to bear fruit, and it soon became an inspiring example to follow for the authorities trying to enhance Swedish farming. The land reform program carried out on Svaneholm thus became a blueprint for the "one shift" enclosure regulation that was established in 1803. This meant that the "one shift" enclosure plan could also be carried out in villages

---

[5] The sources concerning the enclosure process and the landscape changes it involved in Vittskövle are land survey materials, foremost maps and protocols. They are kept in the regional Land Survey Archive in Kristianstad (in Swedish Lantmäteriarkivet i Kristianstad).

where the farms were owned by freeholders, or otherwise had several proprietors, as the regulations comprised a set of rules concerning how the interests of different owners should be handled. Both before and after 1803, Macklean's pioneering project served as an inspiration for other estate owners to carry out radical enclosure programmes on their own domains.

It is quite clear that the enclosure on Vittskövle in the beginning of the 19th century bears the mark of the Svaneholm enclosure. It was commanded and organised by the estate owner, who controlled all of the farms in Vittskövle village except for the vicar's farm. The planning process started in 1806 and was completed in 1815. In the following years, the plan was implemented on the land. In well-preserved maps and protocols, the process of evaluating the land and redistributing it in single land units can be followed.

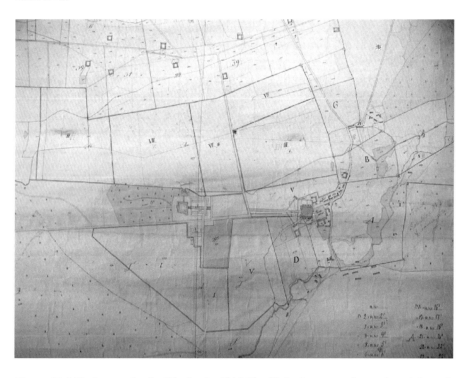

Figure 15-2. Enclosure plan for Vittskövle, 1814 (detail). In the centre, the castle and the park are visible. The plots numbered I-VII are the arable fields run from the estate's home farm. The numbers indicate a seven-year crop rotation introduced at this time. The tenant farms, previously situated in the village core (cf. Figure 1), have been relocated to rectangular plots north of the home farm's fields. The distance between the castle and the church is ca 1.2 km (Lantmäteriarkivet, Kristianstad).

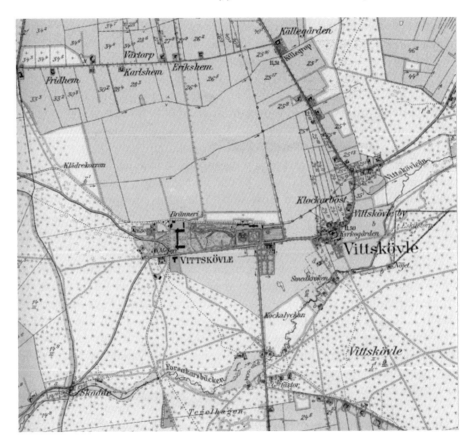

Figure 15-3. Economic map of Vittskövle 1930 (detail). The map shows that the enclosure plan from 1814 was implemented – the settlement pattern was very much the same more than 100 years later (as it is today). New farm buildings were erected at the estate in the late 19th and early 20th centuries (west of the park). The map also represents the modern features of a railroad, a dairy (*mejeri*), and a distillery (*bränneri*), all connected to the estate. The distance between the castle and the church is ca 1.2 km (Ekonomisk karta över Kristianstad län, printed in 1930).

Figure 2 shows a detail of the enclosure map from 1814. In the centre, the castle and the park is visible, and the fields numbered I-VII are the arable lands run from the estate's home farm. The numbers indicates a seven-year crop rotation introduced at this time. Outside the fields of the home farm, the tenant farms are situated. They are allocated along new roads that often bend in sharp angles, and the whole scene displays a rational and efficient impression. The changes meant that the old village of Vittskövle was more or less dissolved: only three farms remained in the village core, while the others were moved out in the landscape. There was no reduction in the number of farms at this time, but rationalisations were carried out in the 1870s and a number of adjoining tenant farms were closed down and demolished. In the centre of their lands, a new large farm with an imposing manor house was erected (Persson 2002). This was a process quite common on the estates in Skåne during this time period.

It should be remembered that the map in Figure 2 is a plan, but that taking a look at a later map shows that the plan was implemented. Figure 3 shows a detail of an economic map from around 1930 covering the whole province of Skåne.

The map shows that obstacles to modern agriculture have been effectively erased from the vast areas making up the estate's arable land. Wetlands have been drained and there are no signs of the old commons. The previous open-field system is totally gone, and the Vittskövle farms are spread out in areas north of the estate according to the enclosure plan. This reflects how the collective work on the village fields during the pre-enclosure situation has been replaced by individual work on each farm's collected grounds. The forest is also represented as a homogenous area in the map, and it is crossed by straight and right-angled roads to make the extraction of its resources efficient. Behind the picture provided by the map lay new ways of handling arable land and forest, meaning, among other things, that the previous customary rights held by the peasants regarding the use of the out-fields and the forest disappeared. The enclosure, the new crop rotations, drainage, and other agriculturally progressive measures can all be seen as aspects of how modernity was mediated in this landscape during the 19th century.

Simultaneously, there are strong traditional traits visible in the map from 1930, as there are in the present-day landscape. No doubt the magnificent estate with its castle and park is this landscape's leading theme. In the map, they lay as an enchanting island within the surrounding open landscape of arable fields. The moat signifies seclusion, even if it has long since lost its practical defending role. The manor house – the castle – has preserved its exterior from the 16th century, and stands out as an authoritative nexus of the surrounding area (Figure 4). The interiors have been renovated through time, and the castle is today the private home of the possessor's family. The roads leading to the manor from all the cardinal points are emphasised by their straight layout and their avenues, and there is no doubt where the centre and the power is situated in this landscape. The park provides variation in the milieu close to the castle, and its paths and watercourses indicate how it could be used for exclusive and recreational purposes. Also, the park has been successively reshaped. It is today quite similar to the picture displayed in the map from 1930, which means it shows some elements of a Renaissance park close to the castle, while the part to the west is designed as an English park.

One very symbolic feature illustrating modernisation is the small railroad leading up to the manor. It was financed by the estate owner and constructed in 1910, and run by locally-produced electricity (Swenson 1993). The railroad reached the storage buildings and barn near the estate by following the straight course of an earlier avenue – a quite expressive illustration of modernisation within a traditional setting.

Figure 15-4. The estate of Vittskövle. The earliest known traces of a noble estate dates back to the 14th century, while the existing building was erected in the 16th century. Photo: T. Germundsson.

The picture sketched so far, using the enclosure plans and the maps from 1815 and 1930, gives an outline of how the estate landscape of Vittskövle was influenced by the processes of modernity. New agricultural technologies and methods were introduced and changed the shape of the landscape. Historical traits and elements of this landscape were, however, still utilised as a means of maintaining a social order of hierarchy and authoritative power. Consequently, the ideological dimensions of this landscape are ambiguous. Prevailing from pre-industrial time is the tendency to "naturalise" the estate's productive landscape and the work put in to it. The park, the well-arranged alleys, the thoroughly-cultivated fields, and the woods and grazing grounds assembled a landscape entity that only vaguely revealed any trace of how it was produced. As in the 17th century England, the combination of elements such as architectonic historicity and pastoral parks was a way of allowing the landscape to mediate a harmony between nature and culture based on ideology (Cosgrove 1998; Olwig 2000). As the enclosure maps and the map from 1930 indicates, the modernisation of the estates during the 19th and early 20th centuries was carried out with deliberate reference to earlier epochs, to a time when the power of an estate owner was of a more concentrated and despotic character than was the case during the period discussed here. As mentioned above, this dimension of estate landscape is often reproduced in preservation documents, popular history, and tourist brochures. There is thus a tendency to forget both the processes and landscape elements that bear witness as to how modernity was mediated on the estates. In the case of enclosure, for instance, it is clear as to how this process totally reshaped the plains of southernmost Sweden during the 19th century – both inside and outside the estates. As A. Pred (1986) has convincingly demonstrated, the changes involved thorough alterations of the peasants' daily lives, and it also strengthened the individual farmer's control over his land. In a broader sense, it was an adaptation to a growing market economy, and the enclosures in Sweden stand out as an important prerequisite of the country's modernisation. It has been seen as a symbol for the disintegration of the traditional peasant society – a turnover from *Gemeinschaft* to *Gesellschaft* (Asplund 1983). On the estates, however, the processes of enclosure were carried out with autocratic power, and – as demonstrated in the case of Vittskövle – they could produce as a result a landscape communicating on behalf of a centralistic power; an alteration of the old order, but still a strongly hierarchical landscape.

With the help of the case of Vittskövle, I will examine in the following paragraphs two more examples that connect economic and social processes during this epoch to place and landscape. Both illustrate how modernity gradually imprinted itself on an estate landscape that bore many manifested traditional traits – and also how "lowly" and obscure landscape elements can reflect significant historical processes. The first example concerns the estate's measures taken to secure its workforce, and the second is a brief note on how the emerging welfare state challenged the landed authority.

3.2   Securing the Estate's Workforce – Renewed Feudalism?

During the 19th and early 20th centuries, the agricultural workforce in the Swedish countryside changed in character. From a situation in which the labour on a farm – beside the work of the farmer's own family – was carried out by farmhands and day-labouring crofters, wage labour became more and more common. On the estate, the system was in principle the same – part of the workforce lived on the estates as farmhands and maids. Depending on the estate's size and organisation, the remaining workforce consisted of people renting farms and crofts on the estate's domains. The rent was paid in kind with labour, and there was a system – often a written contract – that stated how many days the farmer or crofter annually had to work on the estate. A system of semi-wage labour began in Sweden by the end of the 18th century, and flourished during the following century. In Swedish it was called *stat-system*, and it meant that a farmhand's family was hired on a large farm or an estate for a one-year period. They were given a small house or a tiny apartment in a building housing several families. Their wage was paid partly in farm produce, partly in money. The family also had a small plot for their own cultivation. Towards the end of the 19th century, wage labour became more and more common. This was due to changes within agriculture, but also to the industrialisation of the countryside. In villages and growing industrial towns, many people built or rented houses with small gardens, and worked where they could find jobs: on farms and estates as agricultural workers, or as workers in industries and transport (Jonsson 1994).

In brief, one could say that the organisation of labour on farms and estates was moving from a pre-modern to a modern system during this time period. The pre-modern system was characterised by personal bonds and customs. People inhabiting a farm or a croft on the estate's domains "belonged" to the estate, in feudal-like relationships – the estate owner provided housing and farming land in exchange for produce and labour. The growth of a more modern system meant that farmers, crofters, and workers to a higher degree rented their houses and lands from the estates with earned money. Many workers also lived outside the estate domains.

This process of modernisation as sketched above, was, however, not linear or homogenous. Concerning the workforce on the estates in Skåne, there were instances in the late 19th and early 20th centuries of "feudal recoil", to borrow an economic historic expression (Olsson 2001). During this time, many wage labourers worked at the estate; but it was problematic for the estates to match the seasonal demand for manpower through the open labour market. One measure taken then was the estates re-introducing the system of day-work, i.e. the right of occupation was again given in exchange for a specified amount of work days at the estate. This was the case at Vittskövle estate, and we can identify the houses and livelihood positions of occupiers who in the beginning of the 19th century were wage-labourers, while occupiers of the same houses in the early 20th century were day-labourers. In the 1930s, and still today, the landscape comprising the estate and the small crofts is dominated by its hierarchical structure, expressed for instance through the alley leading from the crofts up to the castle (Figure 5).

Figure 15-5. Alley leading to Vittskövle estate from the south. Photo: T. Germundsson.

To proclaim the return of feudalism would, however, represent an explanation focusing more on form than on process. It must be remembered that the main reason for making these changes was to increase the production aimed at the market demands. The ideal role of the estate owner during this period, therefore, turned more and more into being a successful "micro mercantilist", i.e. into selling as much as possible and buying as little as possible. The aim was to make the estate self-sufficient, and at the same time produce as much as possible in order to meet the increasing market demands. To accomplish these ambitions, the process of physically concentrating the estate domains, mentioned above, went hand in hand with connecting the livelihood position on the estate land to work duty; thus expressing a very local and place-bound power over land and life. This was, one might say, a new – maybe even modern – form of feudalism. It is still there to be traced in the landscape, although today concealed by crofts being turned into summerhouses for city people.

3.3   Welfare Policies Changing the Scene

Another expression of how modernity was mediated within the estate milieu can be found using the different ways in which modern society's welfare policies affected social life. One can look at some brief examples of this connected to the village core within the Vittskövle estate domain. This was a place that changed considerably during the time dealt with in this paper. Before enclosure, the village core consisted of more than 20 farms situated close to each other. The church was the natural centre of the village. The social life of the village was to a very large extent enacted within the village core: family life on the farms, village life on the *bygata* (the village green), and of course in and around the church. As the farms in the village were tenant farms under the estate, most of the basic conditions for farming and production were regulated from above. This was also the case for a number of other aspects concerning daily life. Typical for the estate milieus was the patriarchal system, where issues like economy, education, health care, and taking care of the poor were regulated and performed under the conditions of the estate owner, with the aim of shaping a more or less closed social system. The focus of this system was to efficiently run the estate; to possess and control a proper workforce, and to keep up the hierarchical system that made it possible for the owner to maintain influence and power.

During the late 19th and early 20th centuries, this system was challenged in different ways. The authorities noticed the drawbacks of the system in the form of the exclusion of those people not needed within the estate system – they also observed the very poor conditions for many people living on the estates. Especially around and after 1900, the previous more-liberal and *laissez-fair* political majority was overtaken by a socially active and democratic movement, taking partial form as the Social Democratic Party and several other popular political movements (Larsson 1981). Briefly put, one could say that the local, hierarchical, and patriarchal social order on the estates was challenged by a broad, democratic and public social system. Characteristic of Swedish development concerning these processes was that social reforms gradually gained a footing not without harsh political struggle, but seldomly through violent clashes. Peasant uprisings occurred on several occasions during the 19th century, and strikes among agricultural workers were organised at the end of the century. But – especially through well-organised and successful political agitation during the late 19th and early 20th centuries – social reforms were carried out that forced estate owners and other patriarchal employers to modify their businesses towards a new order.

Returning to the village of Vittskövle, traces of these processes can be found. The scene of the village core changed dramatically during the enclosure process. Most of the farms were moved out to their new plots of land, and as has been shown earlier, only a few farms remained in the village. Still, the village core was an important place – the church was situated here, and it was the natural gathering point for the village community and other social events. Following the enclosure in the beginning of the 19th century, the village core was "filled in" with various new houses inhabited by artisans and workers. A later development – reflecting some aspects of the new social order in the Swedish countryside referred to above – can be illustrated with the help of two amateur pictures from the early 20th century.

Figure 15-6. Picture from Vittskövle village in the 1920s. In the background the church is visible, and to the left a grocery shop. The shop was connected to Vittskövle trading cooperative, a local extension of the Swedish Cooperative Wholesale Cooperation. Also, a group of schoolchildren can be seen. The scene shows how the village green – before enclosure surrounded by farmsteads – had become a place visited by members of farming families for certain purposes (shopping, going to school, going to church etc.) instead of being the daily nexus for spontaneous meetings and collective activities concerning agriculture. Photo: Amateur photo from ca 1920, used by kind permission.

Figure 6 is a picture from Vittskövle village in the 1920s. In the background the church is visible, and to the left there is a grocery shop.[6] The shop was connected to Vittskövle trading cooperative, a local extension of the Swedish Cooperative Wholesale Cooperation (*Kooperativa förbundet*) that had safeguarded the consumers' interests since the turn of the last century. The Cooperation was founded in 1899 by G.H. von Koch, one of the leading characters within the liberal political movement working towards

---

[6] The photos and data on Vittskövle village are from *Vittskövle socken genom tiderna, del. 1.* 1993.

social reforms from the late 19th century onward. The Cooperation's aim was to organise the wholesale trade in the interest of the consumers; later on it also started focusing somewhat on industrial activity. In 1900, the Cooperation had 11,000 members, a number that grew to 240,000 in 1920 and 700,000 in 1940 (*Kooperativa förbundet* 1993). Its shops, both in the countryside and in towns, became well-known and often-visited places for many people. In Vittskövle, the cooperation shop stands out as a new and altered way of supplying the village households with their daily needs. Previously this had been done locally, with farms' produce, but also through the estate. The cooperation shop in Vittskövle thus became one way of integrating the village into a broad social movement working on a democratic basis, and thereby challenging the old patriarchal order of the estate.

Figure 15-7. Good templar's lodge in Vittskövle, the 1920s. Photo: Amateur photo from ca 1920; used by kind permission.

Another building in Vittskövle village functioned as a home for elderly people during the first decades of the 20th century. It was founded and run by the local authority – this can also be seen as a representation of how the old system was gradually replaced by a new order, here articulated as a local expression of an emerging welfare state. Yet another building in the village was a small Good Templars' lodge erected around 1925 (Figure 7). The plot where the lodge is situated was donated to the organisation by the estate owner, while volunteers did the building work. It reflects mutual interests based on reducing alcoholism, and can be used to symbolise how the Swedish temperance movement was gaining ground. As well, this is a sign of how modern society mediated the estate milieu, and how Vittskövle can be understood not only as a representation of the landed gentry's great influence, but also as a place of modernity.

The examples mentioned above can be extended and deepened, but my main point here has been to give some concrete pictures of how the village of Vittskövle changed due to the development of Swedish modernity in the 19th and early 20th centuries.

I will now pull these examples together along with the cases of the enclosure and the "renewed feudalism", in order to discuss the historical dimensions of the present estate landscape and how it is understood today.

4.   DISCUSSION

As stated in the beginning of this article, the main theme of the estate landscapes in southernmost Sweden is the grand history of their imposing buildings and magnificent establishments. There is a definite focus on the centuries before the 19th century, and there is seldom any developed discussion concerning the modern history of these units. Such a lacuna in time perspective means that important aspects concerning the interpretation of today's estate landscapes are hidden or forgotten. There is thereby a risk of alienating the estate landscapes, and giving them an aura of past epochs that conceals the formative processes that shaped them during the breakthrough of modernity. By focusing on some aspects of the modern history of Vittskövle estate in Skåne using a landscape perspective, I have aimed to show how the interpretation of the present estate landscape could be widened. This was done by first focussing on the understanding of landscape as a many-facetted concept, founded upon the idea that a landscape is both a concrete, physical phenomenon and a social construction. From this point of view, the landscape of the Vittskövle is ambiguous and open to different interpretations. It is a contested landscape, not the least when it comes to its ideological dimensions. On the one hand, these dimensions can be traced to the conscious efforts of the ruling nobility to express its influence and power over time. In this respect, we can see how buildings, the park, and the landscape layout with its alleys and its designed centrality around the castle speak of a harmony implemented from above, giving room for no other social order than the traditional hierarchical one: "a way of carefully selecting and representing the world so as to give it a particular meaning" (Mitchell 2000: 99-100). On the other hand, we can see how modernity in the form of enclosure, the cooperation, and the welfare ideology challenged this ideological dimension. But just as during earlier periods, the modern landscape as such does not speak clearly about the processes that changed it. There are traces of an increased market economy, of a political democracy, and of a welfare state in the modern historical landscape of Vittskövle. However, careful investigation needs to be undertaken in order to understand it more fully, and to be able to grasp how the modern landscape arrived at its form. I believe that such research should emphasise the work of the people who shaped the landscape, and should consider under what social conditions their work was performed. I furthermore suggest that a conceptual tool to use when analysing the landscape of Vittskövle – and other estate landscapes – can be found in Hetherington's notion of a "heterogeneous landscape of modernity". I will thus end my discussion by reflecting on these ideas.

According to Hetherington (1997), the concept of utopia is central to the understanding of modernity. The vision that utopia actually could be realised lays as a background to modern projects like democracy and the welfare state. The perspective given by Hetherington, however, is that such utopic expressions are never total or without limits. Modern social ordering does not develop out of the eradication of ambivalence. Rather, it implements new spatial relations in which modernity manifests itself as an alternative to what surrounds them. "That process can be said to take place in sites that can be described as heterotopia – places of an altering order that had a significant role within the spatialization processes of modernity" (Hetherington 1997: 70). I believe that the processes and social movements related in this paper via the example of Vittskövle – the enclosure, the co-operation, the Good Templars movement and the welfare state – were practices grounded in more or less outspoken utopian ideals; and as such, characteristic of modernity. We can for instance see relations and functions situated on farms in earlier times – and often within the framework of a patriarchal local society – now enacted in alternative places and in alternative ways. International and national movements interacted with the local life as the landscape of Vittskövle found its form at the crossroads of a modernising estate's tradition-bound impact, and an emerging welfare state's politics. Furthermore, we find members of a traditional and partly-feudal social class implementing modernity at the same time as they were turning to the past in order to justify their goals. Simultaneously, the peasants and other agrarian classes were fighting for their independence, often basing their visions on both modern utopian ideas and customary rights. When these processes unfolded in the estate milieu of Vittskövle, they produced a landscape that is ambiguous and multi-layered – in *heterotopia*, as Hetherington puts it. In this instance, we can see no simple division between old and new, past and present, and thus no culturally monolithic landscape – rather, what's evident is a heterogeneous landscape of modernity.

Consequently, I think it is time for an alternative view regarding estate landscapes. Today, the landscape of Vittskövle estate is the scene of both a documented cultural heritage and a modern agricultural enterprise. As the dominant aspects of the cultural heritage are focused on epochs earlier than the 19th century, there is often a weak connection between the historical dimensions and today's landscape. By paying closer attention to the era of modernity, I think some of the missing links can be identified, and a fuller understanding of the estate landscape can be achieved.

REFERENCES

Ambrius, J. (2001). *Stora boken om skånska slott och herrgårdar*. Västerås: Sportförlaget.

Asplund, J. (1983). *Tid, rum, individ och kollektiv*. Stockholm: Liber förlag.

Bengtsson, S. (1994). *Med K-märkt genom Sverige*. Stockholm: Byggförlaget.

Bjurklint, K. (2001). En fransk riddarborg i Skåne. *Ale 2001, 4*, 18-31.

Cosgrove, D. (1998). *Social Formation and Symbolic Landscape. With a New Introduction.* Madison: University of Wisconsin Press.

Dunér, M. (2002). *När man bevarade Lund: bevarande och modernitet i 1970-talets stadsbyggnadsdebatt.* Rapport R 3: 200. Lund.

Eriksson, E. (1990). *Den moderna stadens födelse: svensk arkitektur 1890-1920.* Stockholm: Ordfront.

Eriksson, E. (2001). *Den moderna staden tar form: arkitektur och debatt 1910-1935.* Stockholm: Ordfront.

Germundsson, T. (1999). Godsen under 1800-talet. In T. Germundsson & P. Schlyter (Eds.), *Atlas över Skåne* (pp. 78-79). Vällingby: SNA.

Germundsson, T. (2001). Adelns geografi: gods och landskap genom tid och rum. In K. Arcadius & K. Sundberg (Eds.), *Skånska godsmiljöer* (pp. 33-52). Lund: Skånes hembygdsförbund.

Germundsson, T. & Lewan, N. (2003). Enskiftet i Skåne 200 år. Reformen 1803 i geografiskt perspektiv. *Ale 2003, 1*, 1-32.

Hagman, O. (2002). *Bilen, naturen och det moderna: om natursynens omvandlingar i det svenska bilsamhället.* Stockholm: KFB.

Hetherington, K. (1997). *The Badlands of Modernity.* London: Routledge.

Johannesson, G. (1977). *Skånes historia.* Lund: Signum.

Johansson, E. (1994). *Skogarnas fria söner: maskulinitet och modernitet i norrländskt skogsarbete.* Stockholm: Nordiska museet.

Jonsson, U. (1994). Från hästkärra till skördetröska: jordbruk och jordbruksarbete under 1900-talet. In G. Broberg, U. Wikander, & K. Åmark (Eds.), *Svensk historia underifrån* (pp. 199-220). Stockholm: Ordfront.

Kooperativa förbundet (1993). Entry in *Nationalencyklopedin.* Stockholm: Bra Böcker.

Kulturminnesvårdsprogram för Skåne (1984). Malmö: Länsstyrelsen i Malmöhus län.

Larsson, J. (1981). En ny nationell effektivitet. In K. Åmark (Ed.), *Teori- och metodproblem i modern svensk historieforskning* (pp. 202-226). Stockholm: LiberFörlag.

Larsson, J. (1994). *Hemmet vi ärvde om folkhemmet, identiteten och den gemensamma framtiden.* Stockholm: Arena.

Lilja, E. (1994). *Modernitet, urbanitet och vardagsliv: om människans förhållande till den byggda miljön i staden, förstaden, grannskapet.* Stockholm: Nordplan.

Lisberg Jensen, E. (2002). *Som man ropar i skogen: modernitet, makt och mångfald i kampen om Njakafjäll och i den svenska skogsbruksdebatten 1970-2000*. Lund Studies in Human Ecology 3. Lund.

Lowenthal, D. (1985). *The Past is a Foreign Country*. Cambridge: Cambridge University Press.

Mitchell, D. (2000). *Cultural Geography. A Critical Introduction*. Oxford: Blackwell.

Morell, M. (2001). *Det svenska jordbrukets historia. Bd 4, Jordbruket i industrisamhället: 1870-1945*. Stockholm: Natur och kultur/LT.

Möller, J. (1985). The Landed Estate and the Landscape. Landownership and the Changing Landscape of Southern Sweden during the 19th and 20th Centuries. *Geografiska Annaler*, 67 B, 45-52.

Möller, J. (1989). *Godsen och den agrara revolutionen – arbetsorganisation, domänstruktur och kulturlandskap på skånska gods under 1800-talet*. Lund: Lund University Press.

Nilsson, S.-Å. (2002). *Slotten och landskapet – skånska kulturmiljöer*. Stockholm: Prisma.

Olsson, M. (2001). *Godset, makten, räntan och byn – fyra agrarhistoriska studier*. Ekonomisk-historiska institutionen, Lunds universitet.

Olsson, M. (2002). *Storgodsdrift. Godsekonomi och arbetsorganisation i Skåne från dansk tid till mitten av 1800-talet*. Stockholm: Almqvist & Wiksell International.

Olwig, K.R. (1996). Recovering the Substantive Nature of Landscape. *Annals of the Association of American Geographers*, 86, 4, 630-53.

Olwig, K.R. (2002). *Landscape, Nature and the Body Politic. From Britain's Renaissance to America's New World*. Madison: The University of Wisconsin Press.

Persson, L.-L. (2002). *Gods och landskap i förändring. Landskapsomvandling och plattgårdsuppkomst på Vittskövle gods*. Manuscript. Department of Economic and Social Geography. Lund University.

Pred, A. (1986). *Place, Practice and Structure. Social and Spatial Transformation in Southern Sweden 1750-1850*. New Jersey: Barnes & Noble Books.

Pred, A. (1995). *Recognizing European Modernities: A Montage of the Present*. London: Routledge.

Rune, J. (2003). *Trafiknostalgi: vykort berättar: en bildkavalkad från 1900-talet*. Saltsjöbaden: Trafik-Nostalgiska Förlaget.

Skansjö, S. (1997). *Skånes historia*. Lund: Historiska media.

Swenson, B. (1993). *Mårtengås och annat. Folklivsskildringar från Skåne*. Degerberga: Ultima Thule bokförlag.

Tykesson, T. (2001). Modern bebyggelse – ett kulturarv. In *Miljötillståndet i Skåne; Skåne i utveckling 2001:48*, 16-17.

Vittskövle socken genom tiderna, del 1. 1993. Studiecirkel, Widtsköfle byalag. Kristianstad: Monitor förlag.

Wahlberg, M. (1996). *Jordens förbannelse: en kulturell studie av moderniseringsprocesser i Norrbotten*. Stockholm: Almqvist & Wiksell International.

Chapter 16

GREENS, COMMONS AND SHIFTING POWER
RELATIONS IN FLANDERS

*The Common Meadows in Semmerzake from the 13th Century to
the 19th Century*

Pieter-Jan Lachaert
Department of Medieval History, Ghent University, Belgium

1.   INTRODUCTION

More than 25 years ago, at the 1975 meeting of PECSRL in Warsaw, the
Belgian scholars Claude and Dussart presented the results of their research
on the *dries*, villages in Inner Flanders (Dussart et al. 1975). At the PECSRL
session in 2000, the late R. Knaepen presented the results of his studies done
concerning the commons of the Campine area, in the eastern part of Belgium
(Knaepen 1994; Knaepen et al. 2002). This topic has remained in focus at
several departments of Ghent University (Belgium), and various inventories
have been drawn up over time (e.g., Thoen 1987; Vinck 1997; Van Der
Haegen 1999). In 2002, an article published by Martina De Moor
concentrated attention on the commons in the Campine area and in West
Flanders (De Moor 2002).

In this paper, I will reconsider the *driesen* and other commons in Inner
Flanders (the portion of the northern part of Belgium called Flanders that is
not subject to marine influences). This theme is dealt with in a dissertation I
am preparing in regards to the landscape history of a specific region located
south-east of Ghent. In this article the case study of the village of
Semmerzake will be presented (Figure 1). This village is situated on the right
bank of the River Scheldt, some 15 km southeast of Ghent. In the near
future, I will present more details and a more quantitative study of the
evolution of the *driesen* and other commons in the entire Land van Aalst

H. Palang et al. (eds.), European Rural Landscapes:
Persistence and Change in a Globalising Environment, 269-287.

region of Belgium, using the fascinating (but as yet largely unpublished) material that my search in the archives has produced (especially SAA, Land van Aalst, 1077 and 4298-4335, and the provincial archives – RAG, Frans provinciaal fonds, Hollands provinciaal fonds en Provinciaal archief 1830-1850). The Land van Aalst is a region between the Scheldt and the River Dender that was incorporated into the County of Flanders in the 11th century.

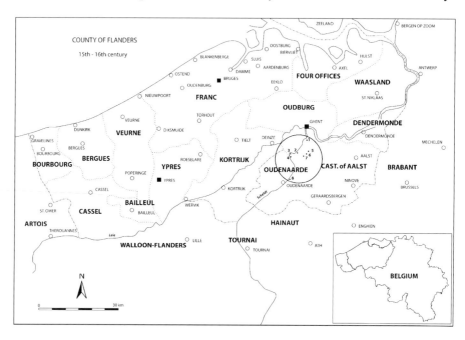

Figure 16-1. Location of Semmerzake and the other villages mentioned in this article, part of the County of Flanders in medieval times and now present-day Belgium, on the east bank of the River Scheldt, south of Ghent: 1. Merelbeke, 2. Melsen, 3. Semmerzake, 4. Gavere, 5. Bottelare, 6. Munte, 7. Baaigem, 8. Ename (figure from Stabel 1997: 5; re-mastered by Wim Van Roy and Marjanne Sevenant).

Although greens – the Dutch word is *driesen* – have already been extensively studied, there remains a great deal of confusion among historians and philologists as to the original meaning of *dries* (see e.g., Van Durme 1998, who gives a good overview of the topic). In my opinion, *dries* had the original meaning of being a common waste or pasture (in this respect I follow the recent toponymical analysis of Van Osta 1990: 437-442; see also Van Durme 1998: 121-123, 149).

The green was used initially for the grazing of cattle and for the gathering of manure and fuel – dung, wood, shrubs etc. (a good overview of the uses of greens that developed later on in Flanders is found in an article by Dussart & Claude 1975).

Later on, the word *dries* developed in two different directions. On the one hand, it signified the parts of a privately-owned holding that were temporarily exposed to a *dries* regime. This meant that these lands remained unplowed and were used temporarily for cattle breeding (Thoen 1986, 1988, 1990, 1993). It is possible that they were subjected to an open access (and a more or less collective) pasturing system. *Dries* was used in this sense from the 9th century until the end of the *Ancien Régime*. On the other hand, from the 12th century onwards the word began to be used to denote a certain form of settlement. This is logical, since the common use of a waste as pasture (which was never cultivated) would attract various kinds of people. Some of them were happy to use this common waste, as it was their only means of subsistence. Some even managed to build their houses on these wastes, and incorporated part of the land in order to cultivate it for their own private use (Thoen 1986).

This is probably how a lot of the so-called *dries* villages and hamlets came into being, and also explains why a lot of them demonstrated a very irregular form – they grew more or less organically, as newcomers built their houses in the neighbourhood of already existing homesteads. Examples such as Ename show us that present-day *dries* hamlets (Natendries, Zwijndries and Ganzendries) are merely small relics of areas that were once much bigger. In Ename, old charters indicate that this area was used for common grazing and gathering purposes as early as the 8th century AD (Tack et al. 1993; Van Durme 1998).

In the discussion above regarding the evolution of dries, one can see that the word evolved from being used for the designation of the natural condition of an area, to being a word used for a specific management-regime and settlement structure.

In this chapter, I also frequently refer to the words *commons* and *Broeken*. The word *commons* (in Dutch *gemene gronden*) in regards to Flanders seems to be used to indicate larger *dries* areas, whereas the word dries itself is limited to more-clearly defined and smaller pieces of land. Still, *commons* is also used to indicate any kind of common land (for an overview of the different kinds of commons in Flanders, see Lindemans 1952: 307-387; De Moor 2002: 117-121).

The Dutch word *broeken* indicates marshy areas with a deficient water management system. These areas were often originally used as a common meadow, but could also have been privately owned (Lindemans 1952; Van Durme 1988).

The juridical status of the commons is often unclear, which has led to numerous disputes. This characteristic has favoured the occurrence and preservation of many archive sources concerning these kinds of lands. Still, in general during the *Ancien Régime*, one can discern two types of rights that were exercised upon the common lands – on one side, there was the *dominium directum* (theoretical land ownership held by the landlord), and on the other side there was the *dominium utile* (rights of use enjoyed by the commoners) (De Moor 2002).

2.   GOALS AND LIMITS OF THE INQUIRY

I have limited my inquiry to one conglomerate of *driesen*, located in Semmerzake. During the *Ancien Régime*, Semmerzake was part of the Land van Aalst region.

The *driesen* in Semmerzake were called the *Broeken*, and were composed of the *Aalbroek*, the *Grooten Broek* and the *Kleinen Broek*. They were located near the River Scheldt, in the alluvial valley at the foot of the village centre. The *Broeken* are remarkably interesting, since there are written sources about them dating from the 13th century onwards. Another interesting feature is that the *Broeken* common pasture regime was maintained over a vast area of about 30 ha up until the 1840s. Even nowadays, the physical environment of the neighbourhood is shaped to a very large extent by these former common meadows.

The location of these meadows and their amplitude allow us to presume that a majority of the people living in Semmerzake depended to a greater or lesser extent on this area for their subsistence. For some reason, these meadows have never been the subject of a serious inquiry before. I think the evolution of this common is fairly typical for this area of Flanders.

Traditional historiography has often considered demography and agricultural economic structures as the prime influences concerning changes in the use and extent of commons, such as the Flemish *dries* villages. In medieval times, a growing population led to reclamations that reduced the extent of the commons and formalised their rights of use. In the 18th and 19th centuries, changing socio-economic circumstances were responsible for attempts to reclaim the remaining uncultivated areas – *driesen* and other commons – with the cultivation of these areas being seen as a means of increasing agricultural output. In this perspective, the unsuccessful attempts to do so in the 18th century and the early 19th century are explained by purely economic and agronomic circumstances, such as the lack of an adequate infrastructure and the lack of manure (e.g., chemical fertilisers); problems that would be overcome during the course of the 19th century, resulting in the final disappearance of these commons by its end (Clicheroux 1957; Verhulst 1990; Vandenbroeke 1975; Tack et al. 1993; Knaepen et al. 2002; De Moor 2002).

Figure 16-2. This is an aerial view of Semmerzake, with the church and village centre in the foreground, and part of the Grooten Broek highlighted. Note also the appearance of the River Scheldt (and some recent river soil deposit) in the background. The physical appearance of the former common, as it appears on the map of 1828, still largely defines the present day position of the surrounding houses. Photo: J. Semey, Department of Archaeology, Ghent University, 45123).

Some critical remarks can be made regarding this traditional schema, as it has been overly influenced by the official discourse of contemporary literature and archive sources. It can be argued that other reasons can also be found in order to explain the change and eventual disappearance of the commons in Flanders.

It is clear that even in medieval times there was a struggle going on between landlords and local communities over the use of the commons: landlords wanted to reclaim these lands, whereas the local community wanted to preserve the traditional uses of them (Billen 1988). In the 18th and 19th centuries, this battle for control continued, but new players entered the field. Regional institutions and the central government discovered the commons.[1] Measures to partition and lease the commons became a key theme in the politics of centralisation advocated by those who held the power. These measures – taken under Austrian, French, Dutch and early Belgian rule – were meant to strengthen the idea of the central state. By taking these measures, the central authorities could make themselves visible

---

[1] In other European states as well, regional and central powers did not take serious interest in the commons before the 18th and 19th centuries. Only in England had the interest in the enclosure of the commons already started a century earlier (see e.g., Hoskins 1955).

to even the lowest strata of society. This is a strategy that has been followed throughout Europe at different times (cf. Brouwer 1995). The fact that the interest of the central authorities in the commons grew only in the 18th and 19th centuries can be explained by the gradual collapse of the traditional structures of power (local landlords and local user-communities) in Flanders at that time. The traditional clerical and seigniorial structures had faded away (a process that had already started at the end of the Middle Ages).

3.   SHIFTING RELATIONS OF POWER AROUND THE COMMON MEADOWS IN SEMMERZAKE

3.1   Lords, Reclamations and Developing Local Communities (11th-18th Centuries)

Little is usually known about the *driesen* and other commons in Belgium before the 13th century. The first charters date from the 11th and 12th centuries, but most of those concerning the commons date from the 13th century (Wauters 1881; Errera 1891; *Table chronologique des preuves*; Billen 1988). According to Billen (1988), the appearance of charters during this period signifies three things:

1. The written rights of use for the commons incorporate the feeling that the commons were limited to a more-restricted area during this period. The lords who granted a community these rights tried to minimise the area of the driesen as much as possible, in order to be able to reclaim as much land as possible. It is no coincidence that this happened in the Great Reclamation Period (11th-13th centuries), during which a huge agricultural expansion took place.

2. At the same time, it must be noted that the granting of rights of use to a community was also a means of recognising this community as a formal collective group. Up until then, lords had only granted individuals rental lands. This recognition gave the community an identity apart from other groups that were not allowed to use the commons. In certain periods, this distinctive identity would create social tensions, not only between the community and outsiders, but also within the authorised community (e.g., did all members of the community have the same rights of use on the commons?).

3. In granting communities rights of use, the lords were taking a risk. Formally speaking, they remained the owners of the common. They kept the dominium directum. In practice, their ownership was very limited, since the users of these lands could take possession of the resources that these lands generated (*dominium utile*). In order to retain some

psychological reminder of his ownership, the lord often asked for some kind of symbolic annual compensation from all the users of the common.

Concerning the *Broeken* in Semmerzake, we possess a 17th century copy of a charter dating from 1271. In this charter, the local lord (the lord of Gavere, who had by then incorporated Semmerzake) admits that a certain predecessor of his had sold the commons to the villagers of Semmerzake. He affirms that these lands must stay uncultivated and in common use by the villagers in perpetuity, and excludes other people from using these lands for grazing cattle.

His reason for granting a new charter is that he wants to create two new lakes on the commons of Semmerzake, and wants to give some compensation to the community for doing this in order to be able to live in peace with them (SAG; Register TT, F 220).

In this charter, we are able to trace the three elements that Billen puts forward. The granting of rights of use for this area is probably a consequence of previous reclamations that had reduced the area of the common. By creating new lakes, this area is further reduced. By selling it to the villagers, the lord excludes outsiders from the use of the common, and recognises the villagers as a formal group with a collective identity. The only element missing is any kind of psychological reminder – no yearly contribution or other symbolic homage to the lord is asked for from the villagers. Nevertheless, we can be sure that even though the charter states that the common has been sold to the villagers, it would still be possible for the lord to take possession of a part of the common, when necessary.

The basic situation can be seen in a court case dating from the 17th century, where it emerges that some parts of the common had been sold to a private person named Louijs Spanoghe. On September 14, 1669, this was the signal for a huge uprising in the local user-community. When the aforementioned tried to enclose the parcels of land he had bought, he was attacked by the women of the village, who hurled abuse at him and even ripped off some of his clothes (RAG, Raad van Vlaanderen, 8569, 8570 and 8595, f2v-3v; De Brouwer 1968; Braekman 1998).

This example helps to show that the battle for control over the commons between the traditional user-community and the lord continued throughout the centuries. The weakness or strength of one of the two parties during a certain period could lead to alterations of the use and extent of the common, especially when the boundaries of the area had never been accurately specified.

It is most likely that, in other periods of time, similar sales of a part of the commons occurred. The hunger for land and housing would have been especially great during periods of a general population increase. It is also certain that some people (especially poor people and people from outside the village) were able to usurp parts of the commons and build houses on them. After a while, the new situation was sometimes accepted by the traditional user-community and the lord. They formalised these kinds of usurpations by letting the "new" residents contribute to the taxes imposed on the village, and by imposing some kind of seigniorial rent on them.

In Semmerzake, from a court case dating from the end of the 18th century, we can see that not all usurpations were welcomed so easily by the traditional local user-community. Sometimes the villagers used extreme violence, and destroyed houses in order to defend their free communal pasturing against newcomers. On April 27, 1776, seven inhabitants of Semmerzake were brought to justice for destroying the newly-built house of one Adriaen Verleijsen, and for using excessive violence against Adriaen and his wife. A person named Pieter Vroije had reacted particularly aggressively, having reportedly declaimed: "I'm going to Adriaen Verleijsen's house – everything living there must die" (SAA; Land van Aalst 13931). Some of Adriaen's neighbours were also attacked. Before the court, one of the arguments used by the aggressors was that they were authorised to act in this manner, according to the stipulations of the charter given to "them" in 1271 (SAA; Land van Aalst 13931).

The above case is a good example of the social tensions that could occur during the existence of a common. Incidents of this kind were commonplace (other examples can be given involving the active participation of the local lords, such as that which occurred in Bottelare (found in RAG, Raad van Vlaanderen, 23593)). By granting the community a charter, the lord had given the exclusive rights of use to a more or less limited number of persons, and excluded others from the use of the commons (cf. Billen 1988).

3.2   Local Customs and Lords Versus a Centralised State (SAA, Land van Aalst, 13931). The Premature Start of a Shifting Power Balance (1769-1778)

In the second half of the 18th century, the attention of the central government in Flanders was increasingly drawn to the issue of the commons. This was a widespread European phenomenon (for studies concerning other parts of Belgium, see Pirenne 1920; Vandenbroeke 1975; for Tuscany, see Rombai 1998; Guarducci et al. 1998; for France, see Vivier 1998, 2002; for England, see Neeson 1993).

This attention resulted from the influence of 18th century physiocrats, and their theories concerning the rationalisation of agriculture and higher agricultural output of the communally-used lands. These thinkers took no account as to the social advantages linked to the traditional use of these lands, especially for the poorest strata in peasant society (Pirenne 1920;

Vandenbroeke 1975). Another reason for the central government's attention was the fact that the commons had until then largely escaped the general taxes in the County of Flanders.

This trend may also be seen as a premature attempt by the government to centralise institutions. In Flanders, however, unlike other parts of present-day Belgium, no resolutions regarding the enclosure of commons were promulgated; merely, a number of preliminary investigations were conducted in the period of 1769-1778 (e.g., an inventory of all *driesen* and commons in the Land van Aalst region). Resolutions were never promulgated because the regional authorities, whose role was to advise on this kind of resolution, were never very enthusiastic to do so. Partly for these reasons, and also because of the resistance of the local lords and local communities, the attempts to enclose and reclaim commons had little effect in Flanders (and in a lot of other Belgian districts as well, even though in some of them resolutions to this effect were published). Local particularism was still strong enough to resist the centralising pull of the central authorities (Vandenbroeke 1975; SAA; Land van Aalst 1077 & 4298-4335).

3.3   *"La propriété c'est la loyauté"*[2]

3.3.1   The French Period (1792/94-1815)

The French conquest of the territory of present-day Belgium signalled the start of a new era in the struggle for control of the *driesen* and commons between the central government, the regional government, the local lords and the local communities. This struggle was to end with the victory of the central government, and the disappearance of most commons. Shortly after the French Revolution, the central government approved several reforms. Seigniorial power over the commons (e.g., the theoretical ownership) was erased by the abolition of all seigniorial dues. Furthermore, a socially motivated partition of the commons among all villagers was proposed (the decrees of August 28, 1792, and June 10, 1793). These actions seemed to strengthen the position of the original user-communities regarding the commons. Notwithstanding the fact that these decrees were never implemented in the Netherlands, they still influenced the situation in Flanders.[3]

---

[2] Paraphrase of the famous quote from Pierre-Joseph Proudhon (1809-1865): "La propriété c'est le vol". I argue that giving the local authorities the property of the commons made them more loyal to the idea of the central state.

[3] See for example the references that are made to these decrees in later periods, for instance in the case of Evergem, *Doorenzeledries* (RAG; Hollands fonds 870/16).

Another French innovation was the introduction of several new institutions, such as *préfets* and village councils appointed by the central government. The village council was given possession of the commons, which resulted in a decline in the authority of the original user-communities. This change of ownership was probably decreed by the central government, in order to "buy" the loyalty of the lowest authorities and gain their support for the dominant ideology of the central state. The profits that derived from this ownership were also later used by the state for propaganda purposes (a better road infrastructure, a better educational system etc.). This process is comparable to similar processes that have taken place in other parts of Europe; for example, the 1976 restitution of the *baldios* to the local authorities in Portugal (see Brouwer 1995), or the situation in Tuscany, where the profits from the partition of the commons in the 18th century were used to create a better infrastructure (Rombai 1998). We can see this latter process taking place in the Land van Aalst region, where the profits of the partition and leasing of the commons were used to build schools (e.g., in Munte (Hollands fonds 870/16)) and to reorganise the road system (e.g., in Semmerzake, with the construction of the Aalbroekstraat towards the neighbouring village of Gavere).

In the beginning of the French period, no attempts were made to change the rights of use on the *driesen* and commons (art. 5, 4th division of the decree of June 10, 1793). At this time, the planned partition of the commons was not implemented in the regions being focused upon in this paper, because a great deal of legal discussions had arisen from attempts in other regions (see e.g., Jones 1991).

Around the revolutionary year XII (ca. 1806), new interest was shown in the partition of the commons, because of the problems foreseen with the coming into being of a new land tax system (*cadastre*). Under this new system, every piece of land would be valued, and on this basis an annual tax would be demanded from its owner. And indeed the commons did cause problems under this new system, since the user-community was not accustomed to paying taxes for them, while the local authorities (who were the formal owners of the commons) were not happy to pay taxes for lands they could not freely "possess".

In some places where the local council allied itself with the policies of the central government, new systems were implemented that reduced the free use of the commons. For example, in another village in the Land van Aalst region called Merelbeke, the *Groote Gemeente* and the *Hoorendriesch* were leased for the first time in 1806 (RAG; Frans fonds 4525/2). In most other places, the resistance of the user-community to leasing operations was still strong enough to prevent them. And even where they were approved, the objections raised by the defenders of the free use of the commons could be strong and intimidating. In Merelbeke, a few days before the date set for public leasing, some materials to make fire were found behind a farm located near the common, with a written note threatening anyone who dared to lease parts of it (RAG; Frans fonds 2536 en 3581).

In Semmerzake, the future prospect of the *cadastre* also produced tensions among the users of the common. In 1807, there was an incident concerning the village council, when one of its members, Jean Franchois Vander Guchte, refused to sign any decisions of the council until the ownership of the commons was given to him or one of his fellow members on the council. This demand could not be approved, of course, and the provincial authorities eventually dismissed the disobedient councillor. Still, friction within the village council persisted. When Mayor Gijselinck undertook a number of initiatives during the period of 1809-1811 to prepare the leasing of the *Broeken* in order to be able to pay the land taxes, he entered into a bitter conflict with the members of his council. He finally had to resign when all of the councillors accused him of corruption and dictatorial behaviour. In 1812, a new mayor Eugène De Smet was appointed (De Potter et al. 1868). The problem of the land tax imposed on the common in Semmerzake remained until after 1815 (RAG; Frans fonds 4207/3).

We may conclude that the rights of the commoners were mostly preserved during this period. This was probably due not only to the strength of the objections made by the representatives of the commoners on the village council, but also to the weakness of the central and provincial authorities. They were not even able to support local representatives who proposed the division and leasing of the commons. In later periods, alliances between some of the local authorities and the provincial and central authorities would prove to be more effective in overcoming the local particularistic objections of the commoners – but this would be a slow process.

3.3.2 The Dutch Period (1815-1830)

After 1815, the territory of present-day Belgium became part of the Dutch Kingdom, which took over kkkmost of the French provincial and local institutions but renamed them. The French *cadastre* (system of land taxation) was also inherited.

In 1824, a new inquiry was set up in order to proceed with the division of the *marken* (a typical Dutch word to indicate commons with a specific management regime, see e.g., Hoppenbrouwers 2002) located in the different provinces of the Netherlands, as proposed by the Minister of the Interior, Education and Public Works (RAG; Hollands fonds 824/1).

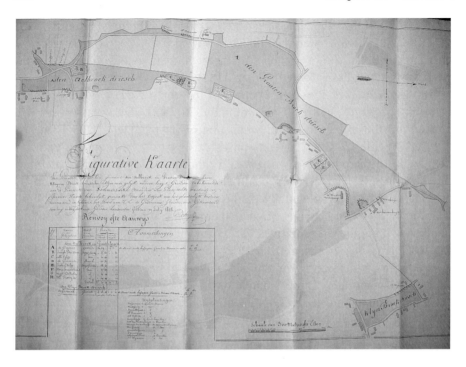

Figure 16-3. A map of the common meadows (left Den Aelbroek Driesch, Den Grooten Broek Driesch, and right Kleijn Broek Driesch) in Semmerzake dating from 1828. The two central white spaces between the Aalbroek and the Grooten Broek are the former lakes that were created by the lord of Gavere around 1271 AD. Parcels of land with a letter in them indicate parts of the common that had been usurped (*uitnemingen*). They are explained in detail in the table in the bottom left corner. It also appears from the map that the surrounding houses were erected next to, or on the common. A lot of ditches and pools could be found on the common, where the pasturing cattle could drink (RAG; Hollands fonds 870/16).

As a result of this inquiry, some commons were divided and leased in the province of East Flanders, mostly around 1826 and 1827. Some examples of this in the Ghent district are the commons in Melsen and Baaigem (*Den Eedt*). In these villages, the village council supported the idea of a division and leasing operation, and the government provided the support and means of establishing it (RAG; Hollands fonds 824/1 and 870/16). In other villages, the local authorities were less enthusiastic. In these cases, the higher authorities exerted some pressure on reluctant local authorities to proceed to a division and leasing operation anyway, but in the end they respected the choice of the village. Local authorities were also supported more than in the previous period in regards to re-establishing their possession on the commons where it had been illegally taken from them by the former village lords, or by people living in the neighbourhood of the commons (RAG; Hollands fonds 870/16).

In Semmerzake, a lot of pressure was exerted to proceed to the division and leasing of the *Broeken*, but the commoners were still able to neutralise this attempt (RAG; Hollands fonds 870/16). However, starting in 1818, taxes were levied on every head of cattle and every horse that pastured on the common, to pay the land taxes imposed by the *cadastre*. This use was further regulated by a police order issued on April 16, 1828, that limited the period of pasturing in ways determined by the local council. The open-access regime that had existed for several centuries came to an end in this way (and the poorest inhabitants could no longer profit from free pasturing in their village). In later periods, this measure would be used against the remaining commoners, through the argument that the use of the commons was being monopolised by a small number of inhabitants, and that not every inhabitant could make use of the possessions of the village (RAG; Modern gemeentearchief Semmerzake 2, f 4v).

Another action that the local council of Semmerzake took, in 1829, was the leasing of parts of the common that had been illegally usurped by commoners to use as their own private houses and lands (*uitnemingen*). The leasing was not at first successful, since no one showed up at the public leasing (RAG; Modern gemeentearchief Semmerzake 2, f 8v). Later on, the usurpations would be legalised by granting long leases to the usurpers – the land was owned by the village, but the house was owned by the occupant (Popp, around 1860, so-called "*bails emphiotiques*").

3.3.3   The Early Belgian Period (1830-1845)

Right from the inception of the autonomous state of Belgium, the partition of the *driesen* and commons was one of the main items on the agenda of the central government. Once again, the intention was to impose central government authority on the lowest levels of society, and to generate more money for the local village council. This money could be used for various purposes that would make the central government more popular and local authorities more loyal to the idea of a central state.

In 1831 and 1835, new inquiries concerning the commons and earlier partition initiatives were held by the Ministry of the Interior (RAG; Provinciaal archief 1830-1850, 824/1 & 798/1). On March 3, 1836, this resulted in the first Belgian law concerning a change in the use of properties held by the villages. New inquiries and a new legal framework exerted pressure on the local authorities to accept the partition and leasing or selling of the commons.

And yet in Semmerzake, the common owned by the village remained a real common, in spite of the fact that the village council – anxious to reject the traditional rights of use for plantation that were exercised by some of the commoners – decided to plant some trees on the common (RAG; Provinciaal archief 1830-1850, 818/5 & RAG; Modern Gemeentearchief Semmerzake 2, f 22r).

In 1836, the traditional common pasturing was still defended by the local authorities against the schemes of the higher authorities to lease the *Broeken* in order to make some more money from lands owned by the village. The local authorities also claimed that the title of ownership that the local landlord granted to them in 1271 did not allow them to countenance these projects.

On March 30, an attack against the common was initiated from a different direction. Some 13 inhabitants wrote a letter to the governor pleading in favour of the leasing of the commons. My further investigations in the cadastral archives have shown that all 13 were living at the extreme east border of Semmerzake (far away from the common), in a separate hamlet called De Steirt (literally meaning "the Tail"), and that they represented the entire population of this hamlet. They said the village council, by refusing the lease of the commons, was acting on behalf of the particular interests of their members and not of the entire community of Semmerzake. Further investigations in the archives of the cadastre have indeed proven that the members of the council were all quite rich owners of lands in Semmerzake, and that their homesteads were located near the *Broeken* (MF; Directie van het Kadaster, Oost-Vlaanderen, 212; Mutatieregister van Semmerzake).

For the first time in the history of the *Broeken* in Semmerzake, the use of the common was being disputed by members that were part (at least in theory) of the group of authorised users of that common. The origins of this historical event should be sought in the fact that the hamlet of De Steirt did not exist (or did not form part of Semmerzake) when the charter of 1271 was granted to the villagers (at that time, the circumscription of Semmerzake was adequate enough, since all villagers lived in the neighbourhood of the common *Broeken* and were capable of using the common).

As a consequence of this dispute, the higher authorities decided that the commons belonged to the whole village, and that the stipulations laid down by the lord in 1271 were no longer unbreakable if the village decided otherwise – but still the village council resisted partition. On February 27, 1837, it decided that a leasing operation was not wanted by the majority of the population, and that the existing traditional use of the "Broeken" was far more convenient and valuable than the higher profits that would be made from leasing it. With this, some representatives of the provincial government accepted that: "I think that, for the moment, we have to renounce the project of allocation of the common land in Semmerzake" (RAG; Provinciaal archief 1830-1850, 815/5).

3.3.4   The End of an "Immemorial" Era (1844-1845)

At last, on July 6, 1844, the division and leasing of the *Broeken* was approved by almost the same village council of Semmerzake that had continued to reject the plan in 1837. The project was approved by a very small majority of three votes against two. The reason for the revision of the view of the council was the growth of the land taxes levied on the *Broeken* in 1844. The normal yearly income from the taxes of pasturing cattle and the sale of the planted trees was no longer adequate to cover the necessary cadastral dues (RAG; Modern Gemeentearchief Semmerzake 2, f 53r-54r). It is notable that the local authorities wanted speedy authorisation from the provincial and national authorities for their proposal, as they feared widespread resistance to their plans (RAG; Provinciaal archief 1830-1850, 815/5).

But still the battle was not won. Opponents organised themselves and sent a petition signed by 28 commoners to the governor of East Flanders in an attempt to prevent this allocation. They referred to the specifications in the charter of 1271 and the decision of 1835. The two leaders of the opposition movement were Augustin De Surgeloose and Louis De Visscher, the two dissenting members of the village council (RAG; Modern Gemeentearchief Semmerzake 2, f 55v; RAG, Provinciaal archief 1830-1850, 815/5). They also hired a lawyer (Eugène Van Acker, jr.) to represent their interests, on the basis of the charter of 1271 (RAG; Provinciaal archief 1830-1850, 815/5, 817/2).

In the autumn of 1844, matters seemed to be going the commoners' way. The actions of their lawyer had caused confusion, and the Minister of the Interior, who had to give his final approval, hesitated. He ordered a new enquiry into the origin of the ownership of the community, and wanted to know specific details about the use of the common up until then (RAG; Provinciaal archief 1830-1850, 815/5). The enquiry was duly conducted by the provincial authorities, and showed that 224 animals were pastured on the *Broeken* every year, and that three out of every four possessors of cattle in the village used the common as a(n) (additional) pasture. It concluded that the income from the common covered expenses at that time, but that the future leasing would turn out to be more profitable, and would even allow the village to abolish personal taxes on all inhabitants.

On the basis of this conclusion, the allocation was finally approved by a decision of the King, represented by the Minister of the Interior, on April 17, 1845. The opposition then embarked upon a number of desperate attempts to cancel the allocation. A hostile revolutionary atmosphere must have reigned in the village. A few days before the date set for allocation, May 6, the mayor reported to the governor "*que notre commune naguerre si paisible est dans une agitation extraordinaire; même l'autorité de la police y est méconnue.*"[4] Augustin De Surgeloose was charged with being the leader of an unauthorised protest and of a number of occupations of the *Broeken* using cattle (RAG; Provinciaal archief 1830-1850, 831/2).

On May 15, 1845, the public allocation finally took place. The common was divided and leased into 72 parcels. All authorities heaved a sigh of relief, and the protest against the division and leasing gradually died out in the following years (RAG; Provinciaal archief 1830-1850, 831/2). The ultimate end came on December 11, 1846, when Dominicus Piens and Bernardus Vander Guchte lost their court case against the village council (De Potter et al. 1868; RAG; Modern Gemeentearchief Semmerzake 2, f 62v; RAB, Eerste Aanleg Gent, Registers der Vonnissen, bundel 97-99, 20 okt., 10, 25 nov., 22 dec. 1845, 5-6 jan. 1846).

The activities of the central authorities had reached their zenith. A year later, in 1847, the reclamation law was also accepted by the parliament, which ended the existence of many of the remaining commons (Clicheroux 1957).

4.   CONCLUSION

Even during the Ancient Regime, hard-fought battles for power took place over the greens and commons. This led almost everywhere to a gradual (but often disputed) decline of communally-used lands (and sometimes their disappearance). This process started during the Middle Ages, and proceeded right through to the end of the 18th century.

In the 19th century, the systematic neglect of traditional management frameworks and specific ownership regulations led to the disappearance of most greens and commons. This was a deliberate policy pursued by the central governments in order to create a central state with uniform ownership regulations, and a centralised organisation structure. In our study area, these political reasons stayed neglected in traditional historiography, which – in the line of the ideology of the physiocrats – placed too much emphasis on rural, economic and demographic reasons as being responsible for altering the traditional use of the commons.

The process of disintegration of the commons started in the French period (1792/94-1815), when for the first time priority was given to ensuring private ownership of the land. Before the French system, *driesen* in Flanders had dual owners: on one side, there was the *dominium directum* (theoretical ownership held by the landlord), and on the other side there was the

---

[4] "Our village that was always a peaceful place has now become extraordinary agitated; even the authority of the police is not respected."

dominium utile (rights of use enjoyed by the commoners). Another severe blow was the introduction of the French *cadastre* (before which taxes were usually not paid for the use of these grounds).

The centralisation of the means of power developed gradually during the first decades of the 19th century. Sometimes commoners were able to resist various attempts at sale or leasing before they finally had to give up, as shown by the fascinating case of Semmerzake.

ACKNOWLEDGEMENT

Many thanks to Prof. Dr. E. Thoen, Dr. A. Verhoeve and Prof. Drs. J. Vervloet for their helpful remarks, and to J. O'Driscoll and E. Lachaert for correcting the first versions of the manuscript.

REFERENCES

Manuscript Sources

Ministerie van Financiën (MF), Directie van het Kadaster, Oost-Vlaanderen, Gent 212. Mutatieregister Semmerzake.
Rijksarchief Beveren (RAB), Eerste aanleg Gent, Registers der vonnissen, bundel 97-99.
Rijksarchief Gent (RAG), Frans provinciaal fonds, 2536, 3581, 4207/3, 4525/2.
Hollands provinciaal fonds, 824/1, 870/16.
Modern Gemeentearchief Semmerzake, 2 (f 4v, f 8v, f 22r, f 53r-54r, f 55v, f 56r, f 62v).
Provinciaal archief 1830-1850, 798/1, 815/5, 817/2, 824/1, 831/2.
Raad van Vlaanderen, 8569, 8570, 8595 (f 2v-3v), 23493.
Stadsarchief Aalst (SAA), Land van Aalst, 1077, 4298-4335, 13931.
Stadsarchief Gent (SAG), Register TT, f 220.

Printed and Secondary Sources

Billen, C. (1988). Jeux de pouvoirs, jeux de profits: remarques à propos de l'histoire des droits d'usage et des biens communaux (XIIe-XIXe siècle). In *Les structures du pouvoir dans les communautés rurales en Belgique et dans les pays limitrophes (XIIe-XIXe siècle).* Bruxelles: Crédit communal de Belgique.
Braekman, W.L. (1998). Privatisering gemeenschappelijke weide. Publiek protest met volkstoeloop te Semmerzake (1669). *Oost-Vlaamse Zanten,* 72 (1), 69-70.
Brouwer, R. (1995). *The Afforestation of the Commons and State Formation in Portugal.* Wageningen: Eburon.
Clicheroux, E. (1957). L'évolution des terrains incultes en Belgique. *Bulletin de l'Institut des recherches économiques et sociales,* 23 (6), 497-524.
De Brouwer, J. (1968). Geburentwist te Semmerzake. *Het Land van Aalst,* 20 (6), 292.
De Moor, M. (2002). Common Land and Common Rights in Flanders. In M. De Moor, L. Shaw-Taylor & P. Warde (Eds.), *The Management of Common Land in North West*

Europe, c. 1500-1850 (pp. 113-141). Comparative Rural History of the North Sea Area, Publication Series, 8. Turnhout: Brepols.

De Moor, M., Shaw-Taylor, L. & Warde, P. (Eds.) (2002). *The Management of Common Land in North West Europe, c. 1500-1850.* Comparative Rural History of the North Sea Area, Publication Series, 8. Turnhout: Brepols.

De Potter, F. & Broeckaert, J. (1868). *Geschiedenis van de gemeenten der provincie Oost-Vlaanderen. Eerste Reeks-Arrondissement Gent. Zevende deel: Semmerzake.* Gent: Annoot-Braeckman.

Dussart, F. & Claude, J. (1975). Les villages de "Dries" en Basse et Moyenne-Belgique. *Tijdschrift van de Belgische Vereniging voor Aardrijkskundige Studies, 44*, 239-294.

Errera, P. (1891). *Les masuirs. Recherches historiques et juridiques sur quelques vestiges des formes anciennes de la propriété en Belgique.* Bruxelles: Weissenbruch.

Guarducci, A. & Rossi, H. (1998). Terres communes et droits d'usage dans la montagne d'Arezzo de l'époque moderne à l'époque contemporaine. In P. Sereno & M.L. Sturani (Eds.), *Rural Landscape between State and Local Communities in Europe. Past and Present. Proceedings of the 16th Session of the Standing European Conference for the Study of the Rural Landscape (Torino, 12-16 September 1994)* (pp. 105-117). Alessandria: dell'Orso.

Hoppenbrouwers, P. (2002). The Use and Management of Commons in the Netherlands. An Overview. In M. De Moor, L. Shaw-Taylor & P. Warde (Eds.), *The Management of Common Land in North West Europe, c. 1500-1850* (pp. 87-112). Comparative Rural History of the North Sea Area, Publication Series, 8. Turnhout: Brepols.

Hoskins, W.G. (1955). *The Making of the English Landscape.* London: Hodder and Stoughton.

Jones, P.M. (1991). The "Agrarian Law": Schemes for Land Redistribution during the French Revolution. *Past and Present, 133*, 96-133.

Knaepen, R. (1994). *Kempische heidorpen tussen Dommel en Neten: historische geografie van het ontginningswezen, de pleinconfiguraties en sociaal-economische ontwikkelingen (met atlas) uit het geografisch pre-industrieel verleden van Mol, Geel (prov. Antwerpen) en Lommel (prov. Limburg).* Proefschrift Universiteit Gent, Vakgroep Geografie.

Knaepen, R. & Antrop, M. (2002). Gemene pleindorpen in de historisch-rurale Kempen van België en Nederland. *Historisch Geografisch Tijdschrift, 20* (1), 22-32

Lindemans, P. (1952). *Geschiedenis van de Landbouw in België, I-II.* Antwerpen: De Sikkel.

Neeson, J.M. (1993). *Commoners, Common Right, Enclosure and Social Change in England (1700-1820).* Cambridge: Cambridge University Press.

Pirenne, H. (1920). *Histoire de Belgique. V. La fin du régime espagnol. Le régime autrichien. La révolution française et la révolution liègeoise.* Bruxelles: Lamertin.

Popp, P.C. (ca. 1860). *Atlas cadastral de Belgique. Province de Flandre Orientale. Arrondissement de Gand. Canton de Oosterzeele. Plan parcellaire de la commune de Semmersaeke avec les mutations.* Bruges: B. Valckenaere.

Rombai, L. (1998). Terres communes et droits locaux en Toscane à l'époque des grands ducs de Lorraine (siècles XVIII-XIX). Libéralisme central et resistances locales. In P. Sereno & M.L. Sturani (Eds.). *Rural Landscape between State and Local Communities in Europe. Past and Present. Proceedings of the 16th session of the Standing European Conference for the Study of the Rural Landscape (Torino, 12-16 September 1994)* (pp. 95-104). Alessandria: dell'Orso.

Stabel, P. (1997). *Dwarfs among Giants: the Flemish Urban Network in the Late Middle Ages.* Leuven: Garant.

Tack, G., Van Den Bremt, P. & Hermy, M. (1993). *Bossen van Vlaanderen. Een historische ecologie.* Leuven: Davidsfonds.

Thoen, E. (1986). Driesch. *Lexikon des Mittelalters, III*, 1399-1400. München: Artemis und Winkler.

Thoen, E. (1987). *Historisch-geografische tekst bij het kaartblad Oosterzele. (Project "Historisch- landschappelijke relictenkaarten van Vlaanderen o.l.v. prof. Dr. A. Verhulst, Dr. A. Verhoeve en dr. F. Verhaeghe")*. Intern rapport, Rijksuniversiteit Gent.

Thoen, E. (1988). *Landbouwekonomie en bevolking in Vlaanderen gedurende de Late Middeleeuwen en het begin van de Moderne Tijden. Testregio: de kasselrijen van Oudenaarde en Aalst, I-II*. Gent: Belgisch Centrum voor Landelijke Geschiedenis.

Thoen, E. (1990). Een model voor de integratie van historische geografie en ekonomische structuren in Binnen Vlaanderen. De historische evolutie van het landschap in de Leiestreek tussen Kortrijk en Gent in de Middeleeuwen. *Jaarboek Heemkring Scheldeveld*, 19, 3-34.

Thoen, E. (1993). Dries versus kouter. De wisselbouw in de Vlaamse landbouw van de Middeleeuwen tot de zestiende eeuw. Bijdrage tot de historische landschapsecologie en geschiedenis van de agrarische techniek. *Jaarboek Heemkring Scheldeveld*, 22, 71-102.

Thoen, E. (2001). A "Commercial Survival Economy" in Evolution. The Flemish Countryside and the Transition to Capitalism (Middle Ages-19th Century). In P. Hoppenbrouwers & J.L. Van Zanden (Eds.), *Peasants into Farmers? The Transformation of Rural Economy and Society in the Low Countries (Middle Ages-19th century) in Light of the Brenner Debate*. (pp. 102-157). Comparative Rural History of the North Sea Area, Publication Series, 4. Turnhout: Brepols.

Vandenbroeke, C. (1975). *Agriculture et alimentation*. Gent-Leuven: Belgisch Centrum voor Landelijke Geschiedenis.

Van Der Haegen, N. (1999). *Inventarisatie en evolutie van de driesen op het kaartblad Gavere-Oosterzele. I-II*. Licentiaatsverhandeling. Mansuscript. Department of Geography, Ghent University.

Van Durme, L. (1988). *Toponymie van Velzeke-Ruddershove en Bochoute*. Gent: Koninklijke Academie voor Nederlandse Taal- en Letterkunde.

Van Durme, L. (1998). Dries, vooral in Centraal- en Zuidoost-Vlaanderen. *Handelingen van de Koninklijke Commissie voor Toponymie en Dialectologie*, 70, 117-212.

Van Osta, W. (1990). *Toponymie van Brasschaat*. Proefschrift, Katholieke Universiteit Leuven, Afdeling Germaanse Filologie.

Verhulst, A. (1990). *Précis d'histoire rurale de la Belgique*. Bruxelles: Ed. de l'Université de Bruxelles.

Vinck, L. (1997). *Driesen en driesachtige ruimten. Inventarisatie en typologie van een cultuurlandschappelijk relict in Centraal-Oost-Vlaanderen*. Licentiaatsverhandeling. Mansucript. Department of Geography, Ghent University.

Vivier, N. (1998). *Propriété communale et identité collective. Les biens communaux en France, 1750-1914*. Paris.

Vivier, N. (2002). The Managment and Use of the Commons in France in the Eighteenth and Nineteenth centuries. In M. De Moor, L. Shaw-Taylor & P. Warde (Eds.), *The Management of Common Land in North West Europe, c. 1500-1850* (pp. 143-171). Comparative Rural History of the North Sea Area, Publication Series, 8. Turnhout: Brepols.

Wauters, A. (1881). *Les bois communaux de Chimay. Recherches historiques sur la nature et l'étendue des droits des communes de Chimay, Saint-Rémy, Beauwelz, Villers-la-Tour*. Bruxelles: Callewaert.

Chapter 17

ENCLOSURE LANDSCAPES IN THE UPLANDS OF ENGLAND AND WALES

John Chapman
Department of Geography, University of Portsmouth, United Kingdom

1. INTRODUCTION

The upland landscapes of England and Wales are a highly distinctive part of the British scene, and are greatly prized today by the general public for their aesthetic qualities, for their wildlife significance, and as areas for informal recreation. The landscapes of areas such as the Lake District, the Yorkshire Dales, and Snowdonia are a major magnet for both British and foreign tourists, giving them considerable economic, as well as emotional, significance. Such landscapes are not, of course, natural. Following the pioneering work of G.W. Dimbleby (1952), and later Ian Simmons (1969), it is recognized that they are a function of past human activities. However, the part played in their evolution by the enclosure movements of the 18th and 19th centuries has been widely ignored. These movements are generally recognized as one of the major formative influences on the English rural landscape, but the academic literature focuses almost entirely on the impact of the Parliamentary part of the movement on the former open-field landscapes of the English Midlands, apart from a few highly localized studies (for example Eyre 1957). It is the purpose of this brief survey to draw attention to the significance of enclosure in moulding upland landscapes, to consider the different landscape outcomes resulting from the different methods used, and to offer some comments on the ways in which the enclosure movements have left a legacy of potential conflict to the 21st century.

In England and Wales, enclosure was, legally speaking, the process whereby an individual assumed sole possession of land previously open to

H. Palang et al. (eds.), European Rural Landscapes:
Persistence and Change in a Globalising Environment, 289-296.
© 2004 *Kluwer Academic Publishers. Printed in the Netherlands.*

some form of communal use or control. During the two centuries under consideration enclosure was carried out in a number of different ways, and clear distinctions must be made between *Parliamentary enclosures* – those carried out under the authority of specific acts of parliament; and the mass of *informal enclosures* – carried out on a small scale by local farmers (see Chapman & Seeliger 2001). Parliamentary enclosures were, by definition, highly formalized, normally large scale, and produced landscapes which were usually very different in kind from the individualistic "intaking" and "encroaching" which characterized the latter. Intermediate between the two, and less significant than either in acreage terms in most uplands, were the formal agreements sometimes reached by the interested parties to divide up a piece of moorland between themselves. These latter might produce results akin to either of the other two forms, though they normally most closely resembled the Parliamentary type.

2.   THE EXTENT OF ENCLOSURE

The physical extent of parliamentary enclosure is far easier to estimate than that of other methods, though even this has many well-known pitfalls (Turner 1978; Chapman & Harris 1982; Chapman 1987). The current author's most recent estimates, based on a ten percent sample of all awards and supported by extensive abstraction of the details from others, suggests that something of the order of 1.7 million acres (688,500 ha) was enclosed by parliamentary means in the upland areas of England, plus a further 0.44 million acres (178,200 ha) in Wales. Not all of this was upland moorland, since many enclosures stretched down into the neighbouring valleys, but it appears that less than 5 percent of the English total, and only a minute fraction of the Welsh, was open field, common meadow, or lowland pasture, so it may be estimated that some 2.1 million acres (850,500 ha) of open moorland was affected in all.

The contribution of the other methods of enclosure during the two centuries under consideration is a matter for conjecture. As has been suggested, the part played by formal agreements was relatively small, though certain local areas made more extensive use of them than others. On the other hand, all the evidence suggests that informal enclosures were of considerable significance, though again with great variation from place to place. Since the commons were technically owned by the lords of the manor, the lord was theoretically in a position to prevent any intaking, but in practice he frequently did not. In many upland parishes there was little likelihood of this intaking seriously infringing the rights of others, and the lord was always able to increase his income by demanding rent for the newly enclosed land. Contemporary rentals and similar documents frequently give evidence of additional fields being added in this way (see Chapman 1961). Ultimately, however, the extent to which either sanctioned extensions or illegal encroachments were permitted to occur was highly dependant on the attitudes of the lord or, often more important, on the vigilance of his steward. Neighbouring parishes might thus show very different patterns. With the utmost caution, it might be suggested that somewhere between a quarter and half a million acres (102,000-204,000 ha) of upland passed into individual occupance in this way. In other words, the various methods of enclosure, taken together, transferred a million hectares or more of uplands from communal control to unfettered individual ownership during the two centuries under consideration.

3.   THE LANDSCAPE IMPLICATIONS

The use which was made of these areas after enclosure, and the type of landscape produced, showed major variations. Individual intakes were almost invariably small, and were characteristically converted at the earliest possible moment to some form of improved use. This was not usually for crops, though oats or barley was sometimes grown, but for the most part the end result was a landscape of fields of permanent grass. Such fields were not only small, but irregular in form, since the individual farmers, working in a difficult environment, tended to follow the contours of the local area for maximum ease when building the stonewalls so characteristic of most of these areas. "Intake Islands" were also a characteristic feature (Eyre 1957). Where a cliff-face bordered the existing enclosed farmland, intaking often took place on flatter ground above the cliff, leaving a break in the cultivated area. Intake islands were also often a means by which the landless were able to establish themselves when the valleys were fully occupied, giving rise to a pattern of isolated farms where none had existed before (see Figure 1).

Such irregularity was, however, not always the outcome of individual intaking. In some areas, notably the Pennine dales in Yorkshire and in the North York Moors, farms tended to run in parallel strips up from the river to the moorland edge. The logic of such an arrangement in a primitive farming community is obvious, since it gave each farm access to a water supply, to drier land above flood level, and direct access from the farm onto the open moor. However, intaking then tended to take a much more regular form, with each farm extending the parallel lines into the moor, since otherwise they were liable to obstruct their neighbours' access, and thus provoke objections.

Figure 17-1. A typical informal enclosure landscape in the uplands: small irregular fields now reverting to moorland. Photo: J. Chapman.

Parliamentary enclosures normally produced a very different pattern. Small allotments might be attached to the edges of existing farms, thus replicating the pattern just described, but for the most part the commissioners responsible for the allocation were dealing with far larger areas, both in total and in terms of the allocations to each individual. Even cottagers with limited common rights might be entitled to allotments of ten acres or more if the moors were large and the settled acreage small. At the other end of the scale, a large landowner might be due for an allotment of several thousand acres. In these circumstances, the commissioners and surveyors tended to lay out large rectangular plots, ignoring any but the most prominent physical features (for example, rivers). Parliamentary landscapes thus tend to display a very high degree of regularity (Figure 2).

While small intakes were almost invariably immediately improved to some degree, the same was by no means true of parliamentary allotments. In many cases the new owner's physical and financial resources were inadequate to allow the immediate absorption of the whole of his allotments, even if he had wished to do so. In practice he may not have done. Though the more enthusiastic agricultural writers may have envisaged vast areas being brought into highly productive cultivation, local landowners were usually both more cautious and more realistic. They were fully aware that the physical character of parts of the moors was such that they were unlikely to get any return from their investment if they attempted to bring them into full-scale cultivation. Most medium–to-large-scale owners therefore tended to reclaim only a portion of their allotments, sometimes experimenting with a small reclamation, and sometimes working to a long-term plan, creating new isolated farms in the moors, which formed the nucleus for future expansion. The remainder of the allotment was then left as open sheep pasture until such time as the owner judged it economically viable to reclaim it.

Figure 17-2. A typical Parliamentary landscape of rectangular fields, with an older less regular landscape in the foreground. Note the Forestry Commission plantations on the enclosures, contrasting with the semi-natural woodland in the foreground. Photo: J. Chapman.

For many larger owners, and even for some of the smaller ones, the evidence is that "improvement" in the sense that this was understood by most contemporary agricultural writers was not necessarily the prime motive for enclosure. Misuse of the moorland commons had become a problem in, for example, parts of the Pennines and South Wales by the mid 19th century, and clarification of the legal right to keep stock on the moors was in itself a considerable benefit (House of Commons 1844). As a consequence, many upland enclosures were only partial, one section of the land being divided up for fencing off, and the remainder being left open but with strictly defined stocking limits assigned to each owner.

Two other major motives emerged in the later 19th century, namely forestry and grouse-shooting. The former was, of course, merely cropping in a different form, but experimentation with introduced conifers proved highly profitable, with long-term consequences, which will be discussed later. The latter, on the other hand, involved neither physical fencing nor any significant change in land-use, and could be readily combined with sheep-farming, offering a double profit from the land. Initially it was the social cachet of owning a grouse moor and being able to invite influential guests for a shooting holiday which motivated many, but certainly by the later twentieth century the economic aspects had come very much to the fore.

In these circumstances, the physical fencing off of the new allotments was neither necessary nor desirable, but parliamentary enclosure in England and Wales traditionally involved the compulsory fencing of the new allotments within a very short time-span, most commonly one year. There thus arose the habit in some upland enclosures of making them permissive, with no time limit on the owner's right to fence off his land (Chapman 1976). Land left unfenced remained open to communal stocking in the proportions defined in the enclosure award, and any owner fencing part of his allotment merely had to reduce his numbers of stock in proportion. This was of considerable help at the time, particularly for the smaller owners, since if they were obliged to lay out large sums of money on fencing they were then likely to be under financial pressure to attempt to recoup their expenditure as rapidly as possible by growing high-value but probably environmentally unsuitable crops. However, it left behind the seeds of future conflict as attitudes to landscape and environment changed.

4.   THE LEGACY OF CONFLICT

Present-day attitudes represent an almost total reversal of those commonly held at the beginning of the 18th century, attitudes which persisted amongst much of the population well into the 19th century. At that time, these areas were seen as a wasted resource, land which, with the application of the evolving technical and scientific knowledge of the period, might be transformed into productive farmland. Contemporaries such as Dr. Johnson and Daniel Defoe saw them as ugly and dangerous, while agricultural writers such as William Cobbett and Arthur Young regarded them as an affront to the country's status as a modern and enterprising nation (see, for example, Defoe 1720: 189-90; Young 1771: 187-8). It was in this context that a major assault on these areas was attempted by means of the enclosure movements, and the legal framework for their future use determined. However, as has just been indicated, formal enclosure did not necessarily involve either the fencing off of the land concerned or any change of land-use.

Over the past 50 years, these issues have been brought to a head. Major clashes have occurred as the interests of an increasingly urbanized population and of the environmentalist movement have challenged the legally established rights of upland landowners. Land which has been long regarded by the general public as open for free access may, as a consequence of the enclosure movements, technically be subject to both fencing off and ploughing up or converting to other uses, at the discretion of the owner, a potentially highly significant alteration to the landscape and environment of the locality.

Two specific aspects may be mentioned. Firstly, the creation of the Forestry Commission after the First World War led to a major extension of conifer plantations on these physically open but legally enclosed uplands, since the land concerned was seen as of little value. However, from the 1950s onwards there has been a growing campaign to restrict such activity, both on the grounds of the threat which it poses to a highly specialized and limited ecosystem and also on the grounds of the limitations imposed on access to walkers. The typical plantation style, with tightly-packed stands of a single alien species, has also attracted much hostile comment for aesthetic reasons. This battle has been largely won by the protestors, as the Forestry Commission has been required to concern itself increasingly with the use of native species planted specifically to enhance the environment, though local disputes continue.

Secondly, the ploughing-up campaigns of the two World Wars and the various forms of financial support offered initially by the British government and more recently by the European Union have encouraged farmers to take in parts of their allotments which would not otherwise have been financially worthwhile, provoking similar hostile reactions on similar grounds. The much-flagged intention of the present British government to introduce "right to roam" legislation aimed at allowing free access for walkers across open land undoubtedly encouraged pre-emptive action by some farmers to fence off and plough up land before this became law. The Countryside and Rights of Way Act was finally passed in 2000, by which time the open areas had been further reduced.

The enclosure movements of the 18th and 19th century thus have a dual significance to the present day. On the one hand, they were responsible for the creation of much of the pattern of hedges and stonewalls which characterize the valleys and hillsides of much of upland England and Wales, and which are such a prized part of the landscape for so many people. Here the clauses contained in the awards and agreements may serve as a significant force for conservation, since they frequently state that such boundaries must be maintained in perpetuity (Chapman 1999). On the other hand, they present a threat to the equally prized open landscapes which characteristically lie above those valleys, for they enshrine the right of the landowner to convert that land to intensive use. In spite of the Countryside and Rights of Way Act, the possibility remains that enclosure acts could be used in certain circumstances to justify further reclamation of these open areas. To the modern conservationist, therefore, the legacy of the enclosure movement in the uplands is very much a double-edged sword.

REFERENCES

Chapman, J. (1961). *Changing Agriculture and the Moorland Edge in the North York Moors, 1750 to 1960.* Manuscript. Masters Dissertation, University of London.

Chapman, J. (1976). Parliamentary Enclosure in the Uplands: the Case of the North York Moors. *Agricultural History Review*, 24, 1-17.

Chapman, J. (1987). The Extent and Nature of Parliamentary Enclosure. *Agricultural History Review*, 35, 25-35.

Chapman, J. (1999). Legal Protection of Hedgerows in England. In G. Setten, T. Semb & R. Torvik (Eds.), *Shaping the Land: vol III, The Future of the Past* (pp. 606-611). Trondheim: Arbeider fra Geografisk Institutt, Universitetet i Trondheim, Ny serie A, no 27.

Chapman, J. & Harris, T.M. (1982). The Accuracy of Enclosure Estimates: Some Evidence from Northern England. *Journal of Historical Geography*, 8, 261-264.

Chapman, J. & Seeliger, S. (2001). *Enclosure, Environment and Landscape in Southern England.* Stroud: Tempus.

Defoe, D. (1928 [1720]). *A Tour in Circuits through England and Wales*, volume 2. London: Dent and Son.

Dimbleby, G.W. (1952). The Historical Status of Moorland in North-East Yorkshire. *New Phytologist*, 51, 349-358.

Eyre, S.R. (1957). The Upward Limit of Enclosure on the East Moor of North Derbyshire. *Transactions of the Institute of British Geographers*, 23, 61-74.

House of Commons Select Committee on Commons' Inclosure, Reports, Committees (1844). vol. 5.

Simmonds, I.G. (1969). Environment and Early Man on Dartmoor. *Proceedings of the Prehistoric Society*, 35, 203-219.

Turner, M.E. (Ed). (1978). *A Domesday of English enclosure acts and awards by W.E. Tate.* Reading: University of Reading Library.

Young, A. (1967 [1771]). *A Six Months Tour Through the North of England.* Vol. 2. New York: Augustus M. Kelley.

Chapter 18

LAND PURCHASE AND THE SURVIVAL OF SWEDISH ETHNICITY IN ESTONIA
Estonian Swedes in Nuckö 1816-1924

Ann Grubbström
Department of Social and Economic Geography, Uppsala University, Sweden

1. INTRODUCTION

There is a growing interest in issues of ethnicity. For example, many people wishing to learn more about their background do so by studying the history of the ethnic group they belong to. For this reason, research focusing on ethnic groups over longer periods of time is of particular interest at present. Despite the importance of the relationship between the opportunity to own land and the survival of an ethnic group, relatively few studies have systematically focused on this question within the field of historical geography. It is the contention of this paper that such a link needs to be established, and that evidence relating to the Swedish minority settled in Estonia points to the value of such an approach.

Following the Vienna Congress of 1815, Central and Eastern Europe came to be dominated by the Russian, Austrian and Ottoman empires. Many ethnic minorities lived within these empires and, with a few exceptions (such as the Jews) the majority had a specific territorial attachment (Lundén 1993). In many cases, territorial minorities of this kind have a long historical tradition in their settlement area (Tägil 1995; Runblom 1996). Such historical tradition encourages solidarity within an ethnic group. The village and the farm bring together different generations and become symbols of ethnic affiliation (cf. Cohen 1995). Ostergren points out that continuous settlement in a specific area is one of the most important conditions for preserving the identity of an ethnic minority (Ostergren 1990). Even though historical conditions experienced by a specific ethnic group may contribute

H. Palang et al. (eds.), European Rural Landscapes:
Persistence and Change in a Globalising Environment, 297-313.
© 2004 *Kluwer Academic Publishers. Printed in the Netherlands.*

to ethnic consciousness, I want to stress that ethnicity is neither primordial nor static. Instead it is in a constant state of change as a result of continually being faced with new circumstances (cf. Calhoun 2001). Relations between individuals within an ethnic minority group, and their relations to both the majority population and to the wider society as a whole, all constitute important factors with regard to the formation of ethnic identity (Nelson 1990; Hylland Eriksen 1997).

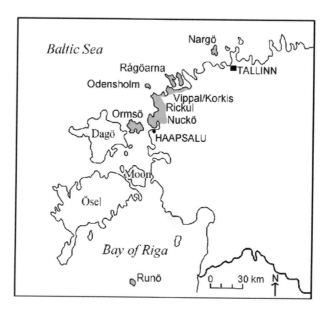

Figure 18-1. The Estonian Swedish settlement during the 1920s and 1930s (Hoppe 1993).

This paper focuses on Swedes in Estonia (Figure 1), or *Estonian Swedes*, during the period 1816-1924. Estonian Swedes have a long history in Estonia. The first Swedish immigrants to settle on the northwest coast of Estonia probably arrived in early medieval times. Over time, the settlers came to form a strong attachment to their territory, and also obtained a privileged legal position, which created conditions favourable to their survival. However, writers at the time and later writers have noted that the size of the Estonian segment of the population increased in the Estonian Swedish villages during the 19th century (Hyrenius 1942; Lagman 1973). The explanations suggested have emphasised both internal migration, with Estonians moving into the Swedish villages, and the fact that Swedes switched ethnic status. It has been assumed that farms in the Swedish villages were purchased by relatively wealthy Estonian peasants, and that this lead to disruptions in the area inhabited by Estonian Swedes (Malmberg 1936; Lagman 1980). A study of German minority groups in Russia and

North America has shown that the opportunity for individual land ownership and a free market for land and labour contributed to the assimilation of a minority group into the majority population. When the opportunity for land ownership was introduced, it influenced this assimilation chiefly by increasing geographical mobility (Waters 1995). Greater mobility may mean more frequent contacts between minority and majority populations, and in-migration may change the general ethnic composition of an area. Ostergren differentiates between *stable villages*, where farms were in most cases kept within the families already living in the village; and *less stable villages*, where farms were purchased by people from other areas (Ostergren 1988).

I have chosen to look in more detail at the parish of Nuckö, which in the 1920s was inhabited by both Swedes and Estonians (Figure 2). My aim is to present a picture of the ethnic composition and settlement patterns of the Swedes and Estonians during the period of the study, 1816-1924. My focus is on the purchases of farms from the estates, primarily at the end of the 19th century, and the question of whether these purchases led to changes in the ethnic composition and settlement patterns of the villages concerned. Finally, I will discuss how the Swedish inhabitants' choice of language and ethnicity were influenced by the increasing size of the Estonian population.

2.   FROM PRIVILEGED TENANT TO FREEHOLDER

The feudal structure survived for a long time in many East European countries, including Estonia. The Estonian peasants had corvée duties and no opportunity to buy land until the second half of the 19th century (Blum 1978). The corvée duties incumbent upon the Swedish peasants were less burdensome than those of the Estonians, since the former group had enjoyed special privileges since the 13th or 14th centuries. These privileges were geographically defined and were only applied within the Estonian Swedish settlement areas (Loit 1996; Berencreutz 1997). Disputes between landlords and the Swedish peasants concerning the interpretation of these privileges were common. Such disputes continued throughout the 17th century and up to the middle of the 19th century (Soom 1956; Jansson 1993).

The Peasant Act of 1856 made land purchases possible for peasants cultivating farms on peasant land, except where this land was located on estates owned by the church. Where a tenant could not afford the purchase price, the landlord had the right to sell the farm to another buyer. The act includes a paragraph relating specifically to the Swedish peasants, stating that their privileges were to be preserved (EB 1860). Research has not yet been able to confirm whether and how these privileges functioned by the mid-19th century. However, it has been shown that the situation of Estonian peasants had been significantly improved by means of legislation and reforms, that the importance of the Swedish privileges had been greatly reduced, and that they appear finally to have become obsolete (Jansson 1993). Legislation was introduced in 1868 to prohibit corvée duties on the estates, and this brought the period of Swedish privilege to a definitive close.

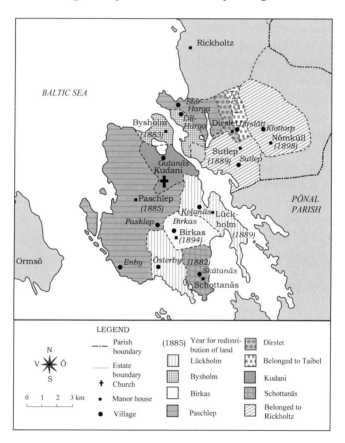

Figure 18-2. Manors, villages pertaining to them, and the year in which land was distributed on each estate investigated in Nuckö parish (SOFI, acc no. 19431, HA, fond 2840).

2.1 Methods and Source Materials

The purchase of land within the Estonian Swedish settlement areas took place between the 1860s and 1919, when the Estonian land reform was introduced. I have employed data covering the period 1816-1924 in order to allow for an analysis of the situation prior to the period of land purchase, and to examine its effects. The most important sources of data for establishing when Estonians moved into Swedish villages, and whether or not they purchased farms, comprise purchase documents, soul revisions (a form of

Russian population register used for the purposes of mobilisation) and church registers.

A soul revision from 1816 and church registers covering the years 1839-1924 have been employed to determine the ethnic status of inhabitants. The existence of the Swedish privileges made it important to establish who was Swedish and who was Estonian. For this reason, Estonian and Swedish families were recorded in separate church registers. This study defines a family as comprising at least two persons with the same surname living on the same farm. The church registers show that an Estonian woman could become Swedish by marrying a Swedish man. An Estonian man, on the other hand, did not become Swedish by virtue of having wed a Swedish woman, during the period when the privileges of Estonian Swedes were applicable. Once the privileges were no longer in place, it was possible for each family themselves to determine which ethnic group they belonged to. On the basis of the church registers, it seems most likely that ethnicity could be chosen freely during the period 1885-1905. There are several examples from this period of switches from Estonian to Swedish ethnicity when the husband was Estonian and the wife Swedish.

According to the church registers, it was quite common for people to switch ethnic status. Such changes could be influenced by many different factors, but the prestige and status of minority groups are often important in regards to the majority population's attitudes towards an ethnic minority (Tandefeldt 1988). In societies where it has been possible to choose one's ethnic status, it has been shown that it is not merely historical factors that play an important role in this decision. Perceptions about success and thoughts about the future are important aspects in choosing one's ethnic status (cf. Östberg 1995). If a person switches ethnic status and language, he or she is usually viewed by others as an assimilated person. However, if ethnicity is studied at a more private or personal level, it is possible to find ethnic feelings and ties which are difficult to detect from the perspective of an outside observer (Hansen & Meyer 1991; Rönnqvist 1999).

Documents relating to the taxation of each manor during the second half of the 19th century have been used to determine whether farms had contracts on peasant land, or whether they were cottage farms or were situated on manor land. Historical maps have also been employed to identify the location of these farms. Since I am interested in examining how the Swedish inhabitants were influenced by the increasing size of the Estonian population, interviews were also conducted with 25 Estonian Swedes who lived in Nuckö (Noarootsi) between the 1910s and the 1930s. These individuals have also provided information passed on to them by older relatives.

3.   THE TRANSITION TO CASH RENTS AND THE CONSOLIDATION OF LAND

The subdivision of land in the villages of Nuckö had been complicated, involving narrow parcels and in most cases common grazing. In order to create well-defined units of land that could be put up for sale, the land had first to be subjected to a process of consolidation.[1] The land on the estates was consolidated between 1871 and 1898 (Figure 2). Generally it took three to four years from the completion of the consolidation of land until the first purchase was realised. There may be many reasons why landlords waited – sometimes for over 30 years from the time of the introduction of the Act of 1856 – before they started selling off farms. One explanation is probably that the estates had to cover their labour requirements (von Rosen 1999). The Act of 1856 stressed that the manors should work towards a transition to cash rents, which meant that estates had to acquire both labour and draught animals. The landlords successively switched over to wage labour, and it was probably advantageous for the estates to make the transition to cash rents and hired labour prior to beginning the sale of land.

3.1   The Ethnic Composition of Nuckö Prior to Land Purchase

In 1816, most of the villages in Nuckö were inhabited exclusively by Swedes. A number, though, such as the villages of Skåtanäs (Tahuküla) and Klottorp (Suur Nõmmküla), already had quite a large proportion of Estonian families at this time (Figure 3). Marriages between the two ethnic groups were relatively rare at this point. When Estonians in Nuckö married a Swedish partner they learned Swedish, and Estonian women that married Swedish men were registered as Swedish. There was a clear social difference between the Swedish and Estonian settlement patterns during the first half of the 19th century. Estonians generally lived close to the estates or as cottagers in the villages, while the majority of the Swedish inhabitants were tenants.

From the mid 19th century until the 1880s, there was a tendency for the numbers of Estonians in the former Swedish villages to rise. The successive shift to cash rent and the final prohibition of corvée duties in 1868 led to a growing need for more labourers on the estates (Siilivask 1985). In most cases, it appears that labourers were recruited from the Estonian population. For the most part, these newcomers became landless estate workers. New dwellings were built close to the manors, as were cottage farms, which

---

[1] The consolidation of land might involve new farms being created, for example, or farms being moved out of the village.

changed the character of the landscape (Kriedte 1983; Kahk & Tarvel 1997). In some villages, the dwellings of the Estonian cottagers were geographically separate from those of the Swedes. Intermarriages, particularly between Swedish men and Estonian women, became more common. These families often chose Estonian ethnicity, particularly in villages where the Estonian share of the population had already been large at the beginning of the century. Where such changes in ethnic status were numerous, they contributed to a less pronounced geographical separation between the two ethnic groups. However, most of the tenants were still Swedes in a majority of the villages.

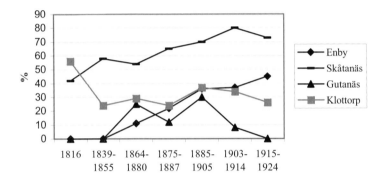

Figure 18-3. Proportion of Estonian families in Skåtanäs, Enby, Klottorp and Gutanäs 1816-1924. Percent[2] (HA, F 1864 and MSC).

3.2   Changes in Ethnic Composition during the Period of Land Purchase

The Nuckö study shows that there were four different factors, often in combination, that led to changes in the ethnic composition of the area during the period of land purchase (Figure 4). In some villages, a considerable number of farms were purchased by Estonian newcomers. This was most evident on the Paschlep (Paslepa) estate. Seven Estonian families came to the village of Enby (Einbi) during the period of land purchase, for example, and five of these bought farms. Prior to the completion of the first land purchase, there had been three Estonian tenants in Enby, two of whom lived on the outskirts of the village. By the end of the period of land purchase, eight farm heads out of 20 were Estonian (Figure 5).

---

[2] Church registers for the period 1839-1905 covered different years for Swedes and Estonians.

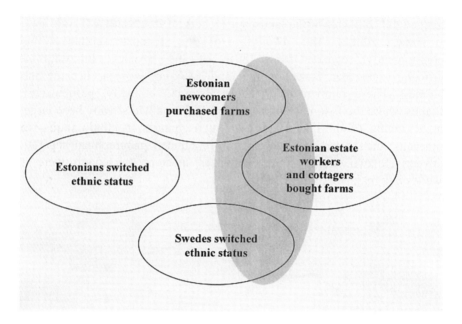

Figure 18-4. Outline of four different factors affecting ethnic composition during the period of land purchase. Shaded area symbolises the fact that a combination of three of these factors was common.

Purchases of land did not lead to any major in-migration of Estonians to the other estates in Nuckö. The land purchase process did, however, provide Estonians who had previously been cottagers or landless estate workers with the opportunity to buy land. In the village of Skåtanäs, for example, Estonian families purchased sixteen of seventeen farms. Seven of these farms were sold to Estonians who had previously been cottagers or estate workers. These farms were situated more towards the fringes of the village and had been established more recently, during the 19th century. Switching ethnic status to become Estonian was a widespread phenomenon in some of the Estonian Swedish villages during the second half of the 19th century, particularly in conjunction with marriage. Three of the Estonian families that bought farms in Skåtanäs had switched ethnic status in connection with their land purchases.

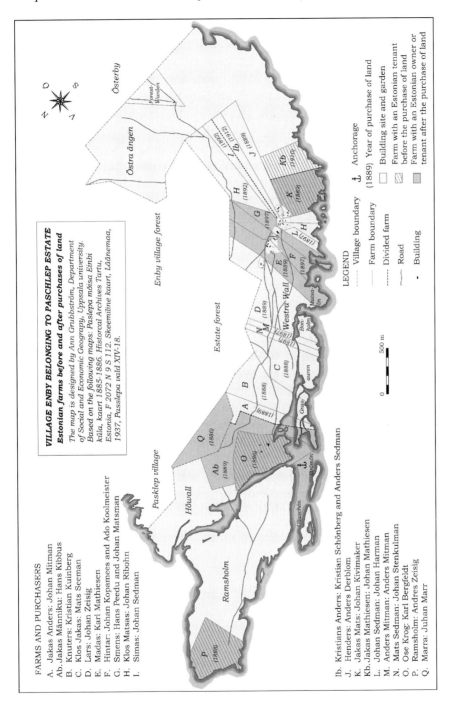

Figure 18-5. Ethnic background of Enby's farm heads prior to and subsequent to land purchase.

There are many possible explanations for the tendency to choose Estonian ethnicity in villages like Skåtanäs. Earlier Estonian in-migration, which then increased during the 19th century, appears to have constituted an important factor – but I would like to underline that land purchase appears likely to have improved the status of the Estonian population. This may have led indirectly to an increased propensity to choose Estonian ethnicity. Another important factor was the loss of the privileged position previously enjoyed by the Swedish peasants. At the same time, a series of laws were introduced during the 1860s which served to increase freedom of movement (Raun 1981; Siilivask 1985). Increases in trade and traveling meant more frequent contacts between Swedes and Estonians. Another important factor was the Estonian national awakening during the period 1860-1880, which involved a general increase in the ethnic awareness of Estonians (Jansen 1985).

Just as in other villages, the size of the Estonian section of the population also increased in Klottorp and Gutanäs (Kudani) towards the end of the 19th century (Figure 3). This was mostly due to an in-migration of Estonians, and to the fact that Swedish men switched ethnic status following marriages to Estonian women. The process of land purchase did not start until 1901 in Klottorp, at which point all of the farms but one were bought by Swedes. Ten farmers in Klottorp did not purchase the farms they leased, but instead continued as tenants. In Gutanäs, there were no purchases of land until the 1920s. Here all the farms were then bought by Swedes. Changes from Estonian to Swedish ethnicity occurred in both villages at the beginning of the 20th century. In Klottorp, all of those who switched to Swedish ethnicity lived in the centre of the village. This meant that Klottorp once again returned to the earlier settlement pattern, with Estonians for the most part living either close to the manor or on outlying farms. The following is an excerpt from an interview with Richard Gineman from Klottorp:

> The outlying farms were Estonian and there were also some Estonian women who had married into the other farms...These outlying farms were a bit on the fringe. There was no animosity, but we just didn't meet that often. People who lived in the village met every day. ... We always spoke Swedish and the Estonian wives had no choice but to learn Swedish (Richard Gineman 1999).

In Gutanäs and Klottorp, families consisting of an Estonian woman and a Swedish man were registered as Swedish. Since the Swedish tenants remained on their farms, it was impossible for the Estonian population to move away from their landless status to become freeholders. This continuity of Swedish settlement improved the chances of ethnic survival, and Gutanäs and Klottorp both became centres for the Swedish cultural awakening that began around 1900. Up to this point, the ethnic identity of Swedes had for the most part been very local in its focus. With the Estonian Swedish cultural awakening, it became possible to manifest a different form of ethnic solidarity, one which included Swedes from across the whole of the Swedish settlement area in Estonia. Those who wanted to preserve their Swedish identity had a common interest in striving for the provision of Swedish education and in arranging cultural events. Many of these Swedish families were nonetheless registered as Estonians by this time, however.

4. SWEDES AND ESTONIANS IN ENBY

In order to provide an example of the ways in which the Swedish inhabitants were influenced by the increasing size of the Estonian population, I chose to look at the village of Enby in greater detail. A considerable number of Estonian families moved into this village, mostly from Dagö (Hiiumaa), in order to buy farms. The interviews conducted provide a picture of the way the Swedish population reacted to the increasing size of the Estonian population. In addition, Enby is also a rather large village,[3] and therefore provides a reasonably good basis for an analysis of ethnic relations.

All the farms in Enby, except for that of the forest warden, were sold to peasants between 1886 and 1912 (Figure 5). The tenants felt forced to buy their farms, even though they lacked the necessary financial resources, since they knew that otherwise the landlord could sell them to someone else (SOFI; Mitman 1946; acc no. 17733; Aman 1992). The documentary evidence indicates that several alternative solutions to this dilemma were employed. For example, two Swedish families left their large tenant farms – which were then purchased by people from the village – and instead bought small former cottages. Such movements provided the opportunity for Estonian families to purchase vacated farms. Three farms were divided into smaller units prior to purchase. This also provided opportunities for Estonians to buy land. The Estonian purchases for the most part related to farms purchased jointly by two families, to farms that had been broken into smaller units, or to farms that had previously been cottager farms. The division of farms meant that new buildings were constructed, which in turn altered the landscape of the village.

The first generation of Estonians, who came to the village in the 1880s, did not learn Swedish. It is possible that they may have been able to understand the language to some extent, but they had no desire to speak it

---

[3] During the period 1839-1924, the village of Enby had between 210 and 399 inhabitants.

(Woman from Enby 2000; Peedu 2000). One of the interviewees had been told by his grandfather that the Estonians who moved to Enby did not feel that they were liked and accepted by the Swedish community (Man from Enby 2001). This may have contributed to the way in which the older Estonians kept to themselves and therefore had little incentive to learn Swedish. Quite soon after moving into the village, the older children from the Estonian families reached marriageable age. All of the eight young men whose marriages are recorded in the church register between the years 1885 and 1905 married an Estonian woman from Enby.[4] Some of the Swedes wanted to prevent the Swedish ethnic group from becoming "mixed". Thus they did not want their children to marry Estonians (Man from Enby 2001). Attitudes of this kind could also be encountered among the Estonians. One woman recollects how her grandmother reacted when she learned that her son was to marry a Swedish woman:

> Father marrying an Estonian Swedish girl wasn't something grandmother approved of. … Grandmother did not think that the Swedes had the same value, those who came from Dagö were better (Woman from Enby 2000).

At first, it is likely that language constituted a barrier between the two ethnic groups. The Estonian newcomers did not understand Swedish, which was the language spoken by all villagers; including those Estonians who had moved into the village earlier in the 19th century. The second generation of Estonians, however, grew up in the village together with Swedes and therefore learned to understand and speak Swedish. Those who married in the 1920s did not consider ethnicity to be an important factor in the choice of marriage partner. It is evident, however, that Swedes changed ethnic status more often in connection with marriage.

Even if a Swedish man or woman formally switched ethnic status, this did not necessarily mean that the person in question gave up his or her Swedish identity. The case of Mathias Ribon, a Swede, and Aet Stenberg, an Estonian born on Dagö, provides us with an example of a mixed marriage. The couple decided to be registered as Estonians. Their son, Johannes Ribon, recounts how the two languages were used in their home:

> Mom and dad spoke Estonian together, even though mom could speak Nuckö Swedish.[5] We children also spoke Estonian. When we spoke with dad we used Swedish. It was like a law (Ribon 1999).

---

[4] References to women's marriage partners are incomplete and unclear and within the limits of this project it has not been possible to reconstruct their marital patterns.

[5] A dialectal form of Swedish spoken by the people of Nuckö.

Parents often thought that speaking Estonian in the home and sending their children to an Estonian school would assist in their children's education (von Rosen 1999). But even though the Ribon family was formally considered Estonian, it was important for Mathias to pass on the Swedish language to his children. Thus there may well be a discrepancy between the formal choice of ethnicity and the language spoken. A minority language often remains in use longest in the private sphere, at home with family, relatives and friends (Tandefeldt 1988). The interview study shows that simply because people switched ethnicity from Swedish to Estonian, it cannot be taken for granted that they had a weak ethnic consciousness as Swedes. Even in villages like Skåtanäs where Estonian was the dominant language, it is possible to find private expressions of identification with a Swedish ethnicity, such as the reading of Swedish literature. The more private a person's ethnic characteristics, the less risk that person runs of being exposed to external pressures – and the greater the likelihood that the person will preserve this ethnic identity (Edwards 1985).

5. DISCUSSION AND CONCLUSION

My investigations in Nuckö indicate that the ethnic composition of a local area may change considerably in different directions within the context of a longer time frame. The settlement patterns of the Swedes and Estonians in Nuckö differed during the first half of the 19th century. The Estonian population consisted for the most part of recently arrived landless estate workers or cottagers, whereas the majority of the Swedes lived on tenant farms in the villages. In combination with the fact that Swedes were a privileged minority, this suggests that in comparison with the Estonians, Swedes enjoyed a relatively high status position during this period. It is evident that one consequence of the successive transition to cash rents following the Peasant Act of 1856 was an increase in the size of the Estonian portion of the population. Estonians accounted for the majority of the estate workers that were hired, and Estonian in-migration can be linked to the establishment of new farms and dwellings in the area. As the Estonian section of the population increased in size, intermarriages between Estonians and Swedes became more common. Swedes still comprised a majority of the tenants on peasant land, however.

Research indicates that the purchase of land is an important factor when it comes to explaining differences in the ethnic composition of different villages. The local landlord was able to control sales of farms, since he was the sole landowner prior to the period of land purchase. It is evident that the policies of different estate owners had a significant impact upon the opportunities available to Estonian Swedes to preserve their ethnicity. Conditions favouring the preservation of Swedish ethnicity and language seem to have been more prevalent in the more stable villages, where the farms were for the most part sold to Swedish tenants. This involved a continuity of Swedish settlement, which according to Ostergren encourages an historical tradition and solidarity within an ethnic group. Estonians who lived in these villages learned to speak Swedish, and even switched ethnic status in the church register.

The purchase of land in other, less stable villages, led to changes in the settlement patterns of Swedes and Estonians respectively. This finding is consistent with those of Waters, who has shown that land purchase led to higher levels of mobility. In the Estonian Swedish villages, this increased mobility manifested itself in two distinct ways. *Firstly,* there was in-migration of Estonians to Estonian Swedish villages in conjunction with land purchase. Some Estonians migrated to the estates in order to make themselves eligible to purchase farms and thus to become freeholders. The case study in Enby shows that as a result of financial considerations, some Swedes were forced to leave their farms or to break them up into smaller units, which provided opportunities for Estonians to buy these farms. Most of the Estonian purchases were of farms of modest size – i.e. former cottager dwellings or units of farms that had already been broken up – or they related to two families pooling resources to buy a farm. Where farms were divided, the person who had cultivated the land up to that point commonly retained the inhabited portion. Thus when the remainder was sold, the new owners had to construct new buildings. It is also likely that new buildings had to be constructed on farms purchased by two families. Thus, to some extent, new settlement patterns may be related to Estonian in-migration. *Secondly,* landless Estonian estate workers and cottagers were given the opportunity to buy farms on the estate where they already lived. Former Swedish peasants who switched ethnic status provide an additional explanation for the increasing numbers of Estonian farm heads. This phenomenon was most common in villages where the Estonian portion of the population was already large at the beginning of the 19th century. The increasing number of Estonians could cause tensions between Swedes and Estonians. Such problems seem to have been concentrated in the period during which the first generation of Estonian newcomers arrived, however, and to have disappeared with the second generation.

Marriages between Estonians and Swedes became more common at the beginning of the 20th century. The fact that Swedes were no longer a privileged minority, in combination with the fact that more Estonians became freeholders, may be assumed to have involved higher status for Estonians in the Estonian Swedish villages. It was common for Swedish men to switch to Estonian ethnicity when they married Estonian women. One of the reasons Swedes switched ethnicity related to perceptions about their children's future. Parents had experienced a trend towards a society with greater mobility, which they thought would require a better command of the Estonian language. It is important to emphasise, however, that even though people may choose to register a change of ethnicity and may even change their use of language, they may nonetheless remain deeply committed to their former ethnicity. Feelings of ethnicity appear to be very resistant and to be able to survive a great deal, even though they may have to become more or less confined to the private sphere.

REFERENCES

Manuscript Sources

Interviews:
Peedu, Artur, born in 1916, interview 31.08.01. Enby;
Gineman, Richard, born in 1930, interview 07.06.99. Klottorp;
Man from Enby, born in 1930, interview 16.09.01. Enby;
von Rosen, Hans, born in 1925, interview 23.08.99. Lyckholm;
Ribon, Johannes, born in 1918, interview 13.08.99. Pasklep and Enby;
Woman from Enby, born in 1907, interview 16-17.03.00. Enby.

Family History Library of the Church of the Latter Day Saints, (MSC), Västerhaninge, Sweden:
Personaalraamatud S:a Catharina, Nuckö parish:
1839-1855, film no. 74201, inv. no. 2030-2031;
1864-1884, film no. 74200, inv. no. 2031-2035;
1879-1914, film no. 74198, inv. no. 2036-2041;
1903-1924, film no. 74199, inv. no. 2041-2048.

Historical archive (HA), Tartu, Estonia:
F 1864 Eestimaa kubermangu revisjonilehtede kollektsioon (1782-1908);
F 2072 Kaardikogu (1664-1965);
F 2486 Eesti maakrediidiselts (1802-1940);
F 2840 Tallinna kinnistusamet (1722-1944).

Institute for Dialectology, Onomastics and Folklore Research (SOFI), Uppsala, Sweden:
Accession number 17 733. Mitman, Johannes, Enby (1946);
Accession number 19 431, Karte von Ehstland 1871. J.H, Schmidt.

Svenska Odlingens Vänner (SOV), Stockholm, Sweden:
Estländsk bondeförordning (EB), (1860). Swedish version, Reval.

Printed and Secondary Sources

Aman, V. (1992). *En bok om Estlands svenskar 4*. Stockholm: Svenska Odlingens Vänner.

Berencreutz, M. (1997). Gods och landbönder i västra Estland. Herravälde, resursutnyttjande och böndernas arbetsbörda under den svenska stormaktstiden. *Kulturgeografiskt seminarium 5/97.* Stockholm: Kulturgeografiska institutionen, Stockholms universitet.

Blum, J. (1978). *The End of the Old Order in Rural Europe.* Princeton: Princeton U.P.

Calhoun, C. (2001). Tradition, but not Mere Inheritance. Symposium on Ethnicity. *Ethnicities,* Vol 1 no 1, 9-17.

Cohen, A.P. (1995). *The Symbolic Construction of Community.* Chichester: Ellis Horwood, London: Tavistock.

Edwards, J. (1985). *Language, Society and Identity.* Oxford: Blackwell.

Hansen, L.-I. & Meyer, T. (1991). The Ethnic Classification in the Late 19th-century Censuses. A Case Study from Southern Troms, Norway. *Acta Borealia,* 2, 13-56.

Hoppe, G. (1993). Jordbruk och landsbygd i Estland: ett långtidsperspektiv med särskild tonvikt på landets svenskbygder. *Östeuropas omvandling, Ymer,* årgång 113, 44-68.

Hylland Eriksen, T. (1997). Ethnicity, Race and Nation. In M. Guibernau & J. Rex (Eds.), *The Ethnicity Reader: Nationalism, Multiculturalism and Migration.* Oxford: Polity.

Hyrenius, H. (1942). *Estlandssvenskarna – Demografiska studier.* Lund: Gleerup.

Jansen, E.A. (1985). On the Economic and Social Determination of the Estonian National Movement. In A. Loit (Ed.), National Movements in the Baltic Countries During the 19th Century. *Studia Baltica Stockholmiensia 2* (pp. 41-57). Stockholm: Almqvist & Wiksell.

Jansson, T. (1993). *Statsmakt och lokalsamhällen: tsarer och baroner, livegna ester och fria svenskar tiderna igenom.* Uppsala: Kungliga Gustav Adolfs Akademien.

Kahk, J. & Tarvel, E. (1997). An Economic History of the Baltic Countries. *Studia Baltica Stockholmiensia 20.* Stockholm: Almqvist & Wiksell International.

Kriedte, P. (1983). *Peasants, Landlords and Merchant Capitalists – Europe and the World Economy 1500-1800.* Cambridge: Cambridge U.P.

Lagman, E. (1973). Estlandssvenskarnas språkförhållanden. Särtryck ur: *En bok om Estlands svenskar 3.* Stockholm: Svenska Odlingens vänner.

Lagman, E. (1980). Ur vår historia. *Kustbon nr 1,* 3-4.

Loit, A. (1996). De estlandssvenska böndernas rättsliga ställning under 1700- och 1800-talen. In M. Engman (Ed.), *Väst möter Öst. Norden och Ryssland genom historien* (pp. 109-124). Stockholm: Carlsson.

Lundén, T. (1993). *Språkens landskap i Europa.* Lund: Studentlitteratur.

Malmberg, T. (1936). Svenskarne i republiken Estland. Off-print *Globen nr 7* 67-69.

Nelson, C.H. (1990). Some Thoughts on the Dynamics of Cultural Transformation. In H. Runblom & D. Blanck (Eds.), Scandinavia Overseas: Patterns of Cultural Transformation in North America and Australia. *Uppsala Multiethnic papers 7* (pp. 111-116). Uppsala: Centre for Multiethnic Research.

Ostergren, R.C. (1988). A Community Transplanted: The Trans-Atlantic Experience of a Swedish Immigrant Settlement in the Upper Middle West, 1835-1915. *Studia multiethnica Upsaliensia 4.* Stockholm: Almqvist & Wiksell International.

Ostergren, R.C. (1990). The Transplanted Swedish Rural Immigrant Community in the Upper Middle West. In H. Runblom & D. Blanck (Eds.), *Scandinavia Overseas. Patterns of Cultural Transformation in North America and Australia* (pp. 18-35). Uppsala: Centre for Multiethnic Research.

Raun, T.U. (1981). The Estonian. In E.C. Thaden (Ed.), *Russification in the Baltic rovinces and Finland 1855-1914.* Princeton, N.J.: Princeton U.P

Runblom, H. (1996). Europa och minoriteterna. In G. Gren-Eklund (Ed.), *Att förstå Europa – mångfald och sammanhang.* Humanistdagarna vid Uppsala universitet 1994 (pp. 297-301). Uppsala: Studentbokhandeln.

Rönnqvist. C. (1999). Scattered Swedes and Single Settlers on Ethnic Identity Reflected in Nationalistic Sentiments, Gender and Class in the 20th Century Canada. In Swedishness Reconsidered. Three Centuries of Swedish American Identities. *Kulturens frontlinjer. Skrifter från forskningsprogrammet kulturgräns Norr 18* (pp. 91-119). Umeå: Institutionen för nordiska språk

Siilivask, K. (1985). Some of the Main Features of the Socio-economic Development of Estonia in the 19th Century. In A. Loit (Ed.), National Movements in the Baltic Countries During the 19th Century. *Studia Baltica Stockholmiensia 2* (pp. 205-214). Stockholm: Almqvist & Wiksell

Soom, A. (1956). De estlandssvenska Ormsö och Nucköböndernas kamp mot det feodala oket under 1600-talet. *Svio Estonica*, XIII, 3-28.

Tandefelt, M. (1988). *Mellan två språk. En fallstudie om språkbevarande och språkbyte i Finland.* Studia Multiethnica Upsaliensia 3. Stockholm: Almqvist & Wiksell International.

Tägil, S. (1995). Ethnic and National Minorities in the Nordic Nation Building Process: Theoretical and Conceptual Premises. In S. Tägil (Ed.), *Ethnicity and Nation Building in the Nordic World* (pp. 1-333). London: Hurst

Waters, T. (1995). Towards a Theory of Ethnic Identity and Migration: The Formation of Ethnic Enclaves by Migrant Germans in Russia and North America. *The International Migration Review*, 29:02 515-541.

Östberg, W. (1995). Land is Coming Up: The Burunge of Central Tanzania and Their Environments. *Stockholm Studies in Social Anthropology*. Stockholm: Almqvist & Wiksell International.

·

Chapter 19

THE DYNAMICS OF PROPERTY RIGHTS IN POST-COMMUNIST EAST GERMANY

Karl Martin Born
Section for Applied Geography, Department of Geographical Sciences, Free University of Berlin, Germany

1. INTRODUCTION

In many respects the transformation process in East Germany proved to be less independent from foreign influences than in other so-called transformation states such as Poland or the Czech Republic. The reunification with West Germany and the adoption of the West German social, economic and legal systems predetermined the process and set the pace for a restructuring process which was interestingly partly orientated towards the future and partly towards the past (Wollmann & Eisen 1995; Reißig 1996; Wiesenthal 1996).

This chapter will give a brief overview over the changes in property rights in rural East Germany after 1990 and will outline the geographical, economic and social impacts of these changes. The results presented stem from an international interdisciplinary research project on property restitution in Germany and Poland.[1] The chapter thus stresses the process of property restitution (Blacksell & Born 2002).

The study area is situated in the far north-east of Germany – a truly peripheral location in relation to Berlin between the Baltic Sea and the Polish border. In order to illustrate the process of property restitution the

---

[1] *Property Restitution and the Post-1989 Transformation Process in Germany and Poland* (Volkswagen Foundation Grant between 1999 and 2001 involving the Department of Geographical Sciences at the University of Plymouth, UK, the Institute for Urban and Regional Sociology at the Humboldt University Berlin, Germany, and the Institute of Sociology at the Jagiellonian University Cracow, Poland).

H. Palang et al. (eds.), European Rural Landscapes:
Persistence and Change in a Globalising Environment, 315-332.
© 2004 *Kluwer Academic Publishers. Printed in the Netherlands.*

small village of Bergholz (Figure 1) and the surrounding district of Löcknitz were selected using statistical data from various privatisation and restitution agencies as well as interviews with experts and people affected. With a population of 334 (2001) and a dominating agricultural sector (94 percent of all employed people in 2001) Bergholz seems a suitable example for the study of privatisation and restitution in rural areas of East Germany. Although Bergholz did not suffer from expropriations ordered by the Soviet Occupation Forces between 1945 and 1949 in the surrounding district of Löcknitz, 28 large estates had been expropriated.

Figure 19-1. Map of Germany. Case study areas.

This chapter focuses on the post-1990 changes and consequences of privatisation in general and furthermore on the various historic decisions which needed to be corrected after 1990. After looking at the legal aspects of these processes the geographical, social and economic consequences of property dynamics will be demonstrated. In a short section the actors' agendas will be taken into consideration in order to assess the social and economic impact of their agendas on restitution and privatisation.

2. RURAL AREAS AFTER UNIFICATION: CHANGES AND CONSEQUENCES

In the public debate the dynamics of property rights after 1990 were mostly discussed in an urban or suburban context as the nationalised property was privatised through privatisation of housing associations or through the return of unlawfully expropriated houses to their original owners. Urban quarters like Prenzlauer Berg in Berlin or suburban areas like Kleinmachnow or Marzahn were subject to these processes which in some cases lead to dramatic demographic and social changes (Blacksell et al. 1996; Czada 1997; Reimann 1997; Glock & Keller 2002).

Table 19-1. Rural areas after unification: changes and consequences. (Bergmann 1992; Thöne 1993; Bork et al. 1995; Albrecht & Albrecht 1996; Clasen & John 1996; Löhr 2002).

Sector	Nature of change	Consequences
Manufacturing and Services	• Privatisation of enterprises • Reorganisation of services	
Agriculture Farm Structure	• Reorganisation of collective farms to new enterprise units • Dispersal of collective farms • Sale of state-owned farms • Re-institution of old farms • Institution of new farms • Unsettled disputes over property rights	• Costs through administrative processes • Costs through criminal activities • Competition for land
Agricultural Production	• Restructuring of collective farms • Introduction of market economy • Introduction of CAP-rules • Loss of market share to West German/European farm products • Responsibility for old debts	• Competition from well-funded West European farms • Adjustment period • Loss of employment • Emigration
Communities	• Responsibility for infrastructure formerly provided by collective farms • Loss of infrastructure: small shops and enterprises • High level of unemployment • Loss of population • Unsettled disputes over property rights to buildings and enterprises	• Over aging of population • Loss of services • Decreasing living conditions

The whole issue of changing property rights in rural areas must be seen within the context of post-unification issues and conflicts. It is therefore necessary to outline the impact of unification and the introduction of the market economy in rural areas before discussing the extent and consequences of changing property rights. Table 1 shows clearly the dramatic changes in many sectors as well as the high transformation costs through the administrative process (e.g., the establishment of administration for private enterprises), criminal activities (e.g., false declarations of bankruptcy or deliberate under-valuation of assets) and the introduction of competition within the agricultural sector.

The reorganisation of agriculture and the privatisation of state enterprises in the secondary sector had major impacts on the development of rural areas after 1990. These transformation processes – all linked to the reinterpretation of private property rights – set the pace for a further peripheralisation of rural areas.

3. THE DYNAMIC OF PROPERTY RIGHTS AFTER 1945

In order to understand the transformation of rural areas and the dimension of changing property rights within this process it is necessary to give a brief summary of the processes which were to be reversed after 1990: Four different processes of more or less forcefully induced ownership changes can be identified while simultaneously many farmers left their farms and fled to the West: the expropriations by the Soviet Occupation Forces, the first redistribution under the Land Reform Act and the second redistribution through the collectivisation reduced private farming to 4.5 percent.

After 1989 the transformation of the agricultural sector can be characterised by the intention to correct the past injustice of expropriation and to adjust the existing structures to the standards of the Common Agricultural Policy of the European Union. This task comprises several sub-tasks (Wiegand 1994):

– The efficiency of the allocation of resources must be increased through newly defined property rights: privatisation and property restitution as future- and past-oriented solutions were implemented.

– Competition within the agricultural sector must be introduced and increased.

– The agricultural market must be liberalised and the rules of the Common Agricultural Policy must be implemented.

The three different approaches: compensation, privatisation and restitution to re-establish property rights in East Germany, must be seen in the light of the following conflicts: The Unification Treaty prohibits the restitution of property expropriated by the Soviet Occupation Forces and offers the payment of compensation to the former owners while all other illegally expropriated land and property should be returned to the former owners according to the Property Restitution Law. However in the case of the so-called *Bodenreform*, the farmsteads' legal status was confirmed although they were located on expropriated land.

Similar to the state-owned industrial complexes the large state farms should be privatised in the same manner by the same institution (*Treuhandanstalt*); the revenue from the sale of former state-owned land should then be used to offset the costs of paying compensation to the owners of the estates expropriated between 1945 and 1949. The large collective farms should be de-collectivised and also transformed into private enterprises.

The transformation of rural areas is, however, not only affected by the direct effect of privatisation and restitution but also by the indirect effects of the agricultural restructuring – The Agricultural Restructuring Law offers different chances for farmers to re-establish and extend their farms by assessing their local roots, their experience and their financial background. Comparing former members of the collective farms and former individual owners, local farmers from the collective farms were clearly advantaged as they benefited from their experience and their share in the collective farms. It was almost impossible for former owners or their heirs to re-establish their farms as they were marginalised from the sale or lease of land (Swinnen et al. 1997).

Table 19-2. Transformation of rural areas after 1945 (Eckart et al. 1994; Wiegand 1994; Albrecht & Albrecht 1996; Hagedorn 1997).

	Scale/Extent	**After 1990**
1933–1945 Expropriations under the Nazi regime (Jews, Anti-Nazi-Activists)	▪ No data available (up to 80% in some urban areas; in rural areas approx. 10%)	▪ Restitution of property through Property Restitution Law
1945–1949 Expropriations under Soviet Occupation	▪ Expropriation of 14,089 farms and agricultural enterprises with 3.3 million ha (mostly farms with more than 100 ha, state-owned land and forests, property of Nazi-activists or anti-communists	▪ Compensation through Compensation Law

1945–1949 Land Reform, Bodenreform	▪ 210,276 new farms
	▪ Enlargement of 125,714 smallholders' farms (under 5 ha) → 2.1 million ha
	▪ New state farms (VEG) → 1.2 million ha
		▪ *Bodenreform*-Land:
		  o Modrow-Law →
		   Confirmation of ownership
		  o Privatisation through Privatisation Agency (Treuhandanstalt)
		▪ State Farms:
		  o Privatisation through Privatisation Agency

1949–1989 Expropriations	▪ Approx. 600,000 ha	▪ Restitution of property through Property Restitution Law (600,000 ha)
1952–1989 Collectivisation (LPG)	▪ 4,530 Collective farms on 82.2 percent of all arable land (1989) ▪ 580 State Farms on 7.5 percent (1989) ▪ Private farming on 4.5 percent (1989)	▪ Decollectivisation through Agricultural Reform Law ▪ Privatisation through Privatisation Agency (1.4 million ha)

Table 2 summarises the dynamics of property rights in East Germany from 1933 to 1989, pointing at the expropriating measures, the scale and extent of these dynamics in property rights and lastly on the political and legal decisions to correct these acts of injustice.

4.   GEOGRAPHICAL IMPACTS

Contrary to urban centres areas characterised by agricultural activities show only little or no direct impact from the restitution process as large-scale farming continued to shape the landscape. It is thus time to turn to the village of Bergholz in order to demonstrate the "resistance" of rural areas towards the impact of restitution policies.

The sequence of the field pattern in Bergholz from 1953 to 1998 (Figure 2) reveals the change in land-use after the collectivisation but it also stresses the continuity in farming after 1990. While collectivisation reduced the number of plots from approx. 1110 to 21 the re-establishment of private farming had only a marginal impact on the actual land-use. Although at least 111 plots were returned to their rightful owners the land-use pattern in 1998 remained almost the same as in 1987. From 1939 onwards the agricultural structure in Bergholz did, however, change dramatically.

In 1939, 72 farms were recorded. Until 1962 the collectivisation created five Agricultural Cooperatives or *Landwirtschaftliche Produktions-genossenschaften* (LPG), a further concentration process – mostly to increase productivity and to eliminate the small LPG – reduced the number of agricultural enterprises to two. After 1990 privatisation recreated four agricultural enterprises from the collective farms. The *Agrarbetrieb Bergholz Kapital Gesellschaft* can be seen as the third successor to the *LPG Neuland* although its land base is smaller – a typical consequence of two earlier liquidations of *Neuland*-success companies. It is not surprising that the four farms in Bergholz work on less land than their socialist predecessors: in general, privatisation increased the number of agricultural enterprises in the surrounding district of Uecker-Randow from 68 in 1989 to 298 in 1999 (in 1970 159 enterprises) In Bergholz, Mr. Pryzibilski purchased land from former LPG-members while Mr. Werth reinstalled his old farm out of the LPG. Although some of the successful restitution claimants had their whole farms with up to 84 ha returned, none of them re-established their farms but sold or leased their land to the existing farms.

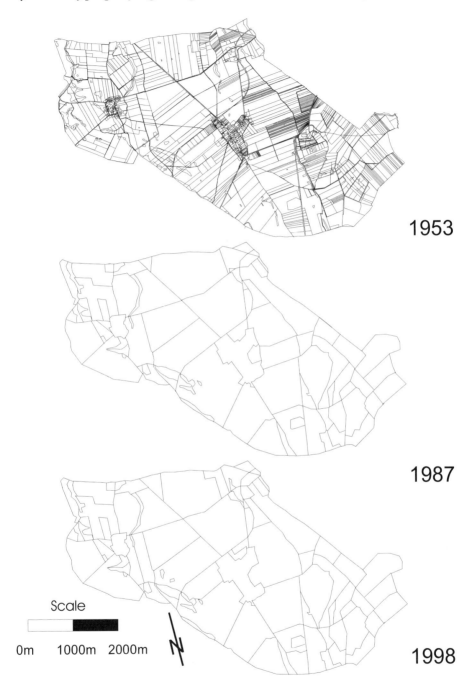

Figure 19-2. Changes in the field pattern in Bergholz (1953-1998) (Landesvermessungsamt Mecklenburg-Vorpommern. Cartography: Department of Geographical Sciences UoP; KMB 2002).

Returned to former owner

Ownership reconfirmed

Claim refused

Claim withdrawn

Claim outstanding

Scale

0   500m   1000m   1500m

Sources: LARoV Pasewalk
Cartography: Department of Geographical Sciences; KMB (2002)

Figure 19-3. Property Restitution in Bergholz, Mecklenburg-Vorpommern.

The map (Figure 3) illustrates the extent of property restitution in Bergholz. The most noticeable feature of the map is surely its cartographic base, the 1954-cadastral map – it illustrates the past-orientation of property restitution, but also visualises the technical difficulty to return a piece of land which is now in the middle of another field. The data from the Office for the Settlement of Disputed Property Rights show that 20 percent of all plots containing 34 percent of the area were affected by restitution claims. Most of the claimants seek the return of their farmsteads and lands. Contrary to the popular belief about the dominance of Western claims the number of claims was equally shared between East and West. In Bergholz the claimants were – compared to the whole of East Germany – outstandingly successful as 57 percent of the claimed area was either returned or the property rights were reconfirmed. For the whole of East Germany only 44 percent of all claims were successful (by December 31, 2002).

4.1   Delay in Investment?

The discussion about positive and negative impacts of privatisation and property restitution has probably been the most controversial issue of the whole unification process as the slow economic recovery of East Germany needed to be explained (Nölkel 1993; Ballhausen 1994). Although for urban areas in the central regions the question whether property restitution slowed the economic development is still being discussed. The situation in peripheral areas seems to be less complex. As privatisation and property restitution led only in a minority of cases to a return of the former owners and the reestablishment of a farm, the spatial impacts were minimal. In the village of Bergholz only 13 houses were affected by property claims – three claims were refused, four of them had been demolished by the German Democratic Republic, four were sold after restitution, and the ownership rights of two were reconfirmed to the occupiers. Given the peripheral location of the village, the reducing number of inhabitants and the non-existing property or rent market in Bergholz, it can be summarised that privatisation processes had little or no impact on any investment. As shown in Figure 2, the impact of privatisation on the land was minimal as the general land-use pattern of large blocks remained. None of the after 1953 destroyed landscape elements, like hedges or *Sölle* were recreated.

4.2 Artificial Blocking of Property Market

Another argument used in the discussion about the spatial impact of property restitution refers to the artificial blocking of the property market while the claim is being processed. As the property market in peripheral areas of East Germany is generally very low, only the sites in the most attractive areas could be affected. In general, houses in rural areas were only marginally affected by privatisation and property restitution as they remained in private hands throughout the GDR-regime. Far more complex is the process of the privatisation of commercial buildings that were brought into the LPG and were later re-privatised – assessing their value offered numerous possibilities for criminal activities.

5. ECONOMIC IMPACTS

Privatisation and re-privatisation in East Germany had their most dramatic impacts on the field of employment. Due to incomparable statistics from the former GDR and Federal Republic of Germany only the number of employees in the agricultural/fisheries sector and the industrial sector for the period from 1989 to 1998 can be shown. Although the total number of employees decreased with 51 percent, agriculture and fisheries lost over 90 percent of their workforce. However, it must be kept in mind that the LPGs offered numerous non-agricultural services like nurseries, pensioners' homes, car repair or veterinary services in rural areas.

6.   AGRICULTURAL ENTERPRISES IN THE DISTRICT OF UECKER-RANDOW

I have already shown the changes in the number of agricultural enterprises but that point deserves further investigation. In the district of Uecker-Randow the agricultural sector changed dramatically after 1989 as the number of agricultural enterprises rose from 68 to 395. At the end of a selection process in 1999 only 258 enterprises remained.

Looking at the agricultural land per farm the non-existing structural change in agriculture becomes obvious – in Uecker-Randow farms over 500 ha work almost 80 percent of the total agricultural land (see Figure 4).

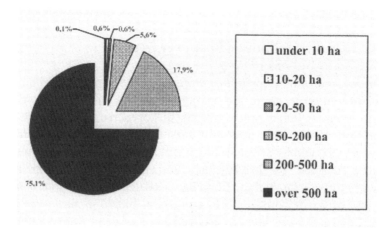

Figure 19-4. Land-use and farm size in the district of Uecker-Randow, 1999 (Statistisches Jahrbuch Mecklenburg-Vorpommern 2000).

This observation leads to the very important point about the dichotomy between the aims of privatisation and restitution on the one hand and economic productivity on the other: privatisation and restitution aim at correcting the de-privatisation of property through collectivisation and expropriation while the restructuring of agriculture aim at creating productive and competitive enterprises. Swinnen et al. (1997) show how the 1992 Agricultural Restructuring Law shaped the agricultural sector in East Germany by differentiating between re-established and newly established farms and calling for local roots and agricultural experience.

Table 19-2. The restructuring of agricultural enterprises – different chances for different enterprises (Beckmann 1997; Swinnen & Mathijs 1997).

Form of operation 1990	Form of operation after 1992	Position for business start and lease
LPG	Successor as legal entity (GmbH, e.G., KG)	Favoured
	Former-LPG-Member with agricultural land: Local re-establisher	Advantaged
	Former-LPG-Member without agricultural land: Local new-establisher	Disadvantaged
LPG and VEG (State Farm)	Non-local new-establisher with little agricultural experience	Disadvantaged
VEG (State Farm)	Former owners:	
	Local re-establisher	Advantaged
	Non-local re-establisher	Advantaged
	Buyers:	
	Local re-establisher	Favoured
	Local new-establisher	Advantaged
	Non-local new-establisher	Disadvantaged

Additionally it must be mentioned that the high leasing rate of three to four leasing contracts per ha gives the local Land Lease Commissions a significant role in the restructuring process. In Bergholz the two major landowners were interested in reducing the number of contracts with different farmers, so that only two leasing contractors remained in the system. Many Land Lease Commissions support the large, capital-weak LPG successor enterprises by providing them with long-term leasing contracts and suitable areas. Table 3 shows how agricultural policies for privatisation and re-privatisation lead to different chances for different enterprises – while successor-companies of the LPG were clearly favoured, former-LPG-members did not enjoy such support. With respect to the privatisation of state farms, former owners were advantaged to buyers. Within the group of buyers, however, local people were advantaged to non-local people. This differentiation in support relates to different levels of compensation to be paid to expropriated people and the practice of the Land Lease Commissions. It thus reflects the different influence of agricultural lobby groups on the formulation of policies.

7.   SOCIAL IMPACTS

Based on the discussion of spatial and economic consequences of property dynamics in rural areas the following analysis of the social consequences will discuss the potential effects for rural areas in general and then demonstrate the impacts for the village of Bergholz.

7.1   Return of Owners

A first consequence of the reorganisation of property rights in East Germany might be the return or the reestablishment of the former owners or their heirs. In the case of Bergholz the claimants for property restitution came from all regions in East and West Germany, as well as from the United States, the United Kingdom and South Africa. The return of former owners would not only lead to a repopulation of the village but would also balance the migrational loss of peripheral areas since 1990. Additionally the village community could be revitalised. Another aspect of a return of former owners would be the additional investment in buildings and property. The reestablishment of private farms from the cooperatives would have similar effects as houses would have been restored and commercial buildings reused.

However, the analysis of our data on Bergholz does not support the above hypothesis, as the activity rate of former owners was very low. Even though the majority of the successful claimants came from outside the region, no former owner returned to Bergholz. It is difficult to assess the impact of privatisation on the demographical situation in peripheral areas but data from rural areas in the land of Brandenburg suggests a moderate return of mostly elderly people leading to a further over-aging of the rural population. It might also be doubted whether village communities could be revitalised as some farmers in Bergholz still see the flight of their neighbours as irresponsible and selfish. During a Round Table Discussion they referred to the animals left behind and the additional workload for them to feed the animals and work on their fields. The reestablishment of one individual farm lead to the return of agriculture into the old village and provided for the restoration of one farmstead. The village community, however, remained untouched with a high proportion of unemployed people and pensioners.

7.2   Resale of Re-Privatised Property

Social consequences might also occur when the re-privatised property is sold either to private owner-occupiers or investors. The impact of these sales ranges from rising rents for tenants and a gradual change in the population to a complete social and demographical restructuring of neighbourhoods. Additionally the ownership structure might be changed from owner-occupiers to absent landlords, a situation in which urban planning and governance becomes increasingly difficult.

Many of these processes are restricted to urban and suburban areas with high numbers of restitution claims and high restitution rates (Glock & Keller 2002). Restitution and privatisation claims in rural areas focus, however, predominantly on agricultural land with only a few houses being affected. As said above for Bergholz no former owner returned and only two houses were sold to new owners. With generally low investment rates in peripheral areas and subsequently only limited development the resale of returned property either to owner-occupiers or investors does not pose a threat to established neighbourhoods and social networks.

7.3 Social Differentiation into Winners and Losers

As in all (re-)distributive processes which are based on the quality of evidence, the population in East Germany can be differentiated into winners and losers from restitution policies. Winners would be the claimants whose claims for restitution or compensation were successful; indirect winners would be those people who would transfer their rights as users of a property into owners of the same property. Amongst the losers the unsuccessful claimants dominate and those whose rights as users of a property are now reduced by a new owner.

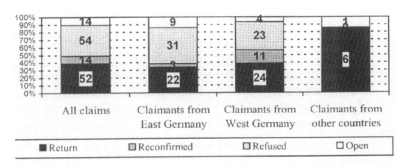

Figure 19-5. Origin of claims and success rates in Ueckermünde (Restitution claims from the Office for the Settlement of Disputed Property Claims Pasewalk 2001).

Whereas for privatisation, no exact and credible data on unfair treatment or even fraud were available, but our investigation in the rural town of Ueckermünde (north of Bergholz) gave some insight into the effect of property restitution. In a detailed investigation of 134 restitution claims in the centre of Ueckermünde it was possible to identify different chances for restitution between the three groups of claimants from East Germany, West Germany and other countries: while only 38 percent of all East German claims were successful, 56 percent of the West German and 86 percent of the claims from other countries led to the return of property or the reconfirmation of property rights (Figure 5).

To explain this pattern it must be stressed that the reason for expropriation and especially the possibility to verify the act of expropriation plays a crucial role in the explanation of the different success rates. Those different success rates stem from the task of the Offices for the Settlement of Disputed Property Rights to identify the claimed piece of land, to verify the expropriation act and to prove the right of the claimant. Most of the unsuccessful claims were denied as the claimant could not prove the expropriation (e.g., the ownership title was lawfully transferred to the community) or his right to claim (e.g., a certificate of inheritance could not be presented).

It is obvious that these differences in the probability to reclaim a property led to a very critical assessment of the principle of property restitution. Even in peripheral areas with fewer restitution claims these differences were perceived and only added to the feeling of being disadvantaged. An indication of a feeling of being disadvantaged might be the remarks of some people in Bergholz who referred to the West German head of the Office for the Settlement of Disputed Property Rights when explaining the success rate of Western claimants.

7.4   Winners/Losers from Restitution in Relation to Winners/Losers from Other Processes

In rural areas these simplistic categories do not fit as the results from privatisation policies interfere with the effects of the whole transformation process. In a society in which the socialist transformation of life was far more advanced than in urban areas privatisation, de-collectivisation and property restitution are seen as one process of unwinding a world and a way of life which many people regarded as better than the present (van Hoven-Iganski 2000). In that context the whole rural population is considered to be losers in the transformation process, as unemployment rose and houses and gardens were under threat from Western claimants. But the winners in the so-called "restitution-lottery" are carefully differentiated into legitimate and less legitimate winners – the legitimacy of a claimant is linked to his affiliation to the community, to the GDR and to those who "fled at night, left their belongings behind them and enjoyed the fine life in the West", as one of the participants in a round table talk put it. Personal success in the West is easily offset against the general hardship of those who stayed.

In addition to the effects of property restitution, other property related processes in rural areas created similar winner-loser-antagonisms that are perceived collectively as a marginalisation of the rural population. A first example refers to the transformation of the collective farms – a process which created winners and losers not only through the redistribution of land and equipment but also through the often criminal activities of managers from the East and the West, e.g., through the deliberate undervaluation of assets and land belonging to those members of the collective farms who wished to leave the collective. Additionally, the Agricultural Restructuring Law made it difficult for members of collective farms to found their own farmstead, as they had to prove broad agricultural knowledge and experience – properties that they might have lost in industrial agriculture. And finally, the still dominating position of the large farms hindered the access of small farmers to the land lease market.

Other winners and losers in rural areas stem from the privatisation of state farms as the initial lease and the subsequent sale of the land advantaged existing large farms and the new farmers amongst the victims of the *Landreform* who enjoy preferential treatment in the sale of their former property (Löhr 2002). However, it is necessary to stress that contrary to the picture in the media the conflict lines between winners and losers in peripheral areas were not exclusively drawn across East-West-dichotomies, but equally across social and professional lines in East Germany.

7.5 Trust in *Rechtsstaat* and Trust in Free Market Economy?

A final dimension of consequences from privatisation policies recurs to central and peripheral areas in the same way – the implementation of laws which deliberately created winners and losers amongst the direct and indirect affected people cannot fulfil one of its original functions as many of the affected people simply lost their newly acknowledged trust in the system of a *Rechtsstaat* or "Rule of Law". Similarly the artificial blocking of the property market and the most hesitating use of the Investment Priority Law was not perceived as a strict application of the elements of a free market economy (Strobl 1994).

8. MAJOR PLAYERS IN THE PROCESS AND THEIR INTERESTS

This analysis of the dynamic of property rights in rural East Germany allows only a first assessment of the social and economic impacts of the relevant institutions. In Table 4 all involved actors are tried to be identified and their tasks and agendas are put in a context with their consequences on the property market, economic development and social cohesion in East Germany.

Table 19-3. Assessment of the social and economic impact of privatisation and reprivatisation institutions in East Germany (*Vermögensgesetz, Treuhandgesetz*; interviews with staff in 2000 and 2001).

Field and Institution	Aim	Consequences
Field: Restitution		
Offices for the Settlement of Disputed Property Rights	▪ Legal clearance	▪ Temporary reduction of the property market ▪ Creation of winners and losers
Agricultural Restructuring (Office for Agriculture)	▪ Agricultural restructuring ▪ Increase of productivity ▪ Implementation of the CAP	▪ Prohibition/aggravation of the re-establishment of farms of the expropriated owners
Field: Privatisation		
Privatisation Agency (*Treuhand*) (Enterprises)	▪ Privatisation	▪ Unemployment
Land Management Corporation (*Bodenverwaltungs- und – verwertungsgesellschaft*)	▪ Priority of sale (until 1995) ▪ Sale and lease	▪ Support of capital-rich agricultural enterprises ▪ Artificial reduction of the production factor "land" through high prices
Building Management Corporation (*Treuhandliegenschaftsgesell schaft*)	▪ Priority of sale (until 1996) ▪ Rent and sale	▪ Support of capital-rich investors through high price policy
Field: Reprivatisation to former owners (SMAD)		
Offices for the Settlement of Disputed Property Rights	▪ Legal clearance	▪ Delay in investment and reestablishment through slow bureaucratic handling
Agricultural Restructuring (Office for Agriculture)	▪ Agricultural restructuring ▪ Increase of productivity ▪ Implementation of the CAP	▪ Preferential treatment for SMAD-Victims
Field: Change of company form		
Agricultural Restructuring (Office for Agriculture)	▪ Agricultural restructuring ▪ Increase of productivity ▪ Implementation of the CAP	▪ Preferential treatment for LPG-Successor enterprises in the form of legal entities

An analysis of the role of the involved institutions reveals their contradictory agendas. While the Offices for the Settlement of Disputed Property Rights try to re-establish the old, mostly small-scale farm structure, the Agricultural Restructuring Offices support large scale farms in order to increase productivity and implement the Common Agricultural Policy of the EU. Similarly, all privatisation agencies suffered from the inherent tension between the task to privatise at the highest price and at the same time to create conditions for the establishment of a functioning competition-based market economy.

8. CONCLUSION

Within the frame of this chapter only some aspects of the spatial, economic and social consequences of the dynamic of property rights in East Germany could be addressed. From this research it did, however, become obvious that the transformation of property rights was as dramatic as the transformation of the political system – with two significant differences. First while the political transformation can be observed in the system of political parties, lobbying organisations etc., the transformation of property rights is far harder to detect – as the agricultural structure of both the shape of the landscape and the enterprises remained identical or similar, private property was not affected as it remained private throughout the GDR-period. A second difference refers to the public debate – contrary to the re-privatisation process in urban areas where many people felt expropriated by the return of the former owners or the arrival of private investors, the transformation process in rural areas did not attract too much attention from the public.

Finally, and in summary, people in rural areas perceive the dynamic of property rights and its consequences as a further element in their own peripherilisation – a process that will, in their own words – only add to the general transformation of rural areas into "homes for the old and the poor".

REFERENCES

Albrecht, G. & Albrecht, W. (1996). Die Entwicklung der Landwirtschaft in Mecklenburg-Vorpommern. Zwischenbemerkung zum Transformationsprozess. In K. Eckart & H. Klüter (Eds.), *Aktuelle sozialökonomische Strukturen, Probleme und Entwicklungsprozesse in Mecklenburg-Vorpommern* (pp. 37-52). Berlin: Duncker und Humblot.

Ballhausen, W. (1994). Die schlimmen Folgen des Rückgabeprinzipes. *Kritische Justiz, 27,* 214-217.

Beckmann, V. (1997). Decollectivisation and Privatisation Policies and Resulting Structural Changes of Agriculture in Eastern Germany. In J. Swinnen, A. Buckwell & E. Mathijs (Eds.), *Agricultural Privatization, Land Reform and Farm Restructuring in Central and Eastern Europe* (pp. 105-160). Aldershot: Ashgate.

Bergmann, T. (1992). The Re-Privatization of Farming in Eastern Germany. *Sociologia Ruralis, 23,* 305-316.

Blacksell, M. & Born, K.M. (2002). Rural Property Restitution in Germany's New Bundesländer: the Case of Bergholz. *Journal of Rural Studies,* 18, 325-338.

Blacksell, M., Born, K.M. & Bohlander, M. (1996). Settlement of Property Claims in the Former East Germany. *Geographical Review,* 86, 198-215.

Bork, H.-R., Dochow, C., Kächele, H., Piorr, H.-P. & Wenkel, K.-O. (1995). *Agrarlandschaftswandel in Nordost-Deutschland unter veränderten Rahmenbedingungen: ökologische und ökonomische Konsequenzen.* Berlin: Ernst und Sohn.

Clasen, R. & John, I. (1996). Der Agrarsektor. Sonderfall der sektoralen Transformation. In H. Wiesenthal (Ed.), *Einheit als Privileg. Vergleichende Perspektiven auf die Transformation Ostdeutschlands* (pp. 188-262). London: Longman.

Czada, R. (1997). Das Prinzip Rückgabe, Die Tragweite des Eigentums. In Deutsches Institut für Fernstudienforschung an der Universität Tübingen, *Deutschland im Umbruch, Studienbrief 4* (pp. 1-40). Tübingen.

Eckart, K. & H.-F. Wollkopf (1994). *Landwirtschaft in Deutschland. Veränderungen der regionalen Agrarstruktur in Deutschland zwischen 1960 und 1992.* Leipzig: Inst. für Länderkunde.

Glock, B. & Keller, C. (2002). Kollektiver Protest ums Eigenheim, individualisierte Konflikte in der Innenstadt. Soziale Folgen der Restitution von Immobilien in Berlin. In C. Hannemann, S. Kabisch & C. Weiske (Eds.), *Neue Länder – Neue Sitten?* (pp. 166-186). Berlin: Schelzky und Jeep.

Hagedorn, K. (1997). The Politics and Policies of Privatisation of Nationalised Land in Eastern Germany. In J.F.M. Swinnen (Ed.), *Political Economy of Agrarian Reform in Central and Eastern Europe* (pp. 197-235). Aldershot: Ashgate.

Löhr, H.C. (2002). *Der Kampf um das Volkseigentum.* Berlin: Duncker und Humblot.

Nölkel, D. (1993). Die Umkehrung des Grundsatzes "Restitution vor Entschädigung" als Instrument zur Förderung von Investitionen in den neuen Bundesländern. *Deutsches Steuerrecht, 31,* 1912-1918.

Reimann, B. (1997). Restitution. Verfahren, Umfang und Folgen des vermögensrechtlichen Grundsatzes für die Stadtentwicklung und Wohnungsversorgung in Ostdeutschland. In U. Schäfer (Ed.), *Städtische Strukturen im Wandel* (pp. 25-37). Opladen: Leske und Budrich.

Reißig, R. (1996). Perspektivenwechsel in der Transformationsforschung: Inhaltliche Umorientierungen, räumliche Erweiterung, theoretische Innovation. In R. Kollmorgen, R. Reißig & J. Weiß (Eds.), *Sozialer Wandel und Akteure in Ostdeutschland* (pp. 245-262). Opladen: Leske und Budrich.

Strobl, B. (1994). *Die Rückgabe von Vermögen in der ehemaligen DDR.* Berlin: Freie Universität.

Swinnen, J., Buckwell, A. & Mathijs, E. (Eds.) (1997). *Agricultural Privatization, Land Reform and Farm Restructuring in Central and Eastern Europe.* Aldershot: Ashgate.

Thöne, K.-F. (1993). *Die agrarstrukturelle Entwicklung in den neuen Bundesländern.* Köln: Verlag Kommunikationsforum.

Van Hoven-Iganski, B. (2000). *Made in the GDR. The changing geographies of Women in the Post-Socialist Rural Society in Mecklenburg-Westpommerania.* Utrecht/Groningen: University of Groningen.

Wiegand, S. (1994). *Landwirtschaft in den neuen Bundesländern. Strukturprobleme und zukünftige Entwicklung.* Kiel: Vauck.

Wiesenthal, H. (1996). Die neuen Bundesländer als Sonderfall der Transformation in den Ländern Ostmitteleuropas. *Aus Politik und Zeitgeschichte. Beilage zur Wochenzeitung "Das Parlament",* 40, 46-54.

Wollmann, H. & Eisen, A. (1995). Transplantation oder Eigenwuchs? In H. Wollmann & J. Meichsner (Eds.), *Transplantation oder Eigenwuchs? Die Transformation der Institutionen in Ostdeutschland. Eine Forschungsdokumentation* (pp. 9-18). Bonn: Informationszentrum Sozialwissenschaften.

Chapter 20

DIFFERENT METHODS FOR THE PROTECTION OF CULTURAL LANDSCAPES

The Example of an Early Industrial Landscape in the Veluwe Region, The Netherlands

Johannes Renes
Faculty of Geosciences, Utrecht University, The Netherlands

1.   INTRODUCTION

In most countries in Europe and Northern America, since the end of the 19th century a growing number of buildings and archaeological objects have been designated as cultural heritage and they are protected as such. Protection of larger areas started also during the end of the 19th century. In this case, the initiative usually came from nature protection quarters. Many of the oldest National Parks are designated mainly for their ecological values. In the course of the 20th century, the interest in cultural landscapes grew. On the one hand, protection of buildings and archaeological sites often proved unsatisfactory when these objects became disparate from their fast changing surrounding landscapes. On the other hand, ecologists realised that most of the landscapes they wanted to protect, were in fact the result of centuries-long human influence. During the second half of the 20th century, protection of cultural landscapes became part of the political agenda in a number of countries. In some countries – most prominently in the United Kingdom – cultural landscapes, including living agrarian landscapes, were designated as national parks. Most of these protected landscapes are situated in regions which are marginal for agriculture, and national park status is often seen as a means to attract tourists and subsidies.

The protection of cultural landscapes brought its own problems and discussions. For archaeological sites, for buildings and, at least to a certain degree, for natural landscapes, the main aim of protection is to preserve the

H. Palang et al. (eds.), European Rural Landscapes:
Persistence and Change in a Globalising Environment, 333-344.

actual state. This aim is also the basis of most Monument Acts and Nature Protection Acts. For living cultural landscapes this is, generally speaking, impossible. These landscapes are continuously under pressure from agrarian and other economic developments. Protection of cultural landscapes is, therefore, a dynamic process for which legal methods such as a Monument Act, are not especially fitting. The legislation and methods of physical planning are better suited to safeguard valuable features within a dynamic landscape.

In the Netherlands, a country whose agricultural systems are the most intensive in Europe, legal protection of cultural landscapes has been extremely problematic. United Nations data on protected landscapes in Europe (Aitchison 1995) show The Netherlands as one of the few countries without protected landscapes. Already in the preparation of the 1962 Monument Act, the possibility of protecting landscape structures, such as roads, field patterns, village types etc., was discussed. However, due to political pressure from landowners, in particular from the farming lobby and their political representatives, this was rejected. The only exception was the possibility of protecting parts of towns and villages comprising ensembles of monuments and their surroundings. In practice, application of such protection was limited. The discussion was duplicated when the Monument Act was revised in 1988, but was again unsuccessful. In the meantime, in 1975, the Ministry of Culture proposed plans to designate a number of agricultural landscapes as *National Landscape Parks*, later renamed *National Landscapes*. These plans could have led to a number of protected landscapes, comparable to the *National Parks* in the United Kingdom. However, the plans were shelved after an intensive and effective lobby by the agricultural sector.

Hence, already during the 1970s landscape protectors turned to the organisations and instruments of physical planning. This seemed an obvious resort as national and regional planning agencies seemed natural allies against the activities of sectoral pressure groups. During this period, the Ministry of Agriculture and the farmers' organisations reached the height of their power and were speeding up the process of complete restructuring of the countryside for agricultural efficiency. It led to a growing public interest in landscape protection, which gave national and regional planning agencies a chance to regain some of their lost influence on the countryside.

Comparison with a number of neighbouring countries shows, that in some of these countries, such as Germany and Switzerland, planning organisations and legislation are the main instruments in protection of cultural landscapes. In many other countries, however, almost all projects for landscape protection focus on a small number of highlights that can be protected by monument or nature conservancy acts (Harteisen et al. 2000). In a third group of countries, such as the United Kingdom, protection by planning is becoming more important in recent years (Grenville 1999).

2.   TWO TYPES OF PROTECTION

As compared to the use of legal protection methods, protection by planning required a new way of thinking. Two diagrams illustrate the differences. The first diagram (Figure 1) shows the process of legal protection. This process starts with an inventory, which is followed by evaluation and selection. This is an almost autonomous process, which mainly involves experts. However, there are three problems:

1. Legal protection is only possible for a limited number of the best examples.
2. Legal protection leaves relatively little room for future development, and is therefore not ideal for cultural landscapes.
3. It is very difficult to apply legal protection to cultural landscapes. The Dutch Monument Act, for example, is designed to protect buildings and archaeological sites, and made no provisions for cultural landscapes (although there seems to be some room for a wider interpretation of the Act).

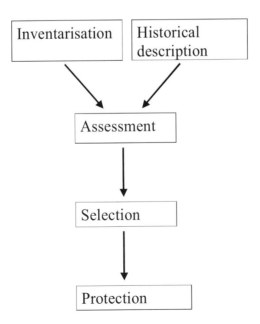

Figure 20-1. Legal protection (Monument Act).

So, for the majority of historical landscape features, the possibilities for protection using the Monument Act are very limited. For these objects and structures, the instruments of physical planning seem to offer better opportunities. Planning never gives the certainty of legal protection, but it is applicable to all objects and, as it leaves more avenues for change, it is better suited for coping with larger structures such as cultural landscapes.

The second diagram (Figure 2) shows that the process of protection by planning is more complex and contextual than legal protection. Instead of belonging to the domain of experts, it is part of a political discussion. There are two main differences between the processes of protection by planning and that of the Monument Act.

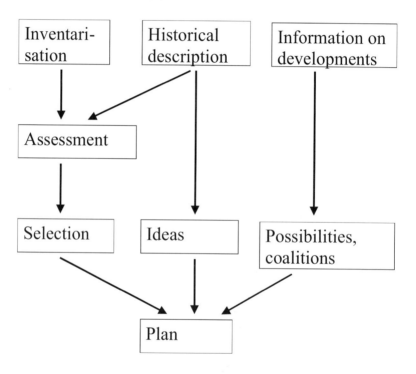

Figure 20-2. Protection by planning.

In the first place, preservation by planning is much more part of a dynamic process. To survive, historic objects need new functions and often new contexts. To preserve, for example, the route of an old mill-leat in an urban context, the watercourse needs a new function such as nature reserve or canoe route. In such situations good ideas are indispensable.

In the second place, it is necessary to obtain insight in external pressures. In a peripheral and uneventful environment, protection can be relatively easy. In contrast, in an environment which is undergoing fast change and which is a recipient of much investment, many historic objects are in danger of destruction. One example is when a rural area becomes part of an urban environment. This situation can also provide opportunities, as funds can become available for restoration. So, in this case protection is not only based on values, but also on possibilities and on creativity. That also means that a larger number of people and organisations are involved. Preservation by planning is dependent on coalitions with potential partners.

Figure 20-3. Location of Veluwe in Holland.

An illustration of the above discussion can be seen in a recent research project on the historical development, present relics and values and future development of an early modern industrial landscape in the Veluwe region.

3. THE PAPERMAKING LANDSCAPE OF THE VELUWE

The Veluwe, a moraine landscape in the Central Netherlands rising up to more than 100 metres above sea level (Figure 3), is nowadays known as the country's most extensive forest. However, it is a region with a complex and multi-layered cultural landscape. Large numbers of tumuli point to a history of occupation that started during the Neolithic. During the Early Middle Ages the central parts of the moraine landscape were the scene of iron production, whereas on the edges agrarian settlements were found. A few centuries later, most of the central region was used as a hunting reserve. During the late Middle Ages, the expanse of heathlands, which had probably existed since prehistoric times, must have grown due to intensive sheep rearing. The number of sheep reached an all-time high during the 16th century, following the growing textile industry in Holland. From the 16th century onwards, on the edges of the moraine landscape paper-mills were built in large numbers, laying the basis for the still existing paper production. During the 19th century the edge of the Veluwe became a popular region for settlement. The owners of the growing number of landed estates bought the large heathlands and started to plant forests, mainly conifers. At present, the central part of the Veluwe is mainly forested; the few remaining heathlands are protected. The main function is recreation. Some of the old settlements around this central area are still agrarian. Many other villages grew into large (sub) urban settlements.

The research project concentrated on the remains of the industrial phase, in particular the paper-mills and the complex system of artificial watercourses that was built to power those mills. The aims of the research project were threefold:

1. To describe the origins and the present relics of this landscape;
2. To evaluate the physical remains;
3. To present ideas for the future of this landscape.

The initiative for the project came from a regional action committee, but funding was from a combination of local and regional governments, regional water boards and large landowners. A small team of three executed the project: a historical geographer did the historical research and edited the report, a field-worker mapped the relics and a landscape architect designed plans for the future. The results were delivered in 2002 (Renes et al. 2002) and a book for the general public is planned for 2004.

3.1   Inventory and Historical Description

The origins of this industrial landscape were rather modest. During the Middle Ages, most of the agrarian villages, which were located in a ring around the moraine landscape, possessed a water-mill, driven by one of the small streams that drained the hills. Around 1500 AD, there were about 20 water-mills, mainly corn-mills in the region.

Somewhere during the 16th century, it was discovered that the moraines contained huge quantities of water that could be tapped to supply the streams. The moraines, ice-pushed ridges dating from the so-called Saalien Ice Age (100-150,000 BC), were built up from older sediments and therefore consist mainly of sand, clay and gravel. The groundwater level follows the shape of the hill. This means that, by digging an almost horizontal ditch into the hill, one will usually find groundwater. After leaving the hill, the water could be kept fairly horizontal until the difference in height is enough to power an overshot water-wheel.

The oldest group of such artificial wells was probably dug in 1517 to power three new water-mills near the village of Apeldoorn. Soon it became clear that this almost inexhaustible water supply could support a large number of new mills. The number grew to some 180 in the middle of the 18th century. Of these 180 mills, 150 were paper-mills; the remaining were corn-mills, a few copper-mills and some other industrial mills (Voorn et al. 1985; Hagens 2000).

Protestant refugees from the southern part of the Low Countries introduced papermaking. In 1591 one of them founded the first paper-mill in the Veluwe region. In a few decades, this developed into a thriving industry. There are some good examples of water systems that were developed on a large scale, and during a short period. Near the village of Loenen, the number of mills rose from two in 1652 to 11 in 1668, made possible by a number of new watercourses. Elsewhere, however, the development of water systems was slow and gradual.

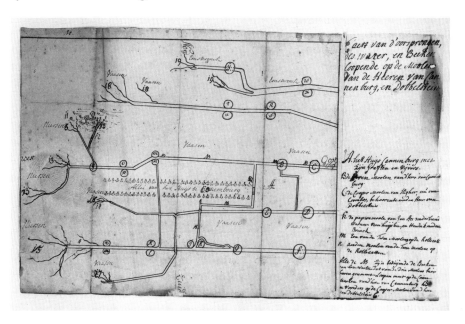

Figure 20-4. An 18th-century map of the mills and watercourses near Vaassen (State Archives Gelderland, Arnhem, Algemene Kaartenverzameling no 1745).

The largest systems are highly complex. An example is shown in Figure 4. This 18th century map shows the system near the village of Vaassen, where the local landlord invested heavily in the water-mills. In the end, this system consisted of four main watercourses, with complex arrangements to divide the water among 20 water-mills. During this period the landlords, who had sold most of the mills earlier, while retaining the water rights, started to buy the mills. In four decades, they managed to become the owners of almost all the water-mills within their territory, thereby showing their determination to maximise their profits from the mills in a period of agrarian crisis (Renes 2002). To this day, this system is almost completely intact.

During the 18th century, papermaking in the Netherlands was concentrated in two regions: the Veluwe and the region along River Zaan (near Amsterdam). In the Zaan region, papermaking was based on windmill technology. The windmills were more powerful, and the companies operated on a larger scale. Compared to this region, the Veluwe had the advantage of large amounts of water in constant quantities and of an extremely high quality. Another advantage was the level of wages, which was substantially lower than in the vicinity of Amsterdam. Both regions supplemented each other.

Table 20-1. Historical development of water-mills and artificial watercourses.

	Water-mills	Artificial watercourses
< 1600	Small number	First experiments (1517?)
1600 – 1740	Very strong growth	Very strong growth
1740 – 1850	Stagnation	Small growth. Additional users: gardens.
1850 – 1950	Decline	Continuing use for industrial processes. New users: laundries, paper factories.
1950 – 1980	Further decline	Replaced by waterworks and sewerage. Neglect.
Since 1980	Restoration of surviving mills and mill-sites	Renewed interest; maintenance; restoration.
Present situation	Few surviving mills; many mill-sites recognisable	System more or less completely preserved (exception: urban areas). Reconstruction of some lost systems is in progress

Table 1 summarises the development of the industrial landscape of the Veluwe region. Three periods can be distinguished: growth until 1740, then stagnation and slow decline until the middle of the 19th century, and a fast decline after 1850. Most of the artificial watercourses must have been constructed during the growth period, although some systems are known to have been developed later. The stagnation in the number of mills during the second half of the 18th century was due to a number of reasons, the most important being foreign competition and the general stagnation of the Dutch economy. In addition, there were growing problems in obtaining rags (old cloths) which were the basic raw material. The stagnating number of mills masks some shifts in the location of the industry. Especially in the southern part, almost all viable sites were occupied early. In the second half of the 18th century, the number of mills on the southern slope diminished, mainly as a result of competing land-uses. At the same time, the industry still grew on the eastern slope of the Veluwe.

Juxtaposed against a gradual decline of the number of water-mills is a survival and even some growth in the number of artificial watercourses as some of these were reused to power fountains and other waterworks in gardens. This different development continued during the 19th and 20th centuries, when most water-mills were closed. The best mills developed into industrial, steam-driven paper-mills; many others were converted into laundries: a growing industry from the 1860s onwards, which could easily use paper-mills after minor adaptations. After some decades, most of these laundries switched to fossil fuels. However, all these industries needed process-water. Therefore, the watercourses survived the mills. Around 1870 a number of new watercourses were even constructed to supply water for a new shipping canal.

It was only during the 1950s and 1960s that most of the watercourses lost their function, in view of many companies which closed down, whereas the surviving companies introduced waterworks and sewerage works. For the watercourses this led to a period of neglect, but already during the end of the 1970s local citizens became interested in the historical and ecological values of the watercourses. They started a campaign for preservation and formed groups of volunteers to carry out maintenance. Also they succeeded in convincing authorities of the values of this system, and during the 1990s the water boards took over the maintenance. In their management plans, the new designation of "historical water" was invented.

Whereas most of the water systems are still intact, only a small number of mills have survived. However, our inventory showed that many mill sites are still recognisable. It is a heritage worth preserving. As a landscape of industrial water-mills the Veluwe is unique for the Netherlands, but is less rare in an international perspective: large concentrations of water-mills can be found elsewhere in Europe, and the remains of large-scale paper-making during the Early Modern period can be found elsewhere too (for example in northern Italy and in north-eastern France). The system of artificial watercourses, however, seems even unique from an international perspective. For systems of artificial watercourses of this type to develop, two things are necessary. In the first place, the geological and geomorphological situation must be favourable. Sandy hills of this size are mainly the result of glaciation, and such hills are found in an east-west band from the Netherlands eastward. In the second place, there must be sufficient demand to enable the development of an industrial landscape based on water power. The combination of these two factors existed only in the Netherlands in the 17th and 18th centuries. So the value of this landscape should be rated as very high.

On a local level, an assessment of the individual water systems was required as an integral element/feature in planning. Therefore, the ca 60 individual systems were compared according to three criteria:

– Wholeness (Is the system recognisable and in a good state of maintenance?);
– Completeness / diversity (Are the main components of the system all there?);
– Size and complexity (A number of very complex systems were ranked higher).

From this assessment it was clear that quite a few systems are still more or less intact.

3.2   The Future of the Industrial Landscape

This landscape heritage poses serious problems for traditional protection methods. In fact only a few surviving mills are protected by the Monument Act. At the moment, efforts are being made to bring the most valuable complex – the system that is shown on the above-mentioned 18th century map – under the Monument Act (which is rarely used for cultural landscapes) or under the new Nature Conservation Act of 1998. The latter introduced the category of *protected scenery*, which however is not implemented yet. The outcome of both procedures is still uncertain.

Table 20-2. Possible coalitions for preserving the historic landscape of watercourses and water-mills (italics: potential partner in preserving historic watercourses).

Potential partner	Past position	Present position
Urban planning	No interest	*Interest in historical values and local identity*
Water Board	Loss of valuable groundwater	*Preservation of historic values*
Nature conservationists	*Special flora and fauna*	*Special flora and fauna* Development of more natural ecosystems
Recreation	Facilities for large-scale tourism	*Growing interest in small-scale facilities (footpaths)* *Growing interest in cultural tourism*
Agriculture	Agrarian production	*Rural development*

The strategy of protection by planning seems to work much better. In this case, it is essential to build coalitions with possible partners. In this respect, the situation has changed for the good during the last few years. Table 2 shows how many of the potential partners have altered their position during recent years. The growing interest in historic landscapes has been one of the main factors.

The support from nature conservation, which was for a long time the main partner of landscape protection interests, is now less evident. Although the artificial streams with their extremely high water quality are the main habitat for a few rare plant and animal species, many ecologists nowadays have other priorities. Some would like to fill all man-made watercourses in order to save the groundwater in the moraines and some even designed plans to change artificial mill-leats into winding, pseudo-natural streams.

On the other hand, most of the old adversaries are now becoming potential partners, following the growing public awareness of, and interest in, landscape heritage. For a long time, in particular the dynamics of urban sprawl and agrarian restructuring have been a threat to the historic landscape. The agrarian sector in this region is now slowly starting to realise that its future lies no longer in intensification, but in widening its scope to include landscape management and small-scale recreation. In both towns in the region (Arnhem and Apeldoorn), the majority of relics have disappeared. Where important relics were preserved, it was in many cases through oversight, in places where there was no competition from alternative land-use types. In recent years, however, this situation changed. Now, the two towns in the region show an active policy in preservation and even reconstruction. In some cases the reasons for this are very practical. A number of watercourses had stopped functioning in the past when industries and waterworks extracted large quantities of water from the hills. As the pumping of drinking water from the region has diminished during recent years, groundwater levels are rising. In one case, this led to flooding of houses and a hospital – reason enough to reconstruct the lost watercourse. But reconstructions now go further than necessary. The town of Apeldoorn has chosen a new local identity as the "town of water-mills and man-made streams" and is at present developing plans to reconstruct the complete system of watercourses.

In this situation, our research can, and probably will, be used in a number of ways. Firstly, it identifies the most valuable systems, which are likely to qualify for a protected status. Secondly, it shows how the system originated, developed and functioned, and thereby gives ideas for new developments. For example, we identified salient features such as the industrial origins and low-profile character. Therefore, we urged the municipalities to give restored or reconstructed watercourses a rather informal design (as opposed to some boulevard-like designs). Lastly, it brings a relatively unknown, but very interesting and unique system to the attention of administrators as well as the general public, and by doing that we enlarge the chances of survival of this unique historic landscape.

4.   CONCLUSION

Protection of cultural landscapes differs from the protection of buildings or archaeological sites. In the case of buildings or archaeological sites, the aim of protection is usually the preservation of the existing situation. For living cultural landscapes, this aim is unrealistic. In this case, protection is a much more complex process, for which the instruments of planning are better suited than legal protection. Protection by planning is essentially a contextual process, in which coalitions between different interests are crucial. This is illustrated by a project to find a new future for an early modern industrial landscape in The Netherlands.

REFERENCES

Aitchison, J. (1995). Cultural Landscapes in Europe: a Geographical Perspective. In B. von Droste, H. Plachter & M. Rössler (Eds.), *Cultural Landscapes of Universal Value* (pp. 272-288). Jena/Stuttgart/New York: Fischer.

Grenville, J. (1999). *Managing the Historic Rural Landscape*. London/New York: Routledge.

Hagens, H. (2000). *Op kracht van stromend water; negen eeuwen watermolens op de Veluwe.* Hengelo: Smit.

Harteisen, U., A. Schmidt & M. Wulf (Eds.) (2000). *Kulturlandschaftsforschung und Umweltplanung. Fachtagung an der Fachhochschule Hildesheim/Holzminden/Göttingen am 9.-10. November 2000 in Göttingen.* Herdecke: CGA.

Renes, J. (2002). Beken in kaart; een vroege topologische kaart van de sprengenbeken bij Vaassen. *Historisch-Geografisch Tijdschrift*, 20, 120-124.

Renes, J., Meijer, J. & de Poel K.R. (2002). *Het Veluwse sprengenlandschap; een cultuurmonument.* Wageningen: Alterra.

Voorn, H., Hollestelle, J. & Cornelissen de Beer, G.D. (1985). De papiermolens in de provincie Gelderland, alsmede in Overijssel en Limburg. *Ver. van Ned. Papier – en Kartonfabrieken. De geschiedenis der Nederlandse papierindustrie 3*. Haarlem.

Chapter 21

THE SIGNIFICANCE OF THE DUTCH HISTORICAL GIS *HISTLAND*

The Example of the Mediaeval Peat Landscapes of Staphorst-Rouveen and Vriezenveen

Chris de Bont
Alterra, Wageningen University and Research Centre, The Netherlands

1. INTRODUCTION

In the Netherlands, there is quite a long tradition of applied historical geography (Vervloet 1994), in which attempts are always made to find a balance between what is at times quick and dirty contract research on the one hand, and a more scientific approach on the other. The outcomes of investigations are reported in rather substantial quantities of text, which are always accompanied by detailed maps. Text and maps together give an idea of the reclamation and habitation history of the Netherlands, which, in the case of applied historical geography, is linked to relics found in the present-day landscape. Until recently, it was not always easy to transform 2000 years of Dutch occupation and habitation history into old-fashioned, hand-drawn maps. With the introduction of a multi-layered GIS system, this problem seems to be solved. The main question is to what extent GIS – and the Dutch historical GIS *Histland* in particular – is able to do justice to the complex and highly dynamic habitation history of many Dutch regions.

In the first part of this article, I will briefly introduce the Dutch historical geographic GIS *Histland*, and discuss some of the aspects of historical geographic research methods; which are of importance in regards to the translation of historical geographic information into our GIS *Histland*. I will describe the way information concerning mediaeval peat reclamation areas in the Netherlands is incorporated into this particular GIS. Subsequently, that

H. Palang et al. (eds.), European Rural Landscapes:
Persistence and Change in a Globalising Environment, 345-358.
© 2004 *Kluwer Academic Publishers. Printed in the Netherlands.*

matter will be explored further by discussing the historical geography of two peculiar peat reclamation areas. Afterwards, the way their reclamation and habitation history is translated into the GIS *Histland* will be looked at, and the parts of their history not "covered" by this GIS system will be examined. I will end with some conclusions.

2.   THE HISTORICAL GEOGRAPHIC GIS *HISTLAND*

A historical geographer can deal with his subject in different ways. Every historical man-made element or structure in the landscape has at least four aspects, namely: location, function, time of origin, and dynamics through time. Some of these relics survived until today: others did not make it and simply faded away. If we want to investigate and describe the genesis of historical elements and structures, we can look back into the past (the so-called "retrospective" method), we can give a more chronological overview, or we can cut the course of history into slices (the so-called "retrogressive" method). It was always problematic to translate this multi-spatio-temporal information into two-dimensional maps. The multi-layered historical geographic GIS *Histland* lays claim to a more modern way of mapping and analysing our data (De Bont et al. 2004).

GIS *Histland* provides – on different scales – an indication of the history of topographical structures in the present-day landscape, and of changes in the parcellation structure between 1850 and 2002 (De Bont 1998). Besides a layer containing relics in the present-day landscape, there are several functional layers. The basic layer gives information regarding the old reclamation structure; as far as it is relevant for analysing and explaining the present-day cultural landscape. A second layer provides information on landscape dynamics between 1850 and 2002. A third layer – one that furnishes an indication of landscape dynamics in the entire Dutch cultural landscape from the Late Middle Ages until 1850 – is still under construction.

The historical geographic basic layer consists of three levels. In *Histland* 11, the Netherlands is subdivided into 11 historical geographic types of landscape, in which the reclamation and habitation history along with the accompanying topographical structures and elements in the landscape form the main subject matter. *Histland* 50 breaks up a subdivision further into 50 sub-types, incorporating a more-detailed view of elements and structures. *Histland* 500 is still under construction, although this approach as tested on one of the larger provinces of the Netherlands (Gelderland) seems to work fairly well (De Bont & Neefjes 2004). A fourth level, which is also still under construction, will give information as to the historical function and temporal aspects of individual relics and relic-ensembles (Plöger 2003).

2.1 Mediaeval Peat Reclamations in the Netherlands and GIS *Histland*

For the western and northern lower parts of the Netherlands, the so-called "mediaeval peat reclamation areas" are of great importance. During one millennium, from the reclamation period onwards, these cultural landscapes developed in a highly dynamic way (De Bont 1994). In finding out in what way GIS *Histland* covers this complex dynamic mixture of reclamation history and historical water management, a rather simplified model of this process has encountered the opportunities GIS *Histland* offers.

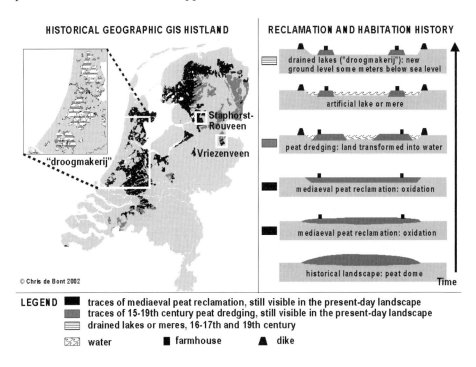

Figure 21-1. 1000 years of reclamation history of Dutch mediaeval peat reclamations (De Bont 2001), in connection with GIS *Histland*.

This model shows the time-bound forms in the landscape that are connected to specific reclamation phases associated with one thousand years of reclamation and habitation history, and which derive from the reclamation techniques and all different forms of water management related to reclamation activities (De Bont 2001). For the sake of argument, I mainly deal with the reclamation of the oligotrophic peat areas. Figure 1 shows these different phases (right column) in relation to the way GIS *Histland* "maps" these features (left column), as far as topographical relics are still to be recognised in the present-day landscape.

The oldest Dutch mediaeval peat reclamations date back as far as the 9th and 10th centuries. Most of the oligotrophic peat areas were reclaimed for agrarian use (mixed farming) by the end of the 14th century. From the 16th century onwards, the oligotrophic peat "domes" had been dredged for fuel purposes. These dredging activities led to a terrible land loss – large areas of arable and pasture land changed into water. From the 17th century onwards, many of those newly-formed lakes were pumped dry (in Dutch *droogmakerij*) and reclaimed. The present-day Dutch landscape still contains many traces of these events, as the left column in Figure 1 demonstrates.

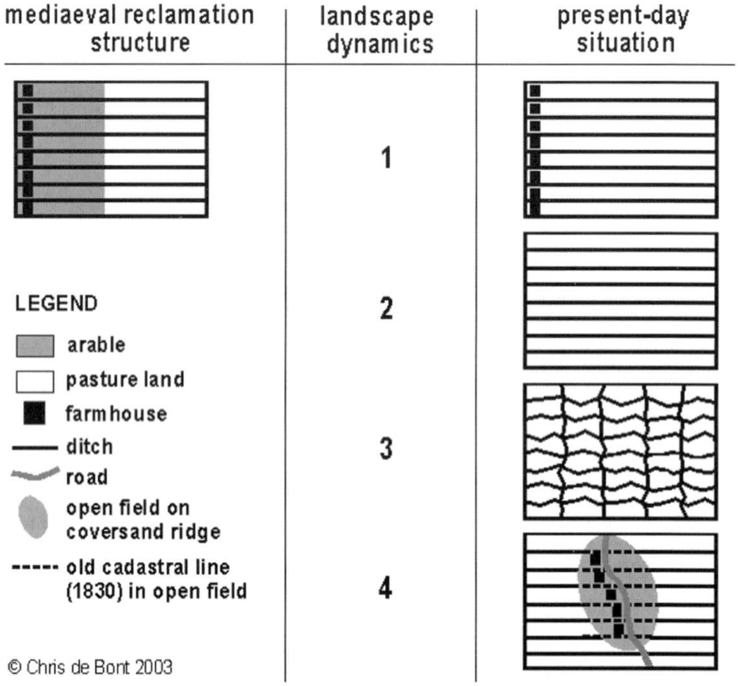

Figure 21-2. Changing topographical structures in the Dutch mediaeval peat reclamation areas (De Bont 1996).

Not all areas of peat reclamation were turned into lakes and reclaimed afterwards. Nevertheless, the topographical structure of those mediaeval peat reclamation areas which were not dug out could change dramatically. These forms of landscape dynamics are incorporated in the Dutch historical GIS *Histland*. Figure 2 provides an idea as to the more specific changes the original mediaeval topographical structure went through, and of the physical forces which initiated those changes. Only four different types of landscape dynamics will be considered here, although more variation did occur (De Bont 1996).

The initial situations in regards to the different peat reclamations were more or less identical. Mediaeval reclamation consisted of a very regular structure of ditches dug in parallel fashion, with the farmhouses situated along a linear reclamation base. Newly-reclaimed peat land was very suitable for mixed farming (Figure 2, left column). Several types of landscape dynamics could occur (Figure 2, right column):

1. In connection with the peat oxidation and setting due to the reclamation activity itself, mixed farming evolved into permanent cattle breeding. Farmers could still live in the area, and the parcellation structure did not change much over the centuries.

2. Some of these changed areas went through another alteration phase. Because of the lowering of the surface of the land due to oxidation and setting, younger marine- or river-clay sediments could be deposited over the reclamation area. For security reasons, farmers had to leave the area to look for higher and drier settlement conditions as nearby as possible. Still, the old parcellation structure remained practically unchanged.

3. After years of oxidation, some of the already-abandoned peat-covered areas appeared to have an older clay soil with some relief under an increasingly thinner peat layer. When this relief came to the surface, the farmers – who were still using the land for cattle breeding (more and more extensively) – had to adjust the old parcellation structure, which led to a more-irregular kind of grid pattern.

4. In some cases during the first centuries of the reclamation, a peat layer still covered the relief of underlying sandy ridges ("cover sands"). Due to oxidation, this sandy relief appeared at the surface. Farmers moved their farmhouses to these sandy ridges and started again with mixed farming. The new arable land was situated on the sandy ridges. It was made fertile with newly-raised layers of plaggen soils.

In large areas in the Netherlands, the present-day landscape still resembles the old mediaeval reclamation structure, and mirrors the mediaeval mind of its creators (De Bont 1992). The GIS *Histland* shows all these different forms of landscape dynamics on a quite-detailed scale; as Figure 3 indicates for an area called the Gueldern Valley, where peat had grown between two ice-pushed moraines. Three out of four stages described above (numbers two to four) are situated rather close to each other, along the coast of the former Zuyder Sea.

Figure 21-3. Different topographical structures (see Figure 2: 2-4) quite close to each other in Gueldern Valley, on a mid-19th century map.

The historical geography of the Dutch mediaeval peat reclamation areas are indicated more implicitly in the Dutch historical GIS *Histland*. Yet historical geography is more than just the specific history of changes in the topographical structure. The remarkable historical geography of two mediaeval peat reclamation areas (Staphorst-Rouveen and Vriezenveen) in the eastern part of the Netherlands (see Figure 1) demonstrates the richness of regional history on the one hand, and to what extent this is embedded into GIS *Histland* on the other. What is missing in GIS *Histland*? And of even more importance, what does it matter?

2.2 Staphorst-Rouveen and Vriezenveen

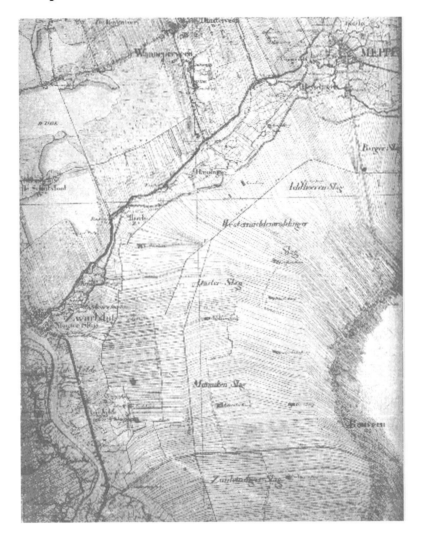

Figure 21-4. The Staphorst-Rouveen area on a mid-19th century map.

The first six centuries of reclamation and habitation history of the rural villages of Staphorst-Rouveen and Vriezenveen are more or less identical. Even on the old topographical map of 1850 (Figure 4), the late-mediaeval peat reclamation area of Staphorst-Rouveen shows a rather systematic curved parcellation structure, which is typical for this type of mediaeval peat reclamation (Vervloet & Bording 1985).

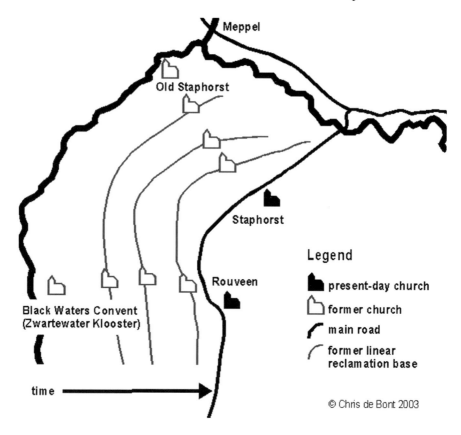

Figure 21-5. Changing occupation structure in the Staphorst-Rouveen area.

On the east side of a bent river system, a large, partly oligotrophic peat area had developed. The oldest farmhouses were situated on small sandy dunes along the river (Figure 5). The two most important settlements were Old Staphorst and Zwartewater Klooster, which was later called Rouveen (in Dutch *rouveen* means rough bogs), both with small mediaeval (parish) churches. In the 12th century, the first reclaimers went from these dunes into the rough bogs, drained them using the well-known Dutch systematic ditch structure, and moved their farms to a new linear reclamation base. Quickly after, both groups of reclaimers built new parish churches there. During the centuries the linear occupation was moved eastwards three times, and each time a new church was built. The churches followed to the moved linear villages within the same lengthened parcel (Figure 6), which started just behind the oldest churches at Old Staphorst, or the Black Waters Convent. During the 17th century, the villages of Staphorst and Rouveen reached their present-day location. Nowadays, the Staphorst-Rouveen area is one of the touristic highlights of the Netherlands; more than 12 km of beautiful old farmhouses in a row. Due to a lack of space within the very long but rather narrow parcels, several generations of farmhouses had to be built one behind the other.

Figure 21-6. The old church of Rouveen within its "own" parcel in an early 20th century photograph. Photo: by kind permission of Stichting Ken Uw. Dorp. Vriezenveen.

Until quite recently, most of the inhabitants of Staphorst-Rouveen were farmers. From the 1960s onward, after some land consolidation and reallocation took place, local economies changed remarkably; although in the renewed parcellation structure, the mediaeval origin is still visible (De Wolde 1980).

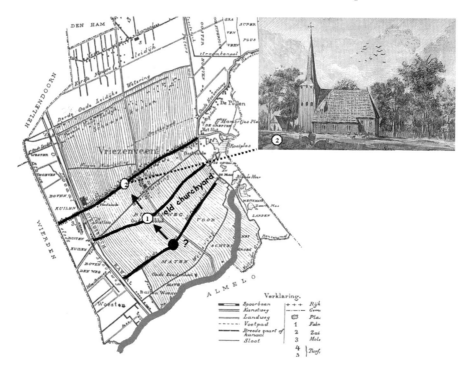

Figure 21-7. Changing occupation structure in the Vriezenveen area.

The historical geography – and more specifically the old parcellation structure of Vriezenveen – resembles to a great extent the reclamation and habitation history of Staphorst-Rouveen. Here also, the whole area was covered with peat. The first reclaimers, who in 1364 were called "free farmers from Friesland" (Entjes 1979), built their first farmhouses at a base line along a natural river course (Figure 7). Because of old local property rights (Dunbar 1859), it is not likely that the first reclaimers built a new church at this first linear reclamation base. After a while they moved northwards into the still-rough bogs. In the middle of the first linear reclamation base and the present-day village along the secondary base line, a field named "old churchyard" indicates the existence of a church there. Halfway through the 17th century, a last leap to the north brought the Vriezenveen farmers to the present-day location of Vriezenveen village, where they built a new wooden church. Here as well the different churches were built in one lengthened parcel. Before a great fire at the beginning of the 20th century, the Vriezenveen village structure almost completely resembled the Staphorst-Rouveen situation; i.e. old farmhouses situated along a very long street, and several generations of farmhouses, one behind the other (Figure 8).

Figure 21-8. Old farmhouses along the youngest linear occupation base in an early 20th century photograph. Photo: by kind permission of Stichting Ken Uw. Dorp. Vriezenveen.

From the 17th century onward, local conditions in both villages started to differ. In the village of Vriezenveen, agricultural conditions had deteriorated rather rapidly, while in the Staphorst-Rouveen area everything went on as usual. In Vriezenveen the more or less fertile peat layer was oxidised almost completely, leaving a sandy soil for the farmers to use in earning a poor living. This harrowing situation led to a quite-unexpected development, which can only be understood in the broader context of North European history. Already in prehistoric times, there were trading routes from the west through the Baltics towards Russia, and even further into the east and south. During the Late Middle Ages, the Hanze merchants appeared, succeeded by Amsterdam merchants during the 16th century who traded with East Prussia and the Baltic States (Van Winter 1983). From the early 18th century onward, former Vriezenveen farmers opened a trade route over land to St. Petersburg and Moscow, almost 2,500 km away from Vriezenveen (Hosmar 1976; Meeuwse 1996). In Dutch they were called *Ruslui*, or Russian folks. They used rather plump *Ruslui*-wagons for their linen trade. In the beginning it was just seasonal trade, but later on the Vriezenveen merchant families stayed in St. Petersburg and Moscow on a more permanent basis, in factories or trading posts.

Figure 21-9. Old merchant house in Vriezenveen in an early 20th century photograph. Photo: by kind permission of Stichting Ken Uw. Dorp. Vriezenveen.

The *Ruslui* still stayed connected with their Vriezenveen hinterland, where they returned for retirement. In their hometown, the wealthy merchant families built beautiful houses. These merchant houses fitted remarkably well into the old agricultural village structure (Figure 9). Trading with Russia was quite lucrative and brought this small community much prosperity for more than three centuries. In the 19th century, Vriezenveen acted as a kind of bank for the whole region. Everything changed after the Russian Revolution in 1917. The linen trade came to an end. Mighty mercantile houses became bankrupt. The village lost its international status and fell back to a local agricultural economy. Just after World War II, land consolidation improved the agricultural situation. The mediaeval parcellation structure was wiped out completely and replaced by a new one, which reflected the new European approach of large-scale agriculture.

2.3   Local History in Relation to GIS *Histland*

From the whole range of historical information in relation to the late Middle Ages until the beginning of the third millennium, GIS *Histland* contains data on the mediaeval peat reclamation – with its topographical and typical village structure on the one hand, and the topographical changes which appeared in the late 20th century on the other. These latest data are nicely incorporated into a GIS *Histland* layer, in which the landscape dynamics between 1850 and 2000 are shown. The more specific information concerns the way the village structure on a micro-scale was related to a typical parcellation structure, which consisted of very small parcels and farmhouses placed within this structure one after another (Staphorst-Rouveen and Vriezenveen). However, the way many old farmhouses were replaced by merchant houses (which in Vriezenveen were incorporated into an originally-agricultural structure), and the unique form of international trading itself, are no part of GIS *Histland* data sets.

3. CONCLUSIONS

The Dutch historical-geographic GIS *Histland* is a powerful tool for understanding the reclamation and habitation history of the Netherlands. In combination with an inventory of old relics in the present-day landscape, it enables a person to on the one hand understand the rather specific regional landscape histories on quite a detailed scale, and on the other hand it gives the landscape planner material for making plans with regards to the history of such landscapes. The rather complicated historical geography of many Dutch regions had to be modeled and simplified in terms of parcellation structure and topographical elements before it could be "stored" into GIS *Histland*. Due to this necessary simplification, parts of the intriguing deeper and subtle features of the regional histories can become lost completely and lead to an impoverishment in the knowledge of a region's history. Simply believing that GIS *Histland* will provide all the relevant answers will mean the end of applied and scientifically-solid historical geography. Analysing old hand-drawn maps always did stimulate the minds of many; in particular historical geographers, since they knew how to read the maps. Experience shows that this is not always the case with planners who are working only with digital maps derived from GIS, and do not have map-reading ability and experience.

The GIS data layers give no answers to intriguing but fundamental questions, such as: To what extent was there a kind of physical determination involved here? Is it beyond science to state that in the case of Vriezenveen, the decline in the soil fertility of an oxidating peat layer on a sandy sub-soil caused (more or less directly) the development of the Russian trade? This answer is not to be found in a geographic information system, but rather in the archives and in discussions amongst historical geographers.

REFERENCES

De Bont, C. (1994). Reclamation Patterns of Peat Areas in the Netherlands as a Mirror of the Mediaeval Mind. In J. Bethemont (Ed.), *L'avenir des paysages ruraux européens entre gestion des héritages et dynamique du changement* (pp. 57-64). Conférence européenne permanente pour l'étude du paysage rural; colloque de Lyon, 9-13 juin 1992. Université Lumière Lyon 2/Université Jean Moulin Lyon 3, Lyon.

De Bont, C. (1996). *Gelre, oud land, nieuwe tekens; beschrijving, waardering en planologische doorwerking van historisch-landschappelijke regionale karakteristieken in de provincie Gelderland.* Repport 421. DLO-Staring Centrum, Wageningen.

De Bont, C. (1998). The Cyclone's Eye: Historical Dynamics and Valuation in the Man Made Landscapes of the Netherlands. In P. Sereno & M.L. Sturani (Eds.), *Rural Landscape between State and Local Communities in Europe: Past and Present* (pp. 203-213). Proceedings of the 1994 meeting of the Permanent European Conference for the study of the Rural Landscape at Torino. Torino: Editzioni dell'Orso.

De Bont, C. (2001). *Delft's Water: Two Thousand Years of Habitation and Water Management in and Around Delft.* Zutphen: Walburg Press.

De Bont, C. & J. Neefjes (2004). *Gelre revisited; historisch-geografische waarden- en kenmerkenkaart van de provincie Gelderland.* Wageningen: Alterra.

De Bont, C., Vervloet, J.A.J. & Dirkx, G.H.P. (2004). *Histlandboek; beknopte handleiding bij het historisch-geografisch GIS Histland,* versie 2.0. Wageningen: Alterra.

De Wolde, J. (1980). *Ontginningen en verkavelingen in de gemeente Staphorst.* Staphorst: Gemeente Staphorst.

Dunbar, G. (1859). *Verhandeling over het graafschap Goor en beschrijving van de heerlijkheid Almelo en Vriezenveen; twee onuitgegeven hoofdstukken van den tegenwoordigen staat van Overijssel. Vereeniging tot beoefening van Overrijsselsch Regt en Geschiedenis.* Zwolle: De Lange Deventer.

Entjes, H. (1979). *Over het ontstaan van Vriezenveen Jaar.* Groningse herdrukken 12bis. Groningen: Satbo.

Hosmar, J. (1976). *Vriezenveense Ruslui in het rijk der tsaren.* Witkam: Enschede.

Meeuwse, K. (1986). *Opkomst en ondergang van de Ruslui.* Utrecht: Bruna.

Plöger, R. (2003). *Inventarisation der Kulturlandschaft mit Hilfe von Geographischen Informationssystemen (GIS). Methodische Untersuchungen für ein Kulturlandschaftskataster.* Doctoral Dissertation. Rheinischen Friedrich-Whilhelms-Universität. Bonn.

Van Winter, J.M. (Ed.) (1983). *The Interaction of Amsterdam and Antwerp with the Baltic Region, 1400-1800.* Papers presented at the Third International Conference of the "Association Internationale de l'Histoire des Mers Nordiques de l'Europe", Utrecht, August 30-September 3 1982. Leiden.

Vervloet, J.A.J. & Bording, J. (1985). *Cultuurhistorisch onderzoek landinrichting Staphorst-Rouveen. Stichting voor Bodemkartering report 1679.* Wageningen.

Vervloet, J.A.J. (1994). *Zum Stand der Angewandten Historischen Geographie in den Niederlanden. Berichte zur deutschen Landeskunde,* 68, 2, 445-458.

Chapter 22

THE FUTURE ROLE OF AGRICULTURE IN RURAL COMMUNITIES
Case Studies in Estonia and Lithuania

Mette Bech Sørensen
Institute of Geography, University of Copenhagen, Denmark

1.   INTRODUCTION

The political and economic transformation that started in the early 1990s has had an important impact on the rural landscapes of Central and Eastern Europe. As a consequence of the radical changes that occurred in the agricultural sector – with the introduction of a market economy, a new ownership structure, and a reorientation of agricultural policy – much land has been taken out of production. Estonian studies (Peterson & Aunap 1998; Mander et al. 2000; Reiljan & Kulu 2002) reveal significant increases in the amount of abandoned land, and decreases of approximately 25 percent in the quantity of arable land. In Lithuania, it is estimated that 10 percent of the agricultural land has been abandoned (MoA 2000). The large-scale collective and state farm structure collapsed and privately-owned entities emerged in the beginning of the 1990s (MoA 2002a). The immediate results were a significant decline in gross agricultural output, and trade deficits in agricultural products. In the Baltic countries, possibilities for employment in rural areas have decreased (Bezemer et al. 2003a) – for example, the number of jobs offered in the rural areas of Estonia has fallen by about 25 percent (Reiljan & Kulu 2002). The collapse of large-scale agricultural production and food processing has not only decreased employment, it has also left the rural areas with inadequate infrastructure – and not very well-adjusted to the emergence of a small-scale farming structure. New family farms have had to adapt to difficult conditions that have come about due to the transition; for example, unsettled land ownership rights, an ill-functioning market for land

H. Palang et al. (eds.), European Rural Landscapes:
Persistence and Change in a Globalising Environment, 359-378.
© *2004 Kluwer Academic Publishers. Printed in the Netherlands.*

and products, a lack of agricultural infrastructure, and frequent changes in terms of agricultural support.

The future development of the agricultural landscape is difficult to foretell. A central question focuses on whether the withdrawal of land from agricultural production in Estonia and Lithuania is going to continue. The central actors in this situation are the farmers, as they are the ones who decide if it is profitable to keep farming the land, and what its future use will be. By looking at different types of existing farms and the farmers' conditions – their motives for farming, perceptions of current problems related to agriculture, and views of the future – it is possible to get some idea of the preconditions important in regards to future agricultural development. This enables an assessment of what factors influence farmers' strategies, and whether farmers in the future will leave more land unused or intensify agricultural production. The farmers' choice of strategy depends on various factors; including personal and socioeconomic motives, and institutional and cultural processes at the local, national and international levels. It is necessary to come to an understanding of the complex interactions between the different influences and levels of involvement, as they set the opportunities for and constraints on agricultural development, thereby playing an important role in regards to pressure on land resources.

Based on preliminary results from an ongoing PhD project, this paper analyses the current agricultural situation in Estonia and Lithuania and discusses the future development of rural areas in these countries. This project focuses on how agriculture adapts to the changes caused by the economic and political transition discussed in preceding paragraphs, and examines the different possibilities in regards to future development. The aim is to go beyond overall statistical findings, and to look at the changes taking place at local levels. Surveys have been carried out with farmers in two rural areas, and are used as a means of illustrating the current agricultural situation and the effects of the transition in rural areas. The aim of the comparative analysis is to find out if the same tendencies can be found in the two countries, and what effects differences between the countries regarding reforms and policies have had at the local level.

2.   APPROACHING AGRICULTURE IN THE BALTIC COUNTRIES

Agricultural restructuring is a central notion in the study of agriculture in the Baltic countries. The word *restructuring* is used to indicate that the study is not concerned with ordinary changes only, which occur in any dynamic and ever-changing society; instead, the focus is on major qualitative changes in social structures and practices. Since these transformations are both interrelated and multidimensional in character, studying restructuring becomes an all-embracing project, as it includes processes in all spheres and on different levels of society (Hoggart & Paniagua 2001). To reach an understanding of these complex and interrelated processes, it is necessary to include a greater integration of insight from different perspectives. The socioeconomic, political, and ecological processes influencing the agricultural system operate on a wide range of spatial and temporal scales. The challenge is to develop an understanding of how macro and micro aspects connect; how, for example, high-level doctrines and decisions influence different situations, where human action and the natural world come into touch with each other.

Inspiration for this paper is derived from Gabriel Bladh (1995), who has defined what he calls a *landscape of action*. The notion is useful because it makes up a framework that enables an integration of physical, individual and social factors acting on both micro and macro levels. The landscape of action is in this study seen as an abstraction of a farmer's decision-making process, in which different structures and procedures interact. The *landscape of action* is a bridge linking the *material landscape* with the *institutional landscape* and the *landscape of meaning*. The different "landscapes" each constitute important aspects regarding the decision-making process; i.e. by obstructing or enabling various decisions. The *material landscape* is the landscape we see and move within, including the different landscape forms and flows of matter and energy. It is an arena for life and different activities. It provides insight concerning local physical conditions, which the farmer has to build his activities on, and which to a large extent condition land use. The *institutional landscape* operates on different levels and covers different institutions, regulating human activities through the use of economic and political tools. These tools include different types of administrative networks, financial and market systems, and different sector interests. The *landscape of meaning* includes individual or group perceptions and attitudes. The *landscape of meaning* involves norm systems and behaviour that are strongly influenced by traditions and culture (Smith 2002). The *landscape of action* is used to link and integrate these immaterial aspects of human behaviour and society with the physical landscape. It is within this *landscape of action* that the farmer has to make his strategic choices (Herslund & Sørensen 1999). The structures set the conditions for a farmer's activity, but how the different structures interact depend on the specific situation.

The farmer's *landscape of action* in the Baltic countries is formed out of the turbulence of political and economical transition and agricultural restructuring. The farmer has to act within a context in which extensive changes can be seen; new institutions and structures are emerging, agriculture is now based on private ownership, and former values are replaced by new. Due to the changed conditions, it might be expected that many different strategies are now evident, as some farmers continue to be influenced by old structures while others adapt more rapidly to the new conditions.

Figure 22-11. Location of study areas.

3.   METHOD

In both Estonia and Lithuania, a rural municipality was selected for an in-depth study of the current agricultural situation. In Estonia, the study was carried out in Olustvere municipality, located in the central part of the country; and in Lithuania, Endrejavas neighbourhood[1] located in the western part of the country was selected for the study (see Figure 1). The Estonian case study took place in 2001 and the Lithuanian case study in 2002.

---

[1] Neighbourhoods (*seniunijos*) are the lowest territorial level in Lithuania. Formerly they were municipalities, but today they have no administrative units (Nordregio 2000).

In each study area, an interview-based questionnaire survey was carried out. Using municipality maps, farmsteads were selected for unannounced visits and interviews. The sample included 146 respondents – 71 in Estonia and 75 in Lithuania. The questionnaire was translated into the respective national languages, and a native speaking field assistant went through the questionnaire with the respondent. There was a possibility for the respondent to explain answers in more detail and to add further questions, when and where required. The interview situations were similar for all respondents. Interviews took place at the respondents' private residences, and only members of the household were present. This method was used to increase the confidence of respondents and to give them the opportunity to answer sincerely regarding questions that often touched upon sensitive political issues.

Though the interviews were carried out under similar conditions, there was variety to be found concerning the answers given in each country. In several of the questions, it was possible to give more than one answer; for example, in regards to the perception of problems related to agriculture. Estonian respondents utilised the possibility of giving several answers more often than Lithuanian respondents did. Thus, no direct comparison can be done on the exact figures between the two countries; but the answers give indications of the relative importance of different problems in the two areas.

Moreover, interviews of people working for agriculture and rural development in local, regional and national administrations and organisations were done. At the national level, representatives of relevant ministries and organisations were interviewed in order to obtain supplementary views on the agricultural situation. All in all, 12 interviews were conducted on the national level in Estonia and 4 in Lithuania, while the number of respondents on the regional/local level was 10 and 6, respectively.

4. NATIONAL AND SUPRANATIONAL SETTING

4.1 National Institutional Landscape

4.1.1 Changes in Farm Structures

The Soviet occupation of Estonia and Lithuania in 1940 brought radical shifts in agricultural development. The land was nationalised, private property was abolished, and in 1947, collective farming was introduced (Pajur 2002). Acts of repression, with large deportations of farmers, left no other choice than to join the collectives (Estonica 2001; MoA 2002a). By the end of the Soviet period, the agricultural structure consisted of large collective and state farms, each typically having some 3,500 ha and 300 employees (European Commission 1995; Davis 1996). These farms were financed through a centralised procurement and delivery price system, which gave the collectives cheap inputs and large credits and subsidies. About five percent of the agricultural land consisted of private household plots, which yielded 30 percent of the potato and 20 percent of the dairy output (Estonica 2001). The major aims of the agricultural policy were to ensure social stability and guarantee the supply of inexpensive food, particularly to the larger cities. In rural areas, self-sufficiency based on individual plots was the rule (OECD 1996).

The dissolution of the Soviet Union in 1991 and the restoration of Estonian and Lithuanian sovereignty brought along a fundamental change in agricultural structure and policy objectives. Strategic objectives now included the introduction of a market economy, privatisation, and the transformation of a system of collective and state farms into an agricultural sector based mainly on individual private farming (OECD 1996). The current farm structure in the Baltic countries reflects this reorientation, and can be described as being in a transitional situation. The current farming structure is comprised of large cooperatives and limited-liability companies, together with numerous small and medium-sized farms (Bager & Oldrup 1997; MoA 2000; MoA 2002a).

4.1.2   The Economic Importance of Agriculture

Agriculture has traditionally played an important role in Estonia and Lithuania not only economically, but also socially and culturally (Unwin 1997; Noorkõiv 2000; LIOAE 2002). After independence, many expected agriculture to regain the social and economic importance it had during the interwar period. However, the transition process has proven to be more difficult and slower than expected, and agriculture has thus far not come to play a leading role in the economic development of Estonia and Lithuania (Knappe 2001). Agriculture has instead experienced a general decrease in terms of its contribution to the GDP and share of employment (Table 1). Agriculture has more importance in terms of GDP contribution and employment in Lithuania than it does in Estonia. The employment in agriculture decreased by 62 percent between 1993 and 2001 in Estonia, while it only decreased by 28 percent in Lithuania. Despite its decreasing economic importance at the national level, agriculture continues to be an important stabilising social and economic factor at the local level. In Lithuania, the share of agricultural employment nationally was 18.9 percent in 2000; in rural areas, agriculture was responsible for 58 percent of employment (MoA 2000).

Table 22-1. Relative share of agriculture in GDP and employment (MoA 2000; 2002b; 2003a; LIOAE 2002).

	1992	1993	1996	1997	1998	1999	2000	2001
Relative share of agriculture in GDP (%)								
Estonia	11.7	9.3	5.8	4.9	4.3	3.7	3.4	3.3
Lithuania		14.2	11.3	10.9	9.5	7.9	6.9	6.9
Employment in agriculture (%)								
Estonia	15.0	13.0	8.1	6.9	6.9	6.2	5.2	5.0
Lithuania		22.4	23.1	20.8	20.6	19.2	18.9	16.1

4.1.3   National Agricultural Policy

Estonia and Lithuania have chosen different strategies concerning the reorganisation of their agricultural sectors. In Estonia, the change from a central planning system to a market economy has been characterised by so-called "chock therapy" (Nørgaard et al. 1999). Comprehensive liberalisation has taken place in the agricultural economy; imports and exports have no limits placed on them, and almost all prices are free (Noorkõiv 2000). Moreover, the extensive agricultural subsidies of the Soviet period have been abolished. Not until 1998 did Estonia introduce direct income support to farmers in relation to the number of animals or amount of cultivated land owned (Valdes et al. 1998; Noorkõiv 2000; Palang et al. 2000). Currently, several support schemes exist. The most important are sources of direct income support for dairy and cereal production, and development support in the form of interest rate and investment subsidies (MoA 2003a). The aim of the state support is to promote rationalisation and specialisation (Hellström 2002).

In Lithuania, the economic transition has been more gradual (Nørgaard et al. 1999). Support for farmers decreased, but was never abolished. Until 1997, production-related subsidies were the major policy tool for market regulation and income support for agricultural producers (Naujokienė & Krivickienė 2000). In 1997, the Rural Support Fund was created, with the goal of reducing the subsidies for agricultural production gradually, and increasing the support of targeted investment programmes – since these would provide greater prospects in regards to supporting long-term structural development concerning the strengthening of the agricultural sector. Due to unfavourable conditions – particularly connected to the economic crises in Russia – the first years did not bring major changes in support policy, and most of the state contributions were still given out as production support (Naujokienė & Krivickienė 2000). The intended decrease in production support has been carried out by changing the criteria for receiving support every year. In 2002, a farmer needed to have five cows to get support, while the year before only three cows were needed (Interview 2002d; MoA 2003b).[2] Moreover, a farm needs to be registered in the national farm register to receive support. Year after year, a decreasing number of farmers fulfill the criteria for support, resulting in mainly large farms qualifying to receive support.

As both Estonia and Lithuania prepare for EU membership, their agricultural policies have been adjusted to suit EU rules and regulations. The EU-financed pre-accession SAPARD programme (Special Assistance Programme for Agriculture and Rural Development) assists in the modernisation of the agricultural sector. Due to delays in the implementation process of the programme, the start was postponed until 2001. At the Estonian and Lithuanian national levels, the programme has received a positive assessment: it helps to modernise agriculture and the processing industry, and adjusts them to EU standards. In 2002, SAPARD support was given to 379 applicants in Estonia (MoA 2003a).

4.1.4   Re-Privatisation of Land

Re-privatisation of land and property has entailed yet another reorganisation of production structures and rural infrastructure. In both Estonia and Lithuania, the re-privatisation process is based on ownership that existed before the Soviet occupation in 1940 (Purju 1996). Persons who owned land in 1940, or their descendants, have the right to demand return, substitution or compensation for the land they lost (Davis 1997; Palang et al. 2000). In Lithuania, all employees of the collective or state farms received the right to privatise two to three ha of land, even if they or their ancestors did not own land before 1940 (OECD 1996; Johannsen & Nielsen 2002). In Estonia, such a rule does not exist: landless persons have to rent or buy land from either the state or private persons. The land market is developing slowly due to unfinished land reforms; how far the process has come varies from county to county. Of the land used for agriculture in Lithuania in 2001, private persons owned 52.8 percent and 47.2 percent was state-owned (Bezemer et al. 2003b). In Estonia, 44.3 percent of all agricultural land was in the ownership of private holders in 2001 (ESA 2002).

In both countries, the re-privatisation process has resulted in numerous farms with small areas of land (Kriščiūnaitė & Uždavinienė 2000; Sepp 2000). According to the agricultural census in 2001 (ESA 2002), there were 68,987 operating farm holdings in Estonia; 67,984 of these were classified as

---

[2] Interview references are organised in four categories: Interview 2001a – survey respondents in Estonia; Interview 2002b – Survey respondents in Lithuania; Interview 2001c – Survey employees in the local administration and organisations in Estonia; Interview 2002d – Survey employees in the local administration and organisations in Lithuania.

individually-owned farms, and 1003 as an enterprise with more than one owner. The average size of farm holdings was 21.2 ha (ESA 2002). For Lithuania, these numbers were 937 agricultural enterprises, and 29,600 family farms with an average area of 11.9 ha (MoA 2000). However, not all farms are recorded in the national registers. In Estonia, there are additionally about 175,000 active household farmers, and there are about 250,000 small households farmers having two to three ha of land in Lithuania (ESA 2002; LIOAE 2002).

5. THE LOCAL LEVEL: THE ESTONIAN AND LITHUANIAN STUDY AREAS

5.1 The Material Landscape

Olustvere municipality is located in the district of Viljandi in the central part of Estonia. Agriculture is widespread in the area, and the soil quality is considered as being a little above average in comparison to the rest of the country (Interview 2001c). The municipality has 1,627 inhabitants (ESA 2003), covers an area of 144 km^2, and consists of 13 so-called "villages". Five of these are regular villages with a central agglomeration, while the other villages have more dispersed settlements.

In Lithuania, the study area was Endrejavas neighbourhood in the district of Klaipeda in the western part of the country. The neighbourhood has about 2,100 inhabitants, and an agricultural potential of around the average for the country (Interview 2002d). The arable land has a good soil quality, though, and extensive meadow and pasture areas dominate (Interview 2002d). Due to large-scale drainage and land amelioration works in the 1970s, many farmsteads were destroyed, with owners subsequently being given new houses in the more centrally located villages. Today two larger villages dominate the settlement structure of the area – however, many farmsteads are found outside these areas.

It is not common for people to commute to the nearby bigger towns of Viljandi (for Olustvere) and Klaipeda (for Endrejavas), located 20 and 40 kilometres away, respectively. According to the inhabitants, this is due to difficulties in finding jobs, low wages for jobs that can be found, and insufficient collective transport. Employees in local service facilities, small businesses and administration, are mainly local people. In both areas, there are considerable areas of unused agricultural land, which is a consequence of the privatisation process. In Olustvere, unutilised agricultural land makes up 20 percent of the agricultural land (ESA 2002). There are different reasons for the large unused land areas. Some of the unused land is owned by people who have obtained it by restitution, but who do not live in the area nor cultivate or rent out the land. Other people feel they are too old to cultivate all their land, and sometimes it is not possible to rent out the land due to a low demand for land – therefore it is left unused.

5.2   The Institutional Landscape

In both areas of study, agriculture dominated the economy during the period of Soviet rule, and it is still the main source of income for rural inhabitants.

5.2.1   Re-Privatisation

In both areas, the land reform of the 1990s has brought important changes to the tenure structure. In the Estonian case, more than 85 percent of the respondents own land, or are in the process of receiving ownership rights of the land. In the Lithuanian case, this share is 92 percent of the households. In both cases, 12 percent of the households only farm on land they rent from either private persons or the state. In Olustvere, this group of farmers is very diverse; half of them are big farmers, with more than 200 ha of land. The respondents in both areas say it is easy and inexpensive to rent land – the rent is often equal to the land tax (Interview 2001a). In Olustvere, 16 percent rent out parts of their land to other private farmers – in Endrejavas, only one percent of respondents rent out land.

The amount of land that each household has the right to use varies considerably. Figure 2 shows the distribution of farms according to the land size in the two municipalities. Despite the large share of small farmers in the Lithuanian case, the numbers in the sample are not fully representative for the area; more people than shown have very small pieces of land. According to the local administration, about 60 to 70 percent of the households cultivate around three ha of land (Interview 2002d). It is deliberate that the share of respondents with larger land is overrepresented in the sample, as it is expected that it is within this group that the most important changes in farming strategy will be found. It appears that farms on average are larger in Olustvere than in Endrejavas – few farms are larger than 40 ha in Endrejavas.

The division of agricultural land by farm size shows differences between the two areas (see Figure 3). In Olustvere, the large farms dominate the land use; farms larger than 50 ha use almost 65 percent of the land used by the respondents, and constitute about 20 percent of the respondents. In Endrejavas, there seems to be a more homogeneous distribution of land between the different farm sizes. However, farms larger than 50 ha occupy

30 percent of the land, but make up only six percent of the sample – and it is significant that farms with less than five ha of land only occupy five percent of the area, but their owners make up 36 percent of the respondents.

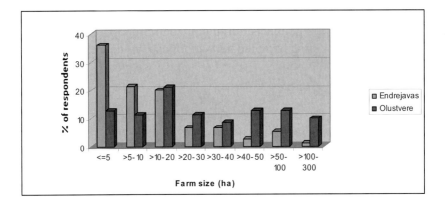

Figure 22-2. Size of land used by respondents in the two case areas: forest land has been excluded, as the interest is on land that can be taken out of production. Estonia – growing season 2001. Lithuania – growing season 2002.

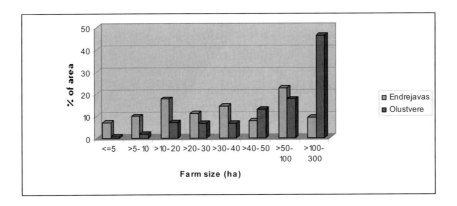

Figure 22-3. Division of land by farm size: forest land has been excluded as the interest is on land that can be taken out of production. Estonia – growing season 2001, Lithuania – growing season 2002.

5.2.2 National Agricultural Support

Many of the farms are not registered as farms in the national farm registers. People do not see any advantages in registering; often their farms are too small to receive national support, or they primarily produce for home consumption (Interview 2002b). In Endrejavas, 20 percent of the respondents had registered their farm, but only 16 percent of the households received national agricultural support. Support for dairy cow farms dominated. In Olustvere, 23 percent of the respondents receive agricultural support, most of them for cereals (75 percent), dairy cows (44 percent), rape (44 percent), and for liming land (25 percent). The majority of those receiving support get it for more than one product.

Almost half of all respondents find the agricultural support too small, and only a few find it satisfactory. The main wish is to get a guaranteed minimum price on all products. This wish is mostly expressed in Endrejavas (43 percent), compared to Olustvere (28 percent). Many farmers complain about unstable prices: they find it difficult to decide what to produce, as the prices vary considerably from year to year. Some of the smaller farmers say that they put more money into their agriculture production than they get out of it (Interview 2002b). Some respondents have stopped producing for sale because of the current low prices. State support is considered to be of great importance for the future development of the farms: 68 percent of the respondents in Olustvere and 78 percent in Endrejavas find that support is crucial for the future development of their farming activities.

5.2.3   EU Pre-Accession Support

Knowledge of the SAPARD programme is widespread, with almost three-quarters having heard of the programme. Of these, nine percent (Endrejavas) and 21 percent (Olustvere) would consider using it for the modernisation of buildings and production equipment. There is a noticeable dissatisfaction and hesitation displayed towards the SAPARD programme. Some have lost their trust in the programme because the start was postponed several times, while others complain about the self-financing element, and that the farmer has to carry all the expenses before they are repaid afterwards. Another point of critique from the respondents is that the target group is larger farmers, and that a lot of paperwork is required to apply for the support (Interview 2001a; Interview 2002b).

5.3   The Landscape of Meaning

It appears as if the rural inhabitants have quite different motivations for their farming activities (Table 2). The reasons are most homogenous in the Lithuanian case, where 79 percent started in agriculture mainly in order to receive an income. The low share of respondents who began in farming because they received restituted family land is in accordance with the fact that many of the people in Lithuania that today live and farm in rural areas did not necessarily own land before the Soviet occupation. This is quite the opposite in the Estonian case, where 39 percent of the respondents give the restitution of family land as a reason for their involvement in farming. Moreover, a large percentage of the Estonian respondents (49 percent) enjoy working in the countryside. This harmonises well with the opinion of a large part of the respondents (77 percent) that rural life constitutes an important part of Estonian self-identity (Interview 2001a). In Endrejavas, being a farmer is not so strongly related to Lithuanian self-identity, and it is more the possibility for generating an income that is emphasised as being important.

Table 22-2. The respondents' motivation for starting to farm (the respondent was allowed to give more than one reason).

Why did you start to farm?	Olustvere (%) N=66	Endrejavas (%) N=75
Got the opportunity to receive family land	39	9
The only possibility to get an income	27	79
Agricultural reform promoted my farming	11	0
Like to work in the countryside	49	8
Like to work with animals	25	3
Want to create a future for my children	17	1
Want to start my own independent business	11	4
Other?	7	0

Table 22-3. Respondents' evaluation of the current situation in relation to the Soviet period.

Sometimes I think it would be better to go back to the situation before 1991	Olustvere (% of respondents) N=67	Endrejavas (% of respondents) N=74
Agree	18	50
Partly agree	36	12
Disagree	46	38

About 70 percent of the Estonian respondents report that they belong to the very poor or poor income groups, and none think they belong to the high income group. These numbers illustrate that respondents find it difficult to reach a satisfactory standard of living. The Lithuanian respondents describe their situation more positively: 31 percent consider themselves to be poor or very poor, and 69 percent say they belong to the middle income group. In spite of this, half of the Lithuanian respondents think it sometimes would be better to go back to the Soviet times (Table 3). This might be a consequence of the dominance of very small farms in the study. They are too small to receive either national or SAPARD support to make investments in modernising the farms.

The Estonian respondents find individual efforts important for creating a good life (Table 4). However, a large share of the Lithuanian respondents disagree (44 percent), and the majority of these also think it would be better to return to conditions in the Soviet times. They are disappointed with the new order – they think they had a better material life before. They had a place to live, inexpensive food, job guarantee, and the social security was much better (Interview 2002b). Others have opposing views and appreciate their freedom: "Today you can see the results and gain the profit of your own work," and "it is necessary to work hard to survive but if you do so, you are able to make a living out of it" (Interview 2002b). These different attitudes are related to differences in personality and mentality, and possibly a respondent's ability to take advantage of the new opportunities. Part of the population finds it difficult to handle the modern requirements of individual initiative brought forth by independence – they prefer the safety net provided by the collective and state farms.

Table 22-4. The respondents' evaluation of people's ability to influence their living conditions.

If a person works very hard, can he/she get a better life for his/her family?	Olustvere (% of respondents) N=65	Endrejavas (% of respondents) N=73
Agree	48	31
Partly agree	46	25
Disagree	6	44

Table 22-5. The respondents' perception of the major problems of farming (the respondent was allowed to give more than one answer).

What are the major problems of farming?	Olustvere (% of respondents) N=60	Endrejavas (% of respondents) N=75
Too low prices for the products	58	96
Lack of market for products	30	28
Lack of support from the state	23	44
Overdue payment for sold production	45	16
Shortage of technology and equipment	43	15
Lack of buildings	22	16
Difficulties in obtaining loans	10	19
Lack of labour	12	8
Shortage of chemicals	7	12
Lack of land	7	8
Lack of knowledge/education	7	5
Laws and restrictions	0	5
Other	7	5

6.   LANDSCAPE OF ACTION – THE CHOICE OF STRATEGY

In both areas, the major problems in relation to farming (Table 5) are the low prices for products and the lack of markets for selling the production. Lack of support is a main problem in Endrejavas and in Olustvere; delayed payments are a widespread problem. Often farmers have to wait for months, especially in regards to milk sold to dairies. Lack of equipment is also regarded as a substantial problem in Olustvere, and is much more frequently mentioned than in Endrejavas. This shortage is connected to the re-privatisation process that was divided into two separate reforms; one redistributing the land and one privatising non-land property from the collective and state farms. Often there is no connection between land and equipment, the result being that some have land but little or no equipment, and vice versa (Interview 2001c). A further problem with equipment is that it often consists of obsolete Soviet technology.

Table 22-6. The respondents' plans for the future (the respondent was allowed to give more than one answer).

The future of the farm?	Olustvere (% of respondents) N=64	Endrejavas (% of respondents) N=75
No changes	59	57
Expand the land	14	9
Specialise the farm	17	11
Set up small business	6	4
Change to organic farming	2	-
Reduce the size of the farm	2	5
Quit farming	13	1
Children will take over	20	28

Despite the differences in the perception of problems between the two areas, there is a high degree of similarity in the respondents' plans for the future (Table 6). More than half of the respondents will continue to farm as they do today, without major changes in the production pattern. The respondents feel insecure, and many do not find the future promising. They continue with agriculture mainly because they cannot find other jobs. Cultivating land and selling part of the production provides income to the household, and makes it self-sufficient in some foodstuffs. Some of the respondents in this group would like to expand their farming operations, but think it is impossible to find the money for necessary investments in more land, buildings and technology. Without extensive assistance from the state (either by more support or price subsidies), they see no possibilities for changing anything in their current farming strategy. About one-fifth of the respondents that will continue their current farming strategy already cultivate more than 50 ha of land. Their reason for not planning substantial changes is that they think their present farming strategy will also be satisfactory in the future. The common denominator for this group of respondents is continuity – they will continue their current agricultural practices and thereby maintain their current utilisation of the land. The impact on the rural landscape will be more or less unaltered.

A rather passive group of respondents are those who will quit farming in the years to come, and those whose children will take over the farm. They do not have any plans for future farming: they will continue as now until they quit or their children take over. Not surprisingly, the majority of the farmers in this group are already pensioners – their average age is 58 years. Many pensioners are more concerned about younger people's future than their own, as they regard their rural life tolerable because of their pension. A minority in this group consists of young people who say they will quit farming if they find a wage job. Reiljan & Kulu (2002) also found that young Estonians often leave farming. The strategies of these respondents might give rise to changed land utilisation – land will be left unused if no land market develops and land becomes more requested. When children take over parents' farmsteads, there is no guarantee that they will cultivate the land. Agriculture is often not attractive for young people, but they will properly keep the land in the hope that the land value will increase in the future. Particularly, the increase in unused land will come from this group of respondents. However, many of the respondents who hope their children will take over their farm have rather small land areas. This means that it is not large parts of the total area that will be left unused, as their shares of the total land only constitute a minor part.

Changes with consequences for the rural landscape will also come from about 20 percent of the respondents, who will either expand their farm lands or specialise their production. Half of those who will expand their land will also specialise the production. The average age in this group is 43 years. These respondents see a future in farming, which makes them more open and responsive to the market and towards the support schemes for investments aimed at modernising their farming. Many of these farmers are those with relatively large farms and those who use a relatively large part of the agricultural land in the case areas.

7.   CONCLUDING REMARKS

The ongoing agricultural restructuring has left several marks in both Estonia and Lithuania. Agriculture is no longer a strongly protected sector with large contributions in terms of GDP and employment. A continuous adaptation to the market economy is taking place, and the aim of state support is to promote rationalisation and specialisation. The privatisation of land has resulted in numerous small holdings; however, the sizes of the holdings differ considerably. There is a higher percentage of small farms in the Lithuanian study area than in the Estonian study area. Due to high unemployment in rural areas, the majority of people living in these regions are involved in agriculture, since they do not have other opportunities for generating an income. Certainly, there are differences between the agricultural situations in the two case areas due to differences in the privatisation policies – but the similarities dominate.

Looking at different influences affecting present land use, it is evident that the *institutional landscape* plays a central role in shaping opportunities and constraints for farmers. Many farmers find liberalisation and the present agricultural policy problematic. Liberalisation has led to increased competition, and the abolishment or decrease of price subsidies – instead, investment support and direct income support is given priority. Privatisation has created a small-scale farming structure. Many farms are not profitable and do not provide a sufficient income with which to support a family. The majority of farmers does not register their farms, and cannot qualify for financial support. They find the criteria for getting support unattainable, and feel that the government only focuses on the larger farms. Smaller farms do not have the financial means for investments: the majority will only make future investments if they are substantially supported by the state. In both countries, the current agricultural policy develops in the opposite direction – concentrating support to farmers with larger cattle herds and land areas.

Important constraints for the farmers are also related to the *landscape of meaning*. Respondents have different motivations for their farming activity. In Lithuania, the possibility for getting an income is a central motivation, while Estonian respondents highlighted the restitution of family land, as well as agriculture being an important part of Estonian identity. Many, particularly among Lithuanian respondents, would like to return to the situation in the Soviet period, and do not think individual hard work can result in better living conditions. This indicates that they find the present rural life hard and that they do not believe that they can improve their economic situation. This defeatist attitude dominates among the owners of small farms. Lack of individual initiative impedes many in developing their farms. Age plays an important role concerning attitude – many farmers worked on large Soviet farms most of their lives as farm workers, being responsible for one specialised function. For many, it has been a difficult jump to start up their own farm and act within a society adjusting to liberalisations and market economy. Elderly people also have a shorter time horizon, which makes investments less attractive for them. Furthermore, they lack physical strength; and as they receive pensions, the pressure for generating an income from farming is not as strong as it is for younger people. They will most probably continue their small-scale farming as long as they find it possible to do so. The younger generation is confronted by the need to generate an income for the household, and the future development of employment possibilities will be decisive in terms of the number of younger people leaving agriculture. As long as there are no other job opportunities, people will stay in agriculture and try to make a living out of it.

The rural farming population can be divided roughly into two groups: one includes approximately three-fourths of the respondents, characterised as having no plans for intensifying or expanding their farming activity; and the other group, with about one-quarter of the respondents, consists of those who have plans for developing their agricultural operations into larger and more-specialised farms. The first group is restricted in a number of ways. The agricultural policy does not focus on small farmers: they cannot qualify for support nor are able to obtain loans for modernisation. These farmers often also have other constraints; their age, for instance, as well as lack of adequate education and initiative. Many small farms will probably be squeezed out of commercial production in the next few years. If they are not able to receive subsidies, it will not be possible for them to compete with those who have a larger production operation, receive subsidies, and have possibilities to obtain credits for investment.

This indicates a continuation of the current land use practice by the large majority of the farming population found in the two study areas. The most important impact will be when they – due to age – stop farming, as the land will be left uncultivated if it is not possible to sell it. Even though they each have only small holdings, this will still influence large areas due to the high percentage of these small farmers. Parts of the abandoned land will be taken over by the second group of farmers, who want to expand their farms. The question is: How much of the abandoned land will they take over? Within the next 20 years, large areas of the land will be abandoned; but there will probably be important regional differences. In areas such as Endrejavas, with a high number of small farmers and only few larger farmers who demand land, this process will most likely be more pronounced than in areas such as Olustvere – where more farmers see a future in agriculture.

The farmers who have the potential to extend their agricultural activities are typically younger and have a different attitude – they are not limited by a lack of education or initiative. They realise that investments are necessary, and that the individual farmer has to take the responsibility to make the best of the transitional situation. They fulfill the requirements of current support programmes, and therefore have the possibility of increasing their competitiveness via specialisation and expansion of production. Nevertheless, they are dependent on continued support, and are therefore strongly influenced by the development of agricultural policy.

In the Baltic countries, farm structure and farming strategies will continue to change in the years to come. Today the farmers are a heterogeneous group, having different attitudes and motives for farming, resulting in a variety of farming strategies. Gradually, the farmers will become more homogeneous, to consist of larger and more-specialised farms due to an increased market orientation, and because many elderly people with small holdings will leave agriculture.

REFERENCES

Bager, T. & Oldrup, H. (1997). *Farm Structure and Farmer Attitudes in Estonia, Latvia and Lithuania.* Esbjerg: South Jutland University Press.

Bezemer, D.J., Stanikunas, D. & Zemeckis, R. (2003a). *Why Are Companies Disappearing in Transitional Agriculture? Evidence from Lithuania.* Economics Working Paper Archive at WUSTL.

Bezemer, D. J., Rutkauskaite, J. & Zemeckis, R. (2003b). *Income Diversity in Rural Lithuania. Benefits, Barriers, and Incentives.* Economics Working Paper Archive at WUSTL.

Bladh, G. (1995). *Finnskogens landskap och människor under fyra sekler – en studie av natur och samhälle i förändring.* Meddelanden från Göteborgs Universitets Institutioner serie B, 87. Göteborg: Handelshögskolan vid Göteborgs Universitet.

Davis, J. (1996). *The Political Economy and Institutional Aspects of the Process of Privatizing Farming and Land Ownership in the Baltic States.* CERT Discussion paper 07/1996, Edinburgh, UK: Heriot-Watt University, Centre for Economic Reform and Transformation (CERT).

Davis, J.R. (1997). Understanding the Process of Decollectivisation and Agricultural Privatisation in Transition Economies: The Distribution of Collective and State Farms in Latvia and Lithuania. *Europe-Asia Studies*, 49 (8), 1409-1432.

ESA – Statistical Office of Estonia. (2002). *2001 Agricultural Census – General data, crop production, Livestock.* Tallinn: Statistical Office of Estonia.

ESA – Statistical Office of Estonia (2003). *Population Census 2000.* From http://www.stat.ee

Estonica (2001). From http://www.estonica.org. February 14, 2001.

European Commission (1995). *Agricultural Situation and Prospects in the Central and Eastern European Countries – Estonia.* Brussels: Directorate-General for Agriculture.

Hellström, K. (2002). *Agricultural Reforms and Policies Reflected in the Farming Landscapes of Hiiumaa from 1850 to 2000.* Doctoral Dissertation. Acta Universitatis Agriculturae Sueciae. Agraria 325. Alnarp: Swedish University of Agricultural Sciences.

Herslund, L. & Sørensen, M.B. (1999). *The Relationship between Agricultural Development and Nature Conservation in a Transition Landscape – A Study of the Farmers in the Nemunas Delta Regional Park in Lithuania.* MSc thesis. Inst. of Geography, Copenhagen University.

Hoggart, K. & Paniagua, A. (2001). What Rural Restructuring? *Journal of Rural Studies*, 17 (1), 41-62.

Johannsen, L. & Nielsen, T.Y.K. (2002). *The Political Economy of Agrarian Reform: Lithuania in Comparative Perspective.* Aarhus: Dept of Political Science, Univ. of Aarhus.

Knappe, E. (2001). Estonia. In S. Goetz, T. Jaksch & R. Siebert (Eds.), *Agricultural Transformation and Land Use in Central and Eastern Europe* (pp. 73-104). Aldershot: Ashgate.

Kriščiūnaitė, L. & Uždavinienė, V. (2000). The Structure of Lithuanian Farms and their Development in Lithuania. In *Agricultural Policy and Rural Development in the Baltic States* (pp. 49-55). Vilnius: Lithuanian Institute of Agrarian Economics.

LIOAE – Lithuanian Institute of Agrarian Economics. (2002). *Agriculture in Lithuania 2001.* Vilnius: Lithuanian Institute of Agrarian Economics.

Mander, Ü., Kull, A., Kuusemets, V. & Tamm, T. (2000). Nutrient Runoff Dynamics in a Rural Catchment: Influence of Land-Use Change, Climatic Fluctuations and Ecotechnological Measures. *Ecological Engineering*, 14 (4), 405-417.

MoA – Ministry of Agriculture of Lithuania. (2000). *Agriculture and Rural Development Plan 2000-2006.* Vilnius: Ministry of Agriculture

MoA – Ministry of Agriculture of Estonia. (2002a). *Estonian Agriculture, Rural Economy and Food Industry*. Tallinn: Ministry of Agriculture.

MoA – Ministry of Agriculture of Estonia. (2002b). *Agriculture and Rural Development – overview 2001/2002*. Tallinn: Ministry of Agriculture.

MoA – Ministry of Agriculture of Estonia. (2003a). *Agriculture and Rural Development – overview 20021/2003*. Tallinn: Ministry of Agriculture.

MoA (2003b). *Agriculture of Lithuania*. Vilnius: Ministry of Agriculture.

Naujokienė, R, & Krivickienė, R. (2000). The Trends of State Financial Support to the Lithuanian Agricultural Sector in 1997-1999. In *Agricultural Policy and Rural Development in the Baltic States* (pp. 33-40). Vilnius: Lithuanian Institute of Agrarian Economics.

Noorkõiv, K. (2000). Estonian Agricultural and Rural Policy Development Strategy. In *Agricultural Policy and Rural Development in the Baltic States* (pp. 15-24). Vilnius: Lithuanian Institute of Agrarian Economics.

Nordregio (2000). *Regions of the Baltic States*. Stockholm: Nordregio.

Nørgaard, O., Johannsen, L., Skak, M. & Sørensen, R.H. (1999). *The Baltic States after Independence*. Cheltenham: Edward Elgar.

OECD (1996). *Review of Agricultural Policies – Latvia*. Paris: OECD.

Pajur, A. (2002). The Baltic States 1914-1939. In U. Vent & I. Kiverik (Eds.), *The History of the Baltic Countries* (pp. 127-196). Tallinn: Avita.

Palang, H., Hiiemäe, O., Sepp, K., Ivask, M. & Mander, Ü. (2000). Predicting the Future of Estonian Agricultural Landscapes: A Scenario Approach. In Ü. Mander & R.H.G. Jongman (Eds.), *Landscape Perspectives of Land Use Changes* (pp. 107-130). Boston: WITPress.

Peterson, U. & Aunap, R. (1998). Changes in Agricultural Land Use in Estonia in the 1990s Detected With Multitemporal LandsatMSS Imagery. *Landscape and Urban Planning*, 41, 193-201.

Purju, A. (1996). *The Political Economy of Privatisation in Estonia*. CERT Discussion paper 02/1996, Edinburgh: Heriot-Watt University, Centre for Economic Reform and Transformation (CERT).

Reiljan, J. & Kulu, L. (2002). *The Development and Competitiveness of Estonian Agriculture prior to joining the European Union*. Tartu: Tartu University Press.

Sepp, M. (2000). Structural Policy and Competitiveness in Estonian Agriculture. In: *Agricultural Policy and Rural Development in the Baltic States* (pp.152-157). Vilnius: Lithuanian Institute of Agrarian Economics.

Smith, A. (2002). Culture/Economy and Spaces of Economic Practice: Positioning Households in Post-Communism. *Transactions, Inst. of British Geographers*, 232-250.

Unwin, T. (1997). Agricultural Restructuring and Integrated Rural Development in Estonia. *Journal of Rural Studies*, 19 (1), 93-112.

Valdes, A., Csaki, C. & Fock, A. (1998). Estonian Agriculture in Efforts to Accede to the European Union. *Post-Soviet Geography and Economics*, 39 (9), 518-548.

Chapter 23

DANISH FARMERS AND THE CULTURAL ENVIRONMENT
Landscape Management with a Cultural Dimension

Per Grau Møller
Centre for Cartographical Documentation, University of Southern Denmark, Denmark

1.   INTRODUCTION – LEGISLATIVE BACKGROUND

Landscape management is a well-known and much-debated issue in so-called "modern welfare" societies. The management of landscapes within these societies is usually in the hands of state or council employees, but is almost always put into practice by local farmers, who have used the landscape for agricultural purposes for thousands of years. For a number of years, state bureaucracies in most of Western Europe have focused on the limited term *nature conservation* in regards to landscape management, and the role of farmers in this respect. So when the term *conservation* has been used, its meaning has usually referred only to the preservation of flora and fauna, or their habitats (Green 1996; Beedell & Rehman 2000; Wilson & Hart 2001). Legislation, registration and strategic research in Denmark concerning landscape management have for decades been directed primarily towards nature conservation; although nobody involved in the process will claim that particular situation to be an ideal one (Agger et al. 1986; Brandt 1994).

For decades, cultural heritage and its preservation has also been an important theme in Danish legislation, mainly in regards to certain visible remnants in the landscape – such as burial mounds from the Bronze Age, ruins and bridges. These remnants are presently covered under the Nature Conservation Act of 1992, aimed at safeguarding natural and cultural heritage in Denmark. Another existing act addresses the preservation of

H. Palang et al. (eds.), European Rural Landscapes:
Persistence and Change in a Globalising Environment, 379-396.
© 2004 *Kluwer Academic Publishers. Printed in the Netherlands.*

buildings, particularly those considered to be of important architectural value.

Since 1994, special attention has been paid to what is referred to as the "third dimension" of the environment – *a cultural dimension* – with the intention of linking the cultural dimension with the two others: one, the traditional protection of nature and culture; and two, environmental protection and pollution control. The term *cultural environment* was presented as a political term by the former minister for the environment in Denmark, but until now it has been lacking a precise definition as to what exactly it consists of. Such a definition is a prerequisite in regards to the management of man-made features and structures found in the landscape, if it is to be done out of consideration for the landscape as history (rather than pre-history). But the problem is that Danish legislation – and especially legislation focusing on public management – is not yet geared towards the actual handling of such features or structures under the concept of cultural environmental management. A *cultural environment* is therefore more a concept than anything that has been actually implemented and made useful for citizens in general – except for the few past designations made by county administrations.

A basic problem – concerning both the specific preservation of sites and their integration into landscape management – is that in Denmark there lacks a definitive survey of landscape sites deemed worthy of preservation; especially for those from historical times (after 1050 AD). For prehistoric periods, there is a Danish national register of all finds that have been made in the landscape (see *Fund og Fortidsminder* 2003). Regarding historical periods, there are only a few national registers that can be viewed in terms of actually playing a role in the management of the cultural environment; none that could be considered as systematic (Møller 2001).

The term *cultural environment* can be understood in two ways. It can either be regarded as a general concept denoting the historical dimension of a cultural landscape (which means everything that is and has been used by humankind); or in terms of a concept that has been used in a particular manner over thousands of years, a concept which has left reminders on the present day landscape. Comprehending this terminology is a necessary step towards understanding the historical dimension of landscape, but it represents difficulties in the administration of the cultural dimension – unless it is accepted that every new development will come to form part of the historical landscape (with the result that any change is acceptable); or, at the other extreme, that every existing element in the landscape is frozen. The other way of handling the cultural environment is to designate certain structures or elements of a preserved historical entity as possessing special values. These values might be historical, may provide an outstanding embodiment of a previous period, or may illustrate the history of the landscape in other ways[1] (see Figure 1). This interpretation allows for the management of the cultural environment at the administrative level, because determined efforts can then be made in selected areas; whereas random development can take place in all other areas.

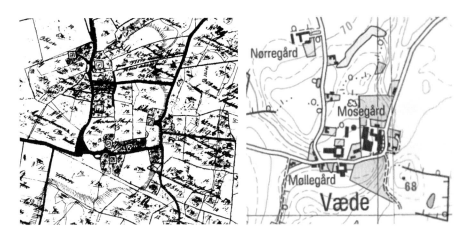

Figure 23-1. This is an example of a Danish village possessing special historical values. The locations of the farms around the enclosure in 1785 (left map) are almost unchanged when compared with the situation in 1996 (right map). The number of farms in the village was 13 in 1511, compared to nine in 1785 – but the structure of the village probably remained unchanged since the late Middle Ages as a small village in a forest region. Enclosure did not change this impression. Only in the last decade has one farm expanded in the south-eastern end, and in 2000 there were three farmers farming more than 30 ha in the village. © Kort & Matrikelstyrelsen G 14-99.

Danish planning legislation provides for the opportunity to make a so-called "local plan" in regards to a larger area. The aim of such a plan might, among others, be to secure the historical significance of buildings, trees, gardens, roads, paths and surroundings. For the time being, an inconvenient weakness of the legislation is that such a plan cannot include areas or buildings used for agricultural production – a serious problem when considering the protection of the cultural environment.

Like other European Union countries, Denmark has for several years used agri-environmental schemes, which can be useful for preserving cultural elements and structures in the landscape. It is, however, characteristic of Danish planning implementation that mainly environmental

---

[1] This corresponds with the terms *completeness* and *authenticity*, which Renes (1999) lists in his overview regarding the evaluation of historical values; but here it is specifically applied concerning the specific registration of historical types.

criteria have been used when it comes to the designation of areas. The use of *MiljøVenligt Jordbrug* (MVJ, agri-environmental support), which supports extensive land-use without the presence of pesticides or artificial fertilizers and with a reduction in the number of grazing animals, can only be exercised for areas registered as being environmentally sensitive (ESA – in Danish, *SFL*). The criteria for inclusion are environmentally-based (such as the presence of drinking water, or de-nitrification) – though in dry areas, certain types of grassland can be designated on the basis of their vegetation. It has not been possible to add a cultural historical dimension to these types of areas, although it may be very relevant in cases pertaining to old meadows or grasslands (Andersen et al. 1998; Møller et al. 2002). In Sweden, a cultural historical dimension has been added as a criterion for the designation of old meadows and pastures, though it has yet to wield much influence (Ihse & Lindahl 2000) – whereas in EU countries like England, Ireland and Austria, cultural landscape and cultural heritage are given much greater consideration in the implementation of agri-environmental schemes (Pedersen et al. 2002).

In Sweden, another initiative has been taken to set up individual plans for the farmers' cultural landscape, taking into consideration nature as well as culture. In Denmark, tests have been undertaken to draw up so-called "voluntary nature plans on farm level", but hitherto they have been implemented only sporadically (Thybirk & Haugaard 2001).

Another guiding principle in the politics of landscape and environment management during the last few decades has been *sustainability*. In a 1987 report from the World Commission on Environment and Development, *sustainability* was defined as "a development that meets the needs of the present without compromising the ability of future generations to meet their own needs" (1987: 43). Sustainability depends on a definition of what "needs" consist of, but according to von Meyer (1999), four fundamental principles can be used as a basis for determining this:
- ecological integrity;
- economic efficiency;
- social equality;
- cultural identity.

The first three principles are well-known, and the ecological integrity dimension in particular is used in connection with many present-day environmental issues. The cultural dimension has hitherto been much neglected. In Denmark, strategic research into how this dimension can be integrated into the process of administering the terrestrial environment has been undertaken (Møller et al. 2002), and the same principles involved in the research are referred to in this text. Denmark, like much of Europe, is a cultural landscape which has had its surface used and shaped by humans over thousands of years. Therefore, much of what makes up cultural identity is connected to landscape, as well as being associated in our minds with features in the landscape; whether they are historical relics, or structures still in use. One of the lessons to be learned is that landscape is only sustainable in relation to cultural identity; that is, sustainable if the most important features (structures or elements) are preserved for the future, so that following generations are able to see the identity of landscape expressed in these features (Møller et al. 2002).

2. METHODOLOGY AND PRESENTATION OF CASE STUDY

What is the role of the modern farmer, how does he or she look upon landscape as cultural environment, and what are the attitudes towards managing the landscape in relation to the attainment of goals as defined by overall political considerations? These were some of the questions raised during interviews with a group of Danish farmers in the year 2000.[2] Several other previous investigations have studied farmers' attitudes to nature conservation (e.g., Primdahl 1999; Wilson & Hart 2001). The present investigation, however, placed a special focus on the management of historical remains found in the landscape, since – despite the need for it having been previously raised on a Nordic level – this specific aspect of management had not previously been researched in regards to Denmark (Gaukstad 2000).

In this context, it is important to know how one important group of agents in the landscape might react in relation to the maintenance of certain elements and structures of the cultural environment. It was a deliberate choice that this group should be professional farmers, defined as owning 30 hectares of land or more, which in Denmark is the maximum a farmer can own without having the requisite agricultural education. This does not mean – as it turns out and as might have been expected – that all those belonging to this group are full-time farmers, whose income derives entirely from their farms. But it is among this group that we can find what can be seen as "the modern farmer" who is expected to base his decisions concerning his production resources – land and buildings – on purely economic considerations.[3] Is he or she willing to consider non-economic arguments in order to preserve (part of) the cultural heritage? The group of full-time

---

[2] For a full documentation of the investigation, see Møller (2002).

[3] Economic arguments are prevalent in the debate from the farmers' perspectives, and the development of fewer and bigger farms shows that economic factors are decisive; however reality might be a little more complex, as a recent investigation which included conversations with farmers shows (Elberg 2002).

farmers cultivates the greatest proportion of the agricultural landscape, and their share is not diminishing, although their number is. A recent Danish investigation has shown that the growing group of small "hobby farmers" demonstrates the same tendency to extend land-use, in the same manner that full-time farmers do (Kristensen 1999). Additionally, it has been stressed that there may exist substantial differences of opinion between an owner and a producer (even if the two are the same person), especially concerning decisions on the management of the countryside. It has also been emphasised that viewing the landscape in terms of an attractive living space provides very strong arguments in favour of landscape management decisions that are not based on production factors only. Therefore, hobby farmers are in fact more active than professional farmers in regards to changing structural parts of the landscape (Primdahl 1999).

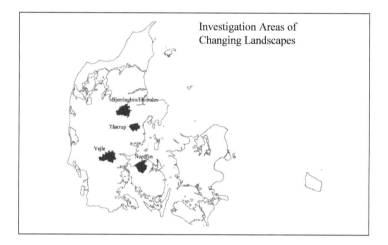

Figure 23-2. The four investigation areas of the strategic research center Changing Landscape (http://www.sdu.dk/Hum/ForandLand/English/Index.htm).

In the spring of 2000, interviews were conducted with farmers in order to throw more light onto these issues. The Danish farmers who were selected by the research centre *Changing Landscapes* resided within four investigation areas in Jutland and Funen counties (Figure 2). From these areas, ten percent of the farmers owning more than 30 ha were selected randomly for interviews.[4] 96 farmers received a letter with a proposed appointment for an interview. Only two-thirds (or 65 in total) were willing to

---

[4] In 16 percent of the villages, one out of every four farmers was selected for an interview. In this temporary cultural historical register, these 45 villages (out of 280) were selected for having a special value concerning preservation – and for providing a potential basis for a discussion with the farmer about conservation values.

participate. The rest refused or were not available for a response. This substantial lack of interest from one-third of this representative selection of farmers can be interpreted as a reluctance to take part in a dialogue with authorities[5] regarding environmental questions in general. This, of course, represents a problem, especially if landscape management is to rely primarily on the attitudes and actions of the local farmers. The intended goal of interviewing farmers who were totally dependent on agriculture did not come to fruition, as it turned out that 17 percent had varying incomes from employment outside the farm (almost all spouses had sources of external income). On average, the farmers were cultivating 90.4 ha, of which they owned 70 ha and rented out 20 ha – the biggest farm was 395 ha (an old manor house). The age of the farmers was on average 48 (ranging from 28 to 73 years) – close to the average age of all Danish farmers, which in the same year was 50. In one-third of the interviews, both spouses participated. In three cases, female farmers were interviewed. Consequently, the majority (a total of 40) was represented by men.[6] But it is notable that two of the three women interviewed were in charge of large-scale pig farms.

The methodology behind the interview was to ask questions from a questionnaire and to subsequently quantify the answers. In the first part, specific factual information pertaining to the individual farm was asked for; the second part consisted of general questions regarding how the farmer would react in relation to features of the cultural environment. After a presentation of some specific historic sites from his farm area and of a digitalised landscape scenario from around the year 1800 (as based on economic maps), the farmer was asked questions on a more "realistic" basis concerning his attitude towards cultural environment in regards to his own farm.

In relation to cultural heritage, a top-down model was used, as the farmers' knowledge of the historical landscape was expected to be sporadic.[7] His knowledge is greatest regarding the times close to the present – i.e. living memory. From older times in particular, it is not expected that he would have much knowledge of the history of structures and elements found in the landscape – he may know stories and legends connected to the local

---

[5] The research project can be interpreted as authoritative, as it was funded from the Danish Strategic Environmental Research Programme, and the opening letter to the farmers was accompanied by official material on cultural environment. A few farmers wanted monetary compensation in order to be interviewed, which of course was refused.

[6] Given that the majority of interviewees were men, I will use the term "he" when referring to a farmer.

[7] A recent Swedish study (Stenseke 2001) investigated the local conception of values in the cultural landscape. It concluded that the locals' main interest is to preserve the landscape as a living, open cultural landscape. But concerning specific natural or cultural values in the landscape, their knowledge is not very extensive or specific. This makes it reasonable to use a top-down method of investigation concerning this point.

area, but they are difficult to incorporate into actual physical planning. Moreover, the farmer will not have had the opportunity to construct priorities based on comparative studies done on other similar cultural environments – the consequence may be that either the historical remains in his fields are very important (to him), or that they are not worth bothering about.

2.1 The Threat Posed by Modern Farmers to the Preservation of Cultural Heritage

The idea that farmers pose threats to features in the landscape is not a very new or original one, as it has been demonstrated for years that their way of cultivating the soil homogenises areas into ploughed fields (or spruce plantations) and diminishes the biodiversity in the landscape (see e.g., Agger et al. 1986; Brandt 1994). The tendency in recent years is, however, for more green elements like hedgerows and small ponds to be added to the landscape (Primdahl 2001). It has also been demonstrated that the number of farm tracks and footpaths has been reduced by 40 percent between 1950 and 1988 (Højring & Caspersen 1999). Facing this background, the farmers were asked about their general attitudes concerning potential actions in the landscape, especially in regards to elements of historical heritage.

Two-thirds of the farmers said they were willing to remove walls or banks and hedges from the landscape, in order to create larger fields more suited to the machinery they used (Table 1). Some consolation concerning this opinion can be derived from the fact that since 1992, it has been illegal in Denmark to remove earthen banks and stone walls from the landscape (even if this does not in fact prevent farmers from doing so, as this investigation indicates – this applies particularly to older farmers (over 55 years old)). This legislation does not apply to hedges. Concerning field tracks and pathways, half of the farmers said they wanted to see them removed if this would aid in production, with the older farmers once again being the keenest to do so. This would have considerable consequences concerning public access to the cultural landscape. The farmers were also asked whether they wanted a burial mound removed from the fields, if it was seen to be hindering more efficient cultivation. The law has preserved all visible mounds since 1937. Only a small minority indicated that they would want them removed – this confirms that the law (in effect for more than two generations) has gained widespread acceptance among present-day farmers, although the sites themselves may date back 100 generations or more. But when it concerns mounds made by their much-more-recent forefathers, the farmers evince a willingness to destroy them (and religious feelings may also influence the attitude towards burial mounds). Furthermore, there is no clear tendency to be noted regarding the type of farm production and the role it plays in the answering of these questions.

Table 23-1. Consideration for elements in the cultural landscape.

Age	Remove walls* (earthen banks or stone walls) or hedges		Remove farm tracks or pathways		Remove preserved burial mounds		Total
	Yes	No	Yes	No	Yes	No	
26-40	11	7	8	10	1	17	18
40-55	17	10	11	16	0	27	27
56-75	19	1	12	8	3	17	20
	47	18	31	34	4	61	65
Production type							
Pig breeding	17	5	12	10	1	21	22
Dairy cattle	12	5	7	10	1	16	17
Cattle	1	2	1	2	1	2	3
Solely plants cultivation	10	3	6	7	1	12	13
Organic farming	1	3	1	3	0	4	4
Other	1	0	1	0	0	1	1
Pigs + dairy cattle	4	0	3	1	0	4	4
Pigs + cattle	1	0	0	1	0	1	1
	47	18	31	34	4	61	65

It should be stressed that the tendency for certain landscape features to be under threat will increase in the near future, as the current agricultural structure demonstrates a clear trend towards farms becoming bigger, and their number fewer. In 1960, there were about 180,000 farms cultivating an average of 18 ha each. In 1990, the number was 80,000, and in 2000 there were 54,500 farms, of which 28,000 farmed less than 30 ha. Furthermore, much of this land is also rented out; and, as mentioned earlier, one owner can express differing and contradictory attitudes from another owner or producer concerning the management of the land (Primdahl 1999, 2001).

From the farmers interviewed, it is possible to find out how much land they have and where this land is located (Table 2 – rented land is also included). In the first place, it was astonishing to see that almost half of the farmers (30 out of 65) only farm land in their own village – they may have bought or rented land from the former common land of the village, but basically they still farm within the economic framework constructed at the time of enclosure 200 years ago.[8] These farms are also somewhat smaller, and their owners somewhat older than the rest. A look at the larger group that farms land outside the village clearly indicates that the further away the

---

[8] Enclosure in Denmark was almost totally completed in the period 1780-1805, based on the main Enclosure Act of 1781, and implemented on a local basis through an interaction between an officially appointed surveyor, the landowner and the farmers.

land, the bigger the farm is; but not necessarily the younger the farmer. This tendency to farm more land bought or rented at a distance is quite critical in regards to the cultural landscape and its cultural heritage, as the farmer cultivating outside of his own village may be more likely to make rationale-based decisions regarding production that have negative effects on valuable elements in the landscape. As well, the knowledge of the farmer in terms of landscape elements and structures is crucial to his way of acting. Certainly, he is not able to possess the same knowledge of land further away as he has of the family farm, where he himself has grown up and has knowledge about the cultural landscape, its elements and the traditional way of cultivating it. 61 percent of the farms had belonged to the same family for more than one generation, but their farms have also been enlarged and will continue to grow in the future.

Table 23-2. Location of the areas farmed by the farmers interviewed.

	Number	Average size (in ha)	Average age	Number of pig-breeding farms
Only in own common village land	30	71.2	51.5	10
Outside own common village land	35	106.9	47	18
In one neighbouring village	13	83.5	50.5	6
In two neighbouring villages	3	63.3	45.7	2
In the same parish	6	85.5	46.7	3
In one neighbouring parish	4	133.25	38.8	1
In several neighbouring parishes	1	168	36	1
Inside 10 kilometres	3	136.7	40.3	2
In neighbouring village and far-flung location	3	159	53	1
In neighbouring parish and far-flung location	2	182.5	50	2

2.2 Modern Farmers' Willingness to Protect Cultural Heritage

Farmers are not just threatening the cultural landscape – they are also developing and creating it, and in this way they can also be seen as part of the management of the future landscape.

First, we can look at buildings belonging to the farms. The big farms need appropriate buildings, either for the machines or for stock. Two-thirds of the farms had actually added new buildings during the last 10 years. But how would they react to already-existing buildings when constructing new ones (Table 3)? Generally, the attitude of most farmers would involve considering the old buildings as much as possible – it is clear that it may be difficult to take into account the size of a building if a pig production system four times larger than the existing one has to be housed on the farm. But when it comes to the pitch of the roof or the choice of building materials, there is a clear tendency for the farmers to consider the existing architecture.

Table 23-3. Considerations for old buildings when erecting new ones.

Age	Consideration – size		Consideration – pitch of the roof		Consideration –choice of building material		Total
	Yes	No	Yes	No	Yes	No	
26-40	13	5	16	2	17	1	18
41-55	27	0	27	0	26	1	27
56-75	19	1	19	1	20	0	20
	59	6	62	3	63	2	65

When it comes to older buildings no longer in use, there is a strong veneration for them from the farmers (Table 4). A great majority (92 percent) wants to keep the buildings, but of these a clear majority (74 percent of all of the farmers) only wants to preserve the buildings if this can be done at no great cost. This means that most buildings will be preserved in the short term, but in the long run they will likely be demolished. Among the youngest generation of farmers, we see the clearest evidence of the tendency to demolish unusable buildings; but several in the younger generation are also willing unconditionally to preserve buildings.

Table 23-4. Consideration for old buildings and newly purchased farm buildings.

Age	Preserving old buildings			Bought farm property				Total
	Positive	Positive, but not at any price	Negative	Keep	Sell	Demolish	No attitude	
26-40	4	11	3	5	13	0	0	18
41-55	7	18	2	5	21	0	1	27
56-75	1	19	0	2	14	0	4	20
	12	48	5	12	48	0	5	65

The current trend of enlarging the size of farms also means that old buildings are bought along with the attractive land (i.e. in order to comply with stipulations regarding land ownership for the purpose of spreading manure from a big pig production). Farmers are well aware of the economic value of such buildings – most of them (three-fourths) want to sell them (along with some of the land) to people willing and wanting to move to the countryside. Another smaller group wants to keep ownership of the buildings, with the aim of renting them to their employees. But the crux of the matter is that nobody wants to demolish the buildings in the first instance – depending, of course, on the condition of the buildings.

The considerations for the buildings correspond very well with a recent study of Danish farms and gardens over the last 200 years (Dragsbo & Ravn 2001). The study shows that Danish farmers still devote considerable consideration to their nearest surroundings. They preserve the old traditional building styles (for instance, many still tend to build around a closed courtyard), and consequently also maintain tradition in regards to the choice of materials, colours of paint, and even the relation between buildings, gardens and plants. In the interviews for the study discussed in this paper, there was evidence of the farmers' same care for the aesthetic quality of their surroundings – they had to look at them every day, so they wanted to be able to appreciate the view. In the ethnological investigation, the researchers also remarked upon some clear regional differences concerning the details of buildings and gardens; but the investigation did not go further into details in regards to these differences.

Table 23-5. Farms, including different types of land-use and their management.

	Number of farms with the type	Grazed	Hay harvest	Unused – overgrown	Ploughed
Meadow	21	18	5	2	1
Bog	11	2		8	1
Grassland	8	6		2	
Tidal meadow	2	1		1	
Heathland	1			1	
Meadow?	4	1		3	
Bog?	3			3	

Concerning the cultural landscape, the farmers were also asked how they treated different types of landscapes (Table 5). It was interesting to note that only one-third of the farms today had meadowland – presumably all farms with roots in the 19th century (or earlier) had access to some meadowland in the village, as it was an essential component of enclosure to allocate all meadowland according to the size of farms. Fewer farms contained bog, grassland, tidal meadow and heathland. Most of the meadows are grazed by cattle, while a smaller number are hay meadows. Several of the bogs are not used, and are thus becoming overgrown. A few of the meadows and bogs are ploughed, even though that is not legal. As can be seen, there is uncertainty for the farmers as to the character of some areas. Although the Danish counties have registered all larger nature areas, they have not provided sufficient information about them (see also Primdahl 2001).

As mentioned, focus was placed on elements and structures in the landscape which are seen to represent particular values – i.e. cultural environments based on a provisional registration of elements and structures found in the investigation areas. On this basis, it was possible to discuss with the farmers and show them specific interesting features of their landscapes. They were then asked some of the same interview questions a second time, now having been told that the element in question was of special (national/regional) value. This produced some interesting results. When first asked, 37 percent of the farmers had said they were willing to respect old demarcations around the farmstead; for example, when constructing a new building, or after buying a neighbouring farm. But upon being told that the demarcation was of special value in the cultural environment (typically a village), a further 22 percent of the farmers indicated they were willing to respect these limits. And the same result occurred in relation to banks or walls in the landscape. Only 22 percent of the farmers had answered the first time that they were willing to respect them – generally, they were willing to respect stone walls, but not earthen banks. But after being told that a certain bank was of a special value, a further 56 percent – for a total of 78 percent of the farmers – were prepared to treat the bank with respect without external pressure being placed on them (i.e. on their own initiative).

Table 23-6. Assessment of consideration for features of the cultural environment.

Age/Action	Respects features independently when informed	Respects features independently if it has elements and structures of *special* value	Respects features only if stated in a local plan for the specific area	Respects features only if it proscribed by law (generally for a type of element)	Respects features only if they are preserved by preservation order (involving economic compensation)	Respects features only if there is annual grant (for instance by CAP grant).	No opinion	TOTAL
26-40	3	9	1	3	0	1	1	18
41-55	8	12	2	3	0	2	0	27
56-75	3	11	2	2	0	1	1	20
	14	32	5	8	0	4	2	65

In a more general fashion, the farmers were asked about their attitude towards cultural environments (Table 6). A large majority (70 percent) said they were willing to take care of the elements and structures, provided they know of their existence. More than half of this number responded that the elements must be of special value, before they would take their preservation into consideration. Regulations handed down "from above" (i.e. governmental action) are not very popular with the farmers – they want to decide for themselves how to treat their farmland. But this particular study demonstrates that they are willing to listen to the knowledge and listing recommendations of experts. They were not asked beforehand in the utmost priority as to whether conservation was dependent on a certain grant of money – of course, it can be assumed that they would not refuse a grant of a certain amount. But this investigation can be interpreted as suggesting that most farmers are willing to cooperate in the preservation of the cultural environment – on the condition that they are given pertinent information in relation to this.

Table 23-7. Historical management of nature areas.

Type of production	Meadow grazed		Hay meadow		Grassland		Coppice wood	
	Yes	No	Yes	No	Yes	No	Yes	No
Cattle	3	1	2	2	2	2	1	3
Dairy cattle	18	3	13	8	16	5	4	17
Pigs	23	5	18	10	19	9	9	19
Other	1		1		1			1
Organic	4		4		4		3	1
Crop only	14	2	12	4	8	8	7	9
All farms	55	10	43	22	44	21	21	44

Concerning the types of land-use, the farmers were also asked about their willingness to manage such areas, if they belonged to their property (Table 7). The answers show a varying willingness to manage such areas – depending upon the type of production involved. An interesting result is that the willingness to manage such areas diminishes in proportion to the time lapsed since the last occurrence of such management in Danish agriculture: for instance, coppice woods disappeared after the Second World War, and coppicing is now an almost unknown method of woodland management, with only one-third of the farmers willing to resume it if deemed appropriate. And it is not surprising that the cattle farmers are the ones most interested in mowing practices.

2.3 The Consideration Given to Cultural Environments by the Younger Generation of Farmers

As already pointed out, there are differences in the attitudes among the generations of farmers. It is almost unambiguous that the younger generation is more willing to preserve parts of, or entire, cultural environments. An explanation for this might be that the younger generation – because of the education it has received – has been taught to place substantial priority on environmental issues. Therefore, the younger farmers are more used to considering ideas related to the preservation of environments, and can easily adapt to the cultural heritage and the cultural landscape. For instance, they were quite unwilling to remove earthen banks from the landscape, in contrast to the older generations of farmers. Another good example regarding the attitude of young modern farmers was that, when asked about their wishes concerning their farms 30 years down the road, a desire for increased size was noticeably absent from their expectations (Table 8). The majority of the young generation was satisfied with and wanted to keep the existing structure, while the older generation (above the age of 56) displayed a tendency to want their farms to become bigger, even if they would not themselves live to experience it. They had started the development towards bigger farms, and could not imagine that this development could stop, or that halting it could be desirable.

Table 23-8. Hopes for the agricultural structure in 30 years.

Age	A. Return to old small farm sizes (before 1965)	B. Retain the present structure	C. Double farm sizes	B in % of A+B+C
26-40	2	13	3	72.2
41-55	1	21	5	77.8
56-75	2	8	10	40
	5	42	18	64.6

3. DISCUSSION AND WIDER PERSPECTIVES

In relation to the future management of the countryside, it is a widely-accepted political goal that it should be sustainable; however, there is usually no precise definition of this term. With special attention paid to cultural identity and more specifically to cultural environments, it can be claimed that the aim is to preserve selected cultural environments (i.e. elements and structures having national or regional values) for the long-term future. There is in all probability agreement as to the general aims involved, but uncertainty exists and discussion is needed as to the number and quality of selected features to be preserved.

In all events, the following preconditions are relevant if such a general goal of sustainability is to be achieved:

1. The authorities should produce a basic registration of cultural environments, especially in regards to the historical period (after 1050 AD);
2. The farmers should understand the contents of cultural environments (the first point mentioned in this list is a precondition of this), and should be willing to take part in conservation, or in better management to effect conservation;
3. Conservation should be economically sustainable for society.

This investigation shows that the farmers are willing to manage cultural environments in co-operation with the authorities, as long as they do not incur substantial costs. But it is very important that the farmer has – or can get access to – the relevant information concerning the cultural environment, so that he knows what he especially has to take care of or manage in some particular way. It has been stressed in other investigations that information is a crucial point for countryside management involving farmers (e.g., Wilson & Hart 2001); furthermore, this information has to be accompanied in most cases by personal advice, in order to establish a dialogue and overcome the perception of farmers that the authorities are setting themselves up as experts. Therefore, the first precondition stated in the list above is essential in order to achieve the desired goals. Unfortunately, the political situation in Denmark has become less favorably-disposed towards a further basic registration of cultural environments since the time of this study's interviews, and the whole task with regards to preserving cultural environments presently seems fraught with difficulty.

This is so much the more aggravating as the two other preconditions seem to have been fulfilled to a large extent. Two-thirds of the farmers are willing to enter into dialogue about the cultural heritage found on their farms, and are willing to preserve essential features. This may prove to be cheaper than if public management by authorities were to take care of the preservation – and would likely be more sustainable in the long run, as there will be a local consciousness instilled in farmers as to what cultural heritage means and how it has to be treated, a knowledge that may be handed on down for future generations.

A realistic way of handling such a goal may be to construct individual plans for the natural and cultural environments of farmlands, using a dialogue between farmers and authorities. This has been done in Sweden for several years with great success, and has also been tested in Denmark concerning aspects related to natural environment. To integrate the cultural dimension would strengthen the management of cultural heritage on the local level, and do it in a sustainable way – economically, ecologically, socially and culturally.

ACKNOWLEDGEMENTS

The author would like to thank John Mason, M.A. for language assistance.

REFERENCES

Agger, P., Brandt, J., Byrnak, E., Jensen, S.M. & Ursin, M. (1986). *Udviklingen i agerlandets småbiotoper i Østdanmark*. Roskilde: Forskningsrapport, Publikationer fra Institut for Geografi, Samfundsanalyse og Datalogi, 48.

Beedell, J. & Rehman, T. (2000). Using Social Psychology Models to Understand Farmers' Conservation Behaviour. *Journal of Rural Studies,* 16, 117-127.

Brandt, J. (1994). Småbiotopernes udvikling i 1980'erne og deres fremtidige status i det åbne land. In J. Brandt, & J. Primdahl (Eds.), *Marginaljorder og landskabet –*

marginaliseringsdebatten 10 år efter (pp. 21-49). Rapport fra et tværfagligt seminar afholdt af Landskabsøkologisk Forening i samarbejde med Institut for Økonomi, Skov og Landskab, KVL, fredag den 25. september 1992 – Forskningsserien nr. 6. Lyngby: Forskningscentret for Skov & Landskab.

Dragsbo, P. & Ravn, H. (2001). *Jeg en gård mig bygge vil – og der skal være have til. En kulturhistorisk-etnologisk undersøgelse af lange linier og regionale kulturforskelle i gårdens landskab: Bygninger, haver og omgivelser ved danske landbrugsejendomme 1900-2000.* Kerteminde: Landbohistorisk Selskab.

Elberg, K. (2002). Nutidige landbrugslivsformer. *Bol og By. Landbohistorisk Tidsskrift 2001:2,* 83-108.

Fund og Fortidsminder (2003). http//:www.dkconline.dk. 11.11. 2003.

Gaukstad, E. (2000). *Jordbrugslandskapets kulturverdier – utfordringer i et tverrsektorielt samarbeid.* Nord 2000:18. København: Nordisk Ministerråd.

Green, B. (1996). *Countryside Conservation: Landscape Ecology, Planning and Management.* 3rd Ed. London: E & FN Spon.

Højring, K. & Caspersen O.H. (1999). *Landbrug og landskabsæstetik.* Udviklingen i landbruget 1950-1995 og dens konsekvenser for landskabets oplevelsesmæssige indhold. Park- og Landskabsserien nr. 25. Hørsholm: Forskningscentret for Skov & Landskab.

Ihse, M. & Lindahl, C. (2000). A Holistic Model for Landscape Ecology in Practice: The Swedish Survey and Management of Ancient Meadows and Pastures. *Landscape and Urban Planning,* 50, 1-3, 59-84.

Kristensen, S. (1999). Agricultural Land Use and Landscape Changes in Rostrup, Denmark: Processes of Intensification and Extensification. *Landscape and Urban Planning,* 46, 1-3, 117-123.

Møller, P.G. (2001). Kulturmiljøregistrering. *Fortid og Nutid,* 1, 3-22.

Møller, P.G. (2002). Det moderne landbrug og kulturmiljøet. *Bol og By. Landbohistorisk Tidsskrift 2001:2,* 41-82.

Møller, P.G., Näsman, U., Ekner, B.D., Höll, A., Myrtue, A. & Sørensen, E.M. (2002). Bæredygtig arealanvendelse – den kulturelle dimension i landskabet. In P.G. Møller, R. Ejrnæs, A. Höll, L. Krogh, & J. Madsen (Eds.), *Foranderlige Landskaber - integration af natur og kultur i forvaltning og forskning* (pp. 144-179). Odense: University Press of Southern Denmark.

Pedersen, K., Jensen, L.B., Post, K., Zink, A.M. & Sørensen, K.D. (2002). *Katalog over EU's landdistriktsstøtteordninger i udvalgte EU-land*e. København: Landboforeningerne & Landbrugsrådet (http://www.lr.dk/planteavl/informationsserier/nyheder/lpnyhed124_eu-stoette.htm).

Primdahl, J. (1999). Agricultural Landscapes as Places of Production and For Living in Owner's Versus Producer's Decision Making and the Implications for Planning. *Landscape and Urban Planning,* 46, 1-3, 143-50.

Primdahl, J. (2001). Landmanden som landskabsforvalter. In T. Hels, K. Nilsson, J.N. Frandsen, B. Fritzbøger & C.R. Olesen (Eds.), *Grænser i landskabet* (pp. 219-228). Odense: University Press of Southern Denmark.

Renes, J. (1999). Evaluating Historical Landscapes. In G. Setten, T. Semb & R. Torvik (Eds.), *Shaping the land, vol. III The future of the past,* Proceedings of the Permanent European Conference for the Study of the Rural Landscape, 18th Session in Røros and Trondheim, Norway, September 7th-11th 1998 (pp. 641-650). Trondheim.

Stenseke, M. (2001). *Landskapets Värden – lokala perspektiv och centrala utgångspunkter.* Choros 2001.1. Göteborg: Kulturgeografiska Institutionen, Handelshögskolan vid Göteborgs Universitet.

Thybirk, K. & Haugaard, H. (2001). *Naturplaner på bedriftsniveau*. København: Danmarks Miljøundersøgelser & Skov- og Naturstyrelsen
(http://www.skovognatur.dk/natur/forskningsartikler/naturplaner_art.htm).

Wilson, G.A. & Hart, K. (2001). Farmer Participation in Agri-Environmental Schemes: Towards Conversation-oriented Thinking? *Sociologia Ruralis,* 41, 2, 254-274.

World Commission on Environment and Development. (1987). *Our Common Future*. Oxford: Oxford University Press.

www.sdu.dk/Hum/ForandLand/English/Index.htm. The Danish Research Centre *Changing Landscapes* 1997-2001 financed by the Danish Strategic Environmental Research Programme.

Chapter 24

THE HUMAN FACTOR IN BIODIVERSITY
Swedish Farmers' Perspectives on Seminatural Grasslands

Marie Stenseke
Department of Human and Economic Geography, Göteborg University, Sweden

1. INTRODUCTION

Landscape management and planning have become significant factors in post-productive rural areas. Through landscape policies, certain ideas about the fashion of the rural landscape are promoted. Policy-making, explicitly concerning the structures and qualities of the agricultural landscape in Sweden, was first initiated in the 1970s as a response to the ongoing abandonment of agricultural land and the degrading of biological values. Ecology and biodiversity developed as major concepts in the 1980s within rural landscape management, thus the planning of agricultural landscapes is very much the domain of the natural sciences (cf. Luz 2000). Agricultural landscapes, however, are not just a question of species, soils, water and climate, but also of culture, being inhabited by people and with a future dependent on human decisions and human activities. Moreover, landscapes vary as do the local and regional contexts, of which the physical features are integrated parts. With common agricultural and rural policies for a large part of Europe, there is an obvious risk that local and regional characteristics will be harmed by such general policies. Thus, landscape planning cannot only take physical facts as a point of departure, but must also deal with the human factor (Pretty 1998; cf. Van den Berg 2000; Bridgewater 2002). There is hence a need to find ways to integrate people and socio-economic aspects within landscape planning (cf. Fry 2001). That is the aim of this chapter.

The concern of this chapter is the maintenance of the seminatural grasslands in Sweden, focusing on the farmers' perspective on these lands. Seminatural grasslands are among the most species-rich landscapes in

H. Palang et al. (eds.), European Rural Landscapes:
Persistence and Change in a Globalising Environment, 397-410.

Sweden (Figure 1). Their biological and cultural values are the outcome of human intervention in ecological systems – human intervention in the form of animal husbandry. Thus, the creation and maintenance of the biological and cultural values have depended, and will continue to depend, on human intervention. Much of the seminatural grasslands in Europe have a long historical heritage with a rich biodiversity and are thus highly valued. This is not so in other parts of the world where animal husbandry is more recent and in fact seen as a threat to native species (Emanuelsson et al. 2001).

Figure 24-1. Swedish seminatural grasslands are lands with rich biodiversity, as a result of centuries of grazing. Photo: U. Emanuelsson.

Throughout the 20th century, the loss of seminatural grasslands have been substantial – in Sweden as in Europe as a whole. In Sweden we do, however, find significantly larger areas in terms of preserved acreage of seminatural grasslands than in most other north European countries.[1] Consequently, the country has a major international responsibility for the care of historical and biodiversity values in these landscapes. By signing the Convention of Biological Diversity, Sweden has pledged to preserve the existing values, and a number of national environmental aims related to seminatural

[1] There are vast areas of grasslands in the United Kingdom, Ireland and Iceland. They are, however, heathlands, which are very different from Swedish meadows and pastures (Gimmingham 1972).

grasslands have been formulated (cf. Rskr 2001/02:36). The main objective is to maintain the area of managed seminatural grasslands, which amounts to 450,000 ha. In order to reach this goal, policy measures have been introduced, mainly in the agri-environmental program within the Common Agriculture Policy (CAP) (SJVFS 2000:132).

Figure 24-2. The biological qualities of the seminatural grasslands in Sweden are dependent on human intervention in the form of animal husbandry. Photo: U. Emanuelsson.

Despite the introduction of agri-environmental schemes, the future of the seminatural grasslands is bleak because of the diminishing number of grazing animals, and the continuous abandonment of agricultural land. Thus, in Sweden unlike the situation in many other parts of Europe, it is not overgrazing but rather the shortage of animals that threatens environmental qualities identified with the grasslands. Furthermore, the diminishing number of people working in agriculture is similarly threatening, since human efforts are needed to maintain the values. Somebody has to look after the animals regularly, provide water and move them to pastures with fresh grass, check the fencing and remove bushes and the old grass that is left when the grazing season is over (Figure 2). The farmers hence play a crucial role in the preservation and enhancement of biodiversity. The development of policy measures thus requires an increased knowledge of how to make farming thrive, in addition to wider understanding of factors and processes in the differing local contexts that are of importance for the continuous management of seminatural grasslands.

2.   AIMS AND OBJECTIVES

This chapter aims at exploring Swedish farmers' perspectives on their seminatural grasslands: What are their experiences of seminatural pasture management, and how do they perceive the possibilities and hindrances for present and future use? In order to capture the local perspectives on these questions, case studies have been carried out with pasture owners and managers within the research project *The human factor in seminatural pasture management.*[2] The project is part of a Swedish multidisciplinary research program on *Management of seminatural grasslands – economy and biodiversity.*[3] The overall goal of the program is to promote farming systems where the use of seminatural grasslands plays an important role. The program is expected to result in good suggestions about future landscape policies for seminatural grasslands. This chapter is thus policy-oriented with the aim of fuelling the discussion by elevating the human factor.

3.   THE STUDY AREAS

Two areas with significant proportions of seminatural grasslands have been investigated, with the aim of developing knowledge about how the seminatural grasslands are integrated in the local context, and farmers' perspectives on these lands. The study areas, 137 square kilometres and 75 square kilometres respectively, include two parishes each. One of them is situated in Mälardalen (the Mälar Valley) in the east of Sweden, and the other one in Västergötland in Southwest Sweden.

Both areas can be characterised as scenic. In the eastern area (in Mälardalen), the mosaic of lakes, forest hills, pastures, fields and farms gives the landscape its character. In the western area (in Västergötland), earlier agrarian periods are still visible, for example in the medieval settlement

---

[2] The project is led by Marie Stenseke, Göteborg University and Ulrich Nitsch, Swedish University of Agricultural Sciences, Ultuna, and financed by MISTRA and Knut and Alice Wallenberg's Foundation. This chapter is based on the results from the first two studies. In continuing the project, comparative studies will be carried out in two more peripheral areas.

[3] There are ten projects within the program: biologists and agronomists, doing research on favourable management systems conduct six of the others. One study concerns the historical perspective, another one possible future and yet another is an inventory of good examples. Further information is accessible on *http//:www.cbm.slu.se.*

structures and the 19th century stone walls. Furthermore, the area is situated on the plateau and slopes of a distinct mountain, giving wide views.[4]

The population density in the areas is approximately 20 inhabitants per km[2]. In both areas, there has been a slight population decline over a fifty years period. The decrease seems, however, to have faded and the areas are quite favourable rural districts for the future, since they are within commuting distance of good labour markets. That said, as in the rest of the country, the number of farm units has constantly been declining during the post-war-period. According to the latest Swedish statistics, there were 78 units in the eastern area and 62 in the western area in 1992, which in both cases means less than a third of the number in 1944. There is, however, a difference in the farm structure, where larger estates are found in the Mälar valley, while small and medium size farms dominate the intermediate areas of Västergötland.[5] Consequently the average farm size (arable land) is notably higher in the eastern area (40 ha) than in the western area (15 ha) (Official statistics of agriculture in Sweden 1993).

In both areas, the amount of seminatural grasslands is the same as it was 50 years ago (5-10 percent). Forested land has slightly increased and covers about 50 percent, while arable land has somewhat decreased, at present covering 25-30 percent (Official statistics of agriculture in Sweden 1946 and 1993). The pastures are used for grazing by cattle or calves, heifers and dairy cows in their dry periods. Moreover, moderate and small herds of sheep are common in the areas. A few farms also have horses.

4. METHODOLOGY

Data has been collected mainly through the use of thematically structured interviews, informally conducted. In the eastern area, 16 farmers were interviewed at their farms in October 2001 and, in the western area, 18 farmers were visited and interviewed in February 2002. Apart from these interviews, county council executives and representatives of the local and regional branches of the Farmers' Union (LRF) have also been contacted for background information.

In order to get a broad picture of the farmers' perspectives there is wittingly a great variety among the respondents chosen and their farm enterprises. All of the interviewed farmers farm some seminatural grassland. As for most farmers in Sweden, the majority of the landowners in both areas

---

[4] With respect to the integrity of the interviewed farmers, the exact locations of the study areas are not revealed here.

[5] Here, small farms are defined as farms with less than 30 ha of arable land, medium sized farms are farms with 30-99 ha of arable land, while large farms manage 100 ha or more.

have an important share of their income from forestry on their own land. Their farms, however, differ in size, kind of agricultural activities and scale of production. The sizes vary from 31 to 595 ha of land. The difference between the two areas with respect to farm structure is reflected in the selection of interviewed farmers. Thus, on the studied farms in the eastern area the median size of arable land is 80 ha, and the median size of seminatural grasslands is 30 ha, while in the western sample of farms the median sizes are 48 and 20 ha respectively. 14 out of the 34 farm units are dairy farms. The other farms raise cattle or sheep. There are also differences in how the farms are operated. Full-time farmers run three quarters of the total sample, while the people operating the others also have a significant off-farm income. Three from the latter group could be categorised as "leisure farmers". Furthermore, six of the farmers in the eastern area and one of the farmers in the western area are tenant farmers. As this is mainly a qualitative study with a limited number of respondents, variations between different categories of farmers cannot be regarded as significant evidence. Thus, these variations will not be discussed and analysed in detail.

At the interviews, the respondents were almost always men, sometimes their wives also took part and even two generations of farmers occasionally represented the farm. During the interviews, notes were taken, and they were also tape-recorded. With regard to the respondents' various abilities to express their ideas, and their underlying motives for answering in one way or another, interpretations of what was expressed are also included in the results. Transcriptions of the interviews were analysed in an iterate process and structured around certain themes. Words and expressions were encoded and sorted into sub-themes, which then constituted the base for the analysis. In the following, some themes central to the aim of this chapter are presented. Citations from the interviews are *italicised*.

5.   FARMERS' PERCEPTIONS AND ASPIRATIONS

5.1   The Seminatural Grasslands

We want more of this kind of land – it gives varied scenery. It provides shadow and shelter for the life-stock.

In general the interviewed farmers have a positive attitude towards their seminatural grasslands. Some say they agriculturally could do without the seminatural grasslands, but because of their affection for these lands, they want to keep them and use them. For many others these lands are necessary parts of the enterprises. As for part-time and "leisure" farmers, the farming activities are to a great extent aimed at managing the landscape, and they adjust the number and kind of cattle to the land resources they have.

The growing attention to environmental aspects and nature protection has stimulated increased activities in the grasslands. During the last ten years, the area of seminatural grasslands has increased at the investigated farms through clearing and restoration. Furthermore, according to the farmers the biological qualities in the existing pastures have improved since they have become more intensively grazed, while the use of fertilisers has ceased. In the eastern area, all but one of the interviewed farmers has restored grazing land, and in the western area, half of them have done some restoration. The areas restored on single farms cover up to 30 ha. Some received payment to cover the costs for restoring grazing lands, but most of the farmers had carried through the cutting and fencing with no subsidies but the extra payment afterwards in the agri-environmental measure for managing pastures. Nowadays, seminatural grasslands are sometimes discussed, especially among farmers. They talk about management strategies, they sometimes compliment somebody's well-kept pasture but most often they comment upon the regulations concerning environmental measures, the visits of executives and other experiences of contacts with the agricultural authorities. As important parts of the scenery in both areas, the pastures are also taken into account when local development strategies have been discussed.

5.2 Values

I enjoy the view every time I pass my pastures and see them well kept and with good fencing. I like the mosaic – the clump of trees in the grassland and the oak hills ... It is as beautiful to me as to the tourist passing by.

The farmers appreciate their seminatural grasslands, not just because of their economic value in terms of agriculture, they see other values too. Many of them mention the unity of flowers, history, beauty and openness as a virtue. Moreover, the view of grazing cattle, sheep and horses is also recognised as a special quality. The farmers want the lands to be well kept, as a signal of vitality and human care.

The alternative is a spruce-curtain. That would mean ugly scenery.

The value of the seminatural grasslands, most often mentioned in the interviews, is their openness. The great appreciation of pieces of open land has to be understood in the Swedish context, where only 8 percent of the land area is agricultural land. The open agricultural land is seen as a sign of a viable countryside, and the forest as a sign of desolation or the abandonment of cultivated land. In general, the interviewed farmers consider themselves to live in very special areas, with certain qualities. The varied scenery is seen as a great asset. When asked about the alternative of farming in more prosperous agricultural areas, the common response from the farmers was that though that would mean better conditions for farming, they prefer the beauty of their mosaic landscape and would therefore not want to move or farm elsewhere. Due to the attention paid and the financial support offered for the land's maintenance, the farmers are quite aware of the biological values within their seminatural grasslands. Most of them have an increasing interest and knowledge in wild flowers, and recognise rare species indicating rich biodiversity. Moreover, many of the farmers have their own ideas about how to maintain or enhance environmental values on their farms, pondering upon how alternative grazing regimes, variations in grazing intensity or the amount of trees and bushes favour different species.

5.3   The Environmental Measures

In general, the farmers are positive to the policy measures that offer payment for the management of seminatural grasslands. Most of the seminatural grasslands in the areas are a part of the Swedish agri-environmental policy scheme for biodiversity (SJVFS 2000), and the managers are paid 1.000-2.400 SEK (120-290 Euro) per hectare for managing the pastures. The measures give better economic conditions for the maintenance of the lands.

It is peculiar that it has to be them against us.

Most of the farmers, however, have complaints about the intricate construction of the schemes within CAP, accompanied by an extensive administration complex. The static structure of the measures and weak incentives for dialogue are also regarded as problematic. Strict prescriptions on the character of the pastures as regards the number of trees and bushes, and the length of the grass at the end of the grazing period, with little possibilities to make local solutions, are commonly questioned. It is further assumed that people with less understanding of agriculture have formulated the regulations of management activities. The static structure undermines part of the freedom of being a farmer, which is considered to be one of the greatest virtues of the profession.

The environmental measures are promoting the development of seminatural pasture deserts.

The farmers commonly disagree with the authorities as to what should be the proper number of trees and bushes in the pastures. The farmers would like to see more vegetation in the grasslands than the regulations in the measures stipulate. Trees and bushes are desirable as shelter for the grazing animals. Moreover, there is also a common opinion that trees and bushes belong to this kind of land-use, and give a more varied landscape character. Some farmers have chosen not to include certain areas in the payment scheme, though that would have rendered them some economic yield. They have chosen to "save" groves in favour of birds and small mammals for the sake of landscape heterogeneity.

The ongoing change of the economic conditions in agriculture, in which direct payments are an increasing part of the farm income, is perceived to bring new uncertainties to the business. Since landscape management measures were first introduced in Sweden in the 1980s, the farmers have experienced a number of changes in the structure and the management regulations in these measures, thus making it difficult to rely too much on the present schemes. Some of them have also been given contradictory instructions and advice from different executives. Consequently, most farmers try to make moderate adjustments, but they hesitate to let the measures be completely decisive, and avoid making greater changes in their farming activities.

5.4 The Future

I see it as my duty to maintain the pastures the way they are.

There is a strong incentive among the farmers to continue maintaining the grasslands. It is commonly considered as a moral duty to maintain the grasslands for their biological values, but even more for keeping the landscape open and well kept. In general, the economic yield is of great importance for their future farming activities on the grasslands. The farmers do not want to be exploited, but to get a fair return for their work.[6] Many of them fear changed economic conditions lurking in the enlargement of the EU. Economy, however, is, not alone decisive. At the very least, most farmers are convinced that the grasslands that are visible from the farmyard will be maintained whatever happens. A couple of farms that plan to change from dairy production to a different kind of farming, try to find solutions that will make use of the grasslands. For those with off-farm incomes, running small farm units, the development of farming systems which demand a limited amount of work-facilitated feeding procedures and light shelter constructions for winter use, is of more interest than a great economic return.

It is a mission in my life, but what will happen when I leave?

---

[6] There is of course no general opinion about what is a "fair return".

There are also other aspects that may threaten the management of pastures in the long run. As regards the farm structure, the farmers expect the enlargement of units to continue. In the eastern area, quite a few farmers worry about future successors on their farms. Even though they have sons and daughters, they doubt that any of them are willing to take over the farm business. In the western area, the farmers are predicting a continuing polarisation of farm units towards large farms on the one-hand and "leisure farms" on the other. Some of the interviewed farmers consider the amalgamation of farms as a threat to the seminatural grasslands. They believe that one farmer can only care for a limited area of seminatural grasslands, and therefore the manual work needed in order to maintain the landscape qualities might be neglected on large farms.

One option, often mentioned by farmers in the western area for small and medium size farms to stay in business, is increased co-operation. The management of the seminatural grazing lands could become more effective if there was more cooperation as regards machinery and grazing. The supervision of the grazing animals could preferably also be co-ordinated. Furthermore, mixed grazing (with various grazing animals), which favours intensive grazing, can be arranged through the collaboration of farms. The efforts of moving animals could also be saved if two or more farms decide on joining their pastures into a larger grazing area.

Other visions for the future concern public access to the seminatural grasslands. Some farmers discussed the possibilities of letting more people experience the pastures. Taking pride in their valuable grasslands, they would like others to enjoy the scenery and the rich biodiversity. That would be inspiring and make their activities more meaningful. Hopefully, it would also make the public more aware of the present conditions in farming and the amount of work needed to care for certain values in the agricultural landscape. In the western area, the farmers were happy to have a common path made by the local government some years ago, stretching over the mountain and into the agricultural landscape. They would like to see more walkers using it. In the eastern area, most of the farmers could not see any signs of appreciation from the public. Commonly, it is only during the visits of the executives of the county council that they get encouraging expressions about their maintenance efforts. A couple of farmers would like to erect signposts to show where the seminatural grasslands are. In line with these ideas, there are also various suggestions about rural tourism as a means of commercialising biodiversity and the attractive scenery. One of the farm units in the eastern area had commercialised the scenery and the biodiversity. There were three houses and cottages for rent on the farm. Furthermore, they offered horseback riding, sauna baths, catering services, and there were also marked trails for visitors that took them out in the fields, forests and pastures.

There are, however, certain exceptions to the willingness towards promoting public access. On one of the investigated farms, a piece of land with a scenic waterfall had been turned into a nature reserve against the will of the landowners. The adverse attitude of that farmer was due to several years of unsatisfactory interaction concerning the practical arrangements during the short spring period when vast amounts of people come by car to see the waterfall.

6. APPLICATION WITHIN LANDSCAPE POLICY

What are the inputs to landscape policy-making that reveal themselves in this study? First, the study shows a great variety among the farms and farmers of the seminatural grasslands – They differ in size, farming activities, intensity and social structure. Since the number of grazing animals in Sweden has to be maintained or increased in order to maintain the seminatural grasslands in the country, suitable solutions have to be developed for large as well as small enterprises. It is important not only to focus on large-scale management methods, but to also search for effective techniques and management methods for small and medium size farms as well as time- and cost-saving solutions for "leisure farmers". It can be noted that the "leisure farmers" in the investigated areas in general farmed grasslands that were less attractive to other farmers. If it were not for their management, however, many of the pastures would otherwise have been out of production.

Having stressed the importance of addressing a variety of farmers that keep grazing animals, three other aspects can be pointed out: *economic efficiency*, *awareness and appreciation among public at large*, and *local recognition*.

6.1 Economic Efficiency

Economies of scale have to be recognised as a central factor in the structural change in agriculture. Amalgamation and the enlargement of farms can therefore be expected to continue in the study areas. Many of the interviewed farmers expressed an interest in increased co-operation as a means of making better use of labour and machinery and to make the maintenance of seminatural grasslands more effective. These attitudes obligate researchers and administrators to explore ways of further stimulating collaboration and to search for forms of agreement and arrangements that have been developed and successfully used in rural areas. Increased co-operation between farmers running small and medium size farms is likely to favour networks and interactions, thus attracting people to live and continue farming in the area. In the small-scale farm structure of the western area, it is thus important to promote time- and cost-saving solutions in the farming systems. Among the larger farm units in the eastern area, the benefits would rather be derived from developing shared knowledge about managing seminatural grasslands and co-ordinated strategies as to how to take care of grasslands that face the risk of being abandoned. Furthermore, for both areas there is an obvious economic potential in attracting more visitors. Promotion could be organised through local community activities, highlighting landscape qualities as an asset.

6.2   Awareness and Appreciation among the Public at Large

Many of the farmers wanted others to experience the values of their pastures. Public interest is considered to be a stimulating sign of appreciation. It makes the efforts to keep the grasslands more meaningful. Increased access to the rural landscape can therefore be said to improve the long-term conditions for the maintenance of seminatural grasslands, in encouraging the farmers to continue their activities and favour a shared understanding of landscape values. In Sweden, there is by tradition *a public right of common access*, which allows people to explore the rural landscape in their own way. However, more can be done to invite people to experience the scenery, the smells, the colours and the serenity of the grasslands. Signs telling where to find valuable grasslands and common paths guiding the way in the countryside would promote visits to these lands. These lands could also be used in education, supporting a long-term increase of the public awareness of the values of grasslands and the context in which they are managed.

6.3   Local Recognition

Landscape planning has much to gain from the involvement of local actors, by making use of their positive attitudes and knowledge. Environmental values are more effectively protected by recognition rather than designation, meaning that sustainable activities are better implemented through awareness, understanding and consensus, rather than through regulation (cf. Meldon & Skehan 1996). The interviews reveal a great interest among the farmers in keeping the rich biodiversity of their grasslands. There is a pride in managing certain values, and also an apparent sense of stewardship. By presenting various ideas as to how to care for the landscape values, many farmers show themselves willing to take a more active part in the process. There is, thus, a potential in getting the local actors more involved in landscape management and planning. By the participation of local actors, landscape design and management methods are likely to be better adjusted to the local context. Promoting local solutions will possibly also reduce the risk of "homogenisation by conservation".

7. CONCLUSIONS

The results from the case studies show that owners and managers of seminatural grasslands in the study areas have in general a positive attitude to these lands. For many of the farmers interviewed the grasslands are integrated parts of their farm units, and, as such, necessary for the ongoing farming activity. The economic values of the pastures are, however, in many cases less important than their values as well-kept and open lands, as signs of a viable countryside. Their beauty, the variety of flowers, and the cultural heritage are also very much appreciated. The farmers do indeed want to see a continuous management of the seminatural grasslands. For that to be fulfilled, the provision of economic support is one condition identified. Most of the farmers, however, are not comfortable with the static structure of the present environmental measures for seminatural grasslands within the CAP, giving little possibilities for farmers to communicate their ideas of how the grasslands better should be designed and managed. Furthermore, the studies indicate that the lack of successors, the ongoing enlargements of farm units and the decrease in the number of farmers are hindrances for the betterment of grasslands.

One conclusion is thus that future common agricultural and rural policies should acknowledge the seminatural grasslands as parts of local "lifeworlds" and give good conditions for the development of many locally adapted solutions. Included in this strategy is also a fruitful interaction between landscape planning and rural development. It has to be recognised, however, that knowledge, awareness and social learning are indispensable ingredients in increased local involvement and responsibility. To be able to adjust to local conditions and allow various local solutions, it is, furthermore, an urgent need to develop flexibility in landscape policies.

The two study areas represent agricultural districts in the more densely populated areas of Sweden, with good labour markets within feasible commuting distances. The number of inhabitants is expected to remain steady or even increase. The seminatural grasslands are significant features in the local environments, and are thus part of the locally perceived image of the areas. As these lands contribute to the acknowledged scenic values of both areas, attracting permanent residents as well as visitors to the areas and giving a potential for rural tourism should be aimed at. The results of this study are likely to be valid in similar areas. A considerable amount of the seminatural grasslands in Sweden are, however, to be found in more remote and sparsely populated areas. In these regions, where distance is also a factor to be reckoned with, the decrease in inhabitants and farmers can be expected to pose greater threats to the future management of the pastures. Hence, other kinds of hindrances and possibilities are at hand, and thus call for other kinds of solutions.

REFERENCES

Berg, L. van den (2000). Negotiated Multifunctionality of Agriculture at the Global Level: Experiences from an FAO-initiated Scoping Project. In J.T. Pierce, S.D. Prager & R.A. Smith (Eds.), *Reshaping of Rural Ecologies, Economies and Communities*. Conference proceedings. Commission of the Sustainability of Rural Systems, International Geographical Union (pp. 223-235). Vancouver: Simon Fraser University.

Bridgewater, P.B. (2002). Biosphere Reserves: Special Places for People and Nature. *Environmental Science and Policy*, 5, 9-12.

Emanuelsson, U., Berg, Å., Svensson, R. & Pehrson, I. (2001). *Management of Seminatural Grasslands – Economy and Biodiversity*. Mistra Programme Plan January 2001. Uppsala.

Fry, G. (2001). Multifunctional Landscapes – Towards Transdisciplinary Research. *Landscape and Urban Planning*, 57, 159-168.

Gimmingham, C.H. (1972). *Ecology of Heathlands*. London: Chapman and Hall.

Luz, F. (2000). Participatory Landscape Ecology – A Basis for Acceptance and Implementation. *Landscape and Urban Planning*, 50, 157-166.

Meldon, J. & Skehan, C. (1996). *Tourism and the Landscape. Landscape Management by Consensus. Sustainable Tourism in Europe's Peripheral Regions*. Dublin: An Taisce and Bord Fáilte.

Official statistics of agriculture in Sweden. *Jordbruksräkningen 1944*. Stockholm: SCB.

Official statistics of agriculture in Sweden. *Lantbruksräkningen 1992*. Örebro: SCB.

Pretty, J. (1998). *The Living Land. Agriculture, Food and Community Regeneration in Rural Europe*. London: Earthscan.

Rskr 2001/02:36 *Svenska miljömål – delmål och åtgärdsstrategier* (Swedish Parliament decision).

SJVFS 2000:132 *Statens jordbruksverks föreskrifter om stöd för miljövänligt jordbruk*. Jönköping: National Board of Agriculture.

Chapter 25

DIVERSITY OF ESTONIAN COASTAL LANDSCAPES: PAST AND FUTURE

Elle Puurmann[1], Urve Ratas[2], Reimo Rivis[2]
[1]*Silma Nature Reserve, Estonia;*
[2]*Institute of Ecology, Tallinn Pedagogical University, Estonia*

1. INTRODUCTION

Estonia is a small country, by the Baltic Sea, with a long shoreline (3,790 km), and thus coastal landscapes comprise a significant part of the country (Figure 1).

Every landscape represents a system of closely interrelated components – abiotic and biotic – in a definite structural complex that is constantly enriched with new features in course of its development. The structure, dynamics and development of the coastal landscapes are closely connected to geologic-geomorphologic features of the area. The ancient topography and lithological compositions of bedrock, as well as accumulation or erosion and distribution of glacial deposits and landforms, have played an important role in contemporary landscape development. Differences in the development of coastal areas were caused by the differences in the land uplift – the highest rate at the present time in north-western Estonia is 2.8 mm per year, but in the geological past it was much higher (Kessel & Miidel 1973). In the postglacial time the development of the landscapes was highly influenced by different phases of the Baltic Sea. The well-developed coastal forms relate to the Ancylus Lake and Litorina Sea transgression. The majority of today's coastal landscapes emerged from the sea in different phases of the Limnea Sea. The development of coastal landscapes is also connected to vegetation, the hydrological regime and climatic conditions as it is transition area from maritime to continental. Based on these features, we may distinguish a series of ecosystems characterising coastal landscapes.

H. Palang et al. (eds.), European Rural Landscapes:
Persistence and Change in a Globalising Environment, 411-424.

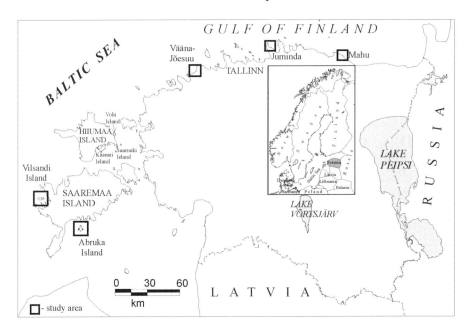

Figure 25-1. Study area and location of sites mentioned in the text.

Being located on the boundaries between terrestrial and marine systems, the coastal landscapes represent a specific structure. The coastal area can be divided into a narrow contemporary shore and inland areas – the territory with ancient coastal formations.

The contemporary coast is the most open and the most remarkable part of coastal landscapes, where soil and vegetation are in the beginning stage of their development. Generally there are two main types of coastal landscapes: areas with active coasts and low coasts with coastal grasslands. A third type is more concerned with constructions. The development of contemporary coast depends mainly on openness to prevailing winds and waves, as well as on their exploitation. Within the main part of coastal landscapes, natural processes have stabilised and are very similar to those of inland areas. Depending on variations in edaphic conditions, different meadow and forest types have developed here. This part of the coastal landscape has been mainly influenced by human activities.

2.   METHODS

Our research on the coastal areas of Estonia started some decades ago. Most of the landscape field studies were carried out between 1995 and 2002.

Monitoring of the coastal landscapes was included in the state environmental monitoring programme in 1996. A total of 26 coastal landscape monitoring areas have been planned (Ratas et al. 1997). Each monitoring area is an area bordering the shoreline and covering the territory of recent coasts and ancient coastal formations.

The width of the studied area (up to 10 km^2) towards the inland is generally 3 km. A great majority of islands in the Estonian shallow coastal sea have an area less than 10 km^2, all belonging to the coastal landscape.

Different study methods have been used for landscape diversity investigations. Maps of localities on a scale of 1:10,000 served as a basis for our research (Ratas et al. 1997). By *locality* we understand the natural territorial complex connected with the mesorelief and mechanical composition of the deposits. The parameters of abiotic environment are under study that act as circumstances of habitat diversity on an area. Moisture regime, nutrient availability, acidity and salinity are revelant abiotic factors for existence of plant-soil systems. The landscape fragmentation within the localities is expressed by vegetation site type. Synthetic maps of landscape units and vegetation indicate the landscape pattern at the present time. Landscape studies enable us to find the optimal ways of using and protecting these areas.

To characterise landscape diversity, the following quantitative parameters have been used: the number of landscape units (localities), number of landscape types, edge index (1) (Forman 1997) as well as Pielou index (2) (Pielou 1975), Patton index (3) (Davis 1986).

1) $I = \Sigma L / A$

2) $D = H / \ln s$

3) $\Sigma T = 4\pi Ai / Pi$

where L — length of edge (m) between different landscape units, A — total area of study site (ha), H — Shannons diversity, s — number of landscape types, Ai — area (m^2) of landscape unit, Pi — perimeter (m) of landscape unit. Edge index shows how mach edges different landscapes has. Patton index characterise shapes of landscape units, and Pielou index shows that some landscape type is dominating or not.

The methods of landscape profiles, showing cross-sections of different landscape units, have been used for studying the landscape structures and mutual relationships between their components. In addition to vegetation units, lists of vascular plants, moss and lichens were compiled to gauge various aspects of biodiversity. In several study areas, the vegetation has been studied in detail, using quadrates of 1 x 1 m in meadow communities and 10 x 10 m in shrub and woodland communities.

On the basis of maps from different periods, the land cover units were distinguished and the trends of the changes in landscapes were analysed. By this retrospective method, the changes in land cover can be followed back to the beginning of the 20th century (in some cases even earlier). The trends of landscape changes are revealed by computer processing of maps. An analysis of the trend of landscape changes would contribute to our understanding of the degree of tolerance and resistance of landscapes to exploitation.

3.   DIVERSITY OF COASTAL LANDSCAPE

3.1   Landscape Structure in Relation to Natural Factors

The diversity of the coastal landscape depends mainly on the variation of geological features, especially deposits (substrate), the climate and hydrological conditions (the distance from sea), the development stages of ecosystems and the influence of human activity. The time factor is great significance, determining the general development of landscape on Estonian coast in the uplifting area.

The coastal landscapes are remarkably diverse, as the development of nature proceeds on the borderline of land and sea. The development of the landscape diversity of contemporary shores is greatly affected by the sea. Determinative role in biodiversity development is the duration and extent of seawater influence. The low-lying coastal area is characterised by a flat sodded shore where seashore meadows with halophilous vegetation on Salic Fluvisols are developed. They are more widely distributed in West Estonia. Seashore meadows support a considerable diversity at the community level, despite the communities themselves not being especially species-rich.

Characteristic features on active beaches are the erosion and sediment movement, where there is no overall soil cover and pioneer communities represent vegetation. The substrate (sand, gravel, pebble) influences development of the biotic component of coastal ecosystems through creating the conditions for plant colonisation, nutrition and water exchange. Biodiversity is evident in the existence of different communities and these areas are habitats for many rare and endangered species.

The last decades have witnessed a remarkable activation of geological processes. Storms cause great damage on sandy beaches, especially in the autumn-winter period when the water level is relatively high and the coastal sea does not turn into ice. At the same time, due to wave erosion of sandy and gravel beaches, some landscape units have been destroyed (Rivis et al. 2002; Orviku et al. 2003).

Seawater does not have a direct impact on inland areas of coastal landscapes. As a rule, farther from the shore, the landscapes become older and the natural processes have stabilised and there is conditionality between the ancient coastal landscape and the contemporary shore type. This can be explained by the existence of quite similar natural conditions in our coastal areas during the last few centuries. For example, sandy shore followed by dune landscape and sodded shore followed by till or sandy plain. Sharp borders are on cliff coast where areas on the bank are several or even tens of meters higher than the contemporary shore and where terrestrial processes have come into play much earlier and have prevailed significantly longer. The distribution of landscape units is primarily based on their location on the landform and character of the substrate, as well as the water regime. In studied areas usually 7-9 types of localities and 12-14 vegetation site types are distinguished. The edge index is very variable (50-150 m per ha).

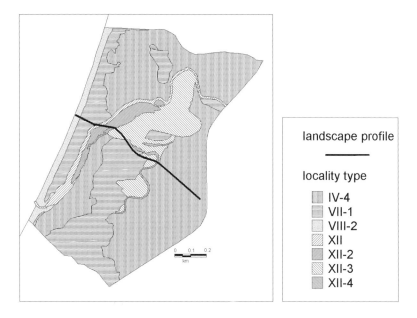

landscape profile

locality type

IV-4
VII-1
VIII-2
XII
XII-2
XII-3
XII-4

Figure 25-2. Landscape map of the Vääna-Jõesuu study site (for abbreviations see Table 1).

In places where the topography rises steeply from the sea, the landscape units are often represented by belts parallel to the shoreline, and the landscape structure is usually simple. This can be illustrated by the Vääna landscape profile, where the dune landscape structure is jointed by the mouth of the Vääna River (Figures 2, 3).

In flat areas, the coastal landscape has been formed through the linkage of small islands. The geomorphology of the islands causes differences in their landscape structure. On islands which have been developed by joining several small islands (Vilsandi Island, Saarnaki Islet etc.), the landscape structure is much more complicated than on islands which have risen up from sea as one area from the beginning (Abruka Island, Vohilaid Islet etc.) (Table 2, Figures 4, 5).

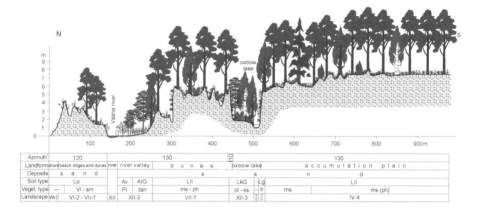

Figure 25-3. Landscape profile of the Vääna-Jõesuu study site.

Table 25-1. Abbreviations of landscape types used in the text and figures.

	Locality types	Vegetation types		
I-1	alvar	heath forests:	**sm**	*Cladina* site type
II-1	limestone plain covered with deposits (0.3-0.6 m)	dry forests:	**ph**	*Vaccinium vitis-idea* site type
			ms	*Vaccinium myrtillus* site type
II-2	abraded moraine plain	floodplain forests:	**tan**	*Carex* and *Filipendula* site type
IV-4	sandy plain	paludified forests:	**tr**	*Carex* site type
VI-1	ancient pebbly beach ridges	peatland forest:	**ks, ld**	minerotrophic swamp forest
VI-2	ancient sandy beach ridges	***Vegetation of dunes and coastal areas:***		
VI-3	ancient shingle beach ridges	**Vl**		white dune site type
VII-1	dunes	**Pi**		*Phragmites australis*
VIII-2	sandy seashore	Soils:		
VIII-3	sodden low shore (on gravely sand)	**Lo**		Cambic Arenosol
		LkI, Lg, LkG		Cambic Podzol
VIII-4	sodden low shore (on bedrock)	**Av, ArG**		Salic Fluvisol on coastal territories
VIII-5	sodden low shore (on till)			
XI-1	lake			
XII	river			
XII-2	floodplain			
XII-3	river terrace			

The area and height of Abruka and Vilsandi Islands are similar. On Vilsandi Island the numbers of landscape types and landscape units are higher than on Abruka Island – which means that the landscape structure of

Vilsandi Island is much more complicated. Also, the lower value of the Pielou index and the higher values of Patton and edge indexes indicate the same.

Table 25-2. Quantitative parameters of landscape structure of Vilsandi and Abruka Islands.

Island	Area (km²)	Max height (m)	Landscape types	Landscape units	Pielou index (%)	Patton index	Edge index (m/ha)
Abruka	9.4	7.6	5	34	93	0.18	71
Vilsandi	9.2	7.4	9	15	86	0.25	115

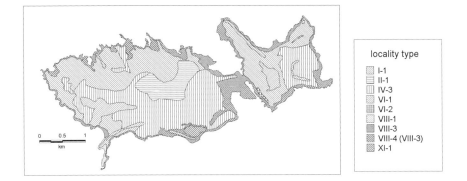

locality type
- I-1
- II-1
- IV-3
- VI-1
- VI-2
- VIII-1
- VIII-3
- VIII-4 (VIII-3)
- XI-1

Figure 25-4. Landscape map of Vilsandi Island.

3.2 Landscape Diversity in Relation to Land-Use

The natural (geological, topographic and edaphic) conditions and the ensuing suitability of land for exploitation are of crucial importance in determining the use of different landscapes. Some effects of human impact have been transient, whereas others have endured for centuries (Lepart & Debussche 1992). Landscape change due to human settlement is manifested foremost in changes in the vegetation, a component of the landscape which can easily be altered. In most cases human activities affect directly the plant cover and landscape pattern.

The traditional resource uses on coasts have included reed cutting, grazing and mowing, and to a lesser extent forestry. Often the shores have already been influenced by human activities from the very beginning of landscape development. The activities of coastal people (fishery, navigation, military activity) are also specific, and affect the diversity of the landscape pattern. Human activity brings along the appearance of new land cover units, making the landscape more varied, but intensive land use can completely destroy some units (Ratas et al. 1996). Therefore, beside natural richness,

coastal landscapes represent also a unique type of cultural heritage, in terms
of ancient villages and peasant architecture, as well as semi-natural meadow
habitats like wooded meadows and coastal pastures.

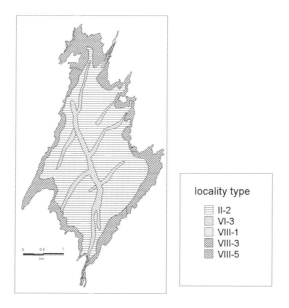

Figure 25-5. Landscape map of Abruka Island.

Generally two landscape types in the coastal area can be distinguished,
based on different combinations of land use and vegetation structure: these
are agricultural landscape and woodland.

3.2.1   Agricultural Landscape

Owing to the occurrence of naturally more fertile soils, there was a larger
agricultural area in the western part of Estonia, compared to the coastal areas
of northern Estonia. The cultivation of coastal areas started on higher
elevations, where edaphic conditions are more suitable for crop growing and
where soils on till as well as coastal deposits usually prevail nowadays. The
majority of fields have been in the same place for centuries, due to lack of
new land naturally suitable for cultivation. This is vividly illustrated by the
distribution of fields on Kassari Island, which has been one of the most
densely settled areas of Hiiumaa County through time (Kokovkin & Loodla
1998; Hellström 2002). On small islands the fields were often very small and
surrounded by stone fences to keep away animals. The low-lying meadow
areas along the shores have been shaped by haymaking and by cattle, sheep
and horse breeding for centuries. The open areas were interspersed with

shrubberies and woods. The most intensive exploitation of coastal landscapes took place from the middle of the 19th century to 1940 (Ratas & Puurmann 1995).

Generally comparative analyses of land cover maps of coastal areas from different times show that the share of the area devoted to agricultural uses has steadily decreased during the last 60 years. Meadows being replaced by reed beds, shrubberies or woodlands and overgrowth is the main threat to the biodiversity of coastal areas (Figures 6, 7). Major changes have been caused mostly by abandoning traditional land-use.

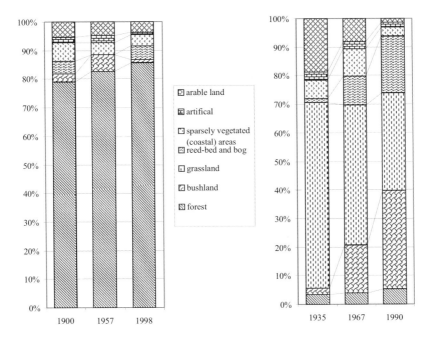

Figure 25-6. Land cover changes in agricultural landscape (Mahu study site) (left) and in woodland (Juminda Peninsula study site) (right).

The most species-rich plant communities in Estonia have been found in wooded meadows. The maximum number of vascular plant species on wooded meadows is 74 species in a 1 x 1 m plot, normally 30 (Kukk & Kull 1998). During the first half of the 20th century the wooded meadows represented a landscape type connected with Estonian peasant culture and were widespread in western Estonia.

The species richness of certain plant communities is affected by different land uses. Comparative analyses of average species richness show that mowed seashore meadows are the most species-rich, followed by grazed meadows and then former pastures (Märtson 1996). After cessation of traditional management activities, the older and more diverse plant communities are replaced by more species-poor ones. The influence of former grazing can be reflected in species composition even after ten years.

Afforestation is one of the most serious threats to open landscapes in the coastal area. For example, at the end of the 1980s the greater part of the *Empetrum* heaths on the small islands of North Estonia were planted with pine, and thus heaths have become subject to overgrowth. This is a great threat to the biodiversity of coastal areas generally, as in North Estonia the oligotrophic heath with *Empetrum nigrum* and *E. hermaphroditum* is on the southern boundary of its distribution area (Kukk & Ploompuu 1992).

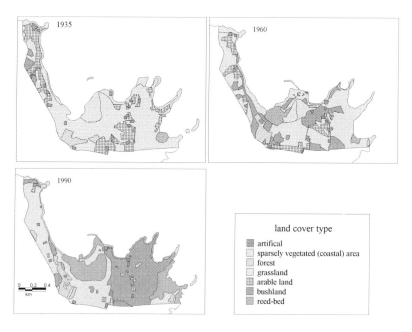

Figure 25-7 Land cover maps of Mahu Islet study site.

The visual exploration of landscapes is clearly linked to the landscape pattern. Usually land type diversity is contrast between forest and open fields and grain size define the possibilities of visual exploration in agricultural landscapes (Dramstad et al. 2001). The share of open and closed landscapes in Estonia has changed a great deal over the years, as has the area of different units. The main reasons for changes have been land reforms. Single households, peasants, and small farms have favored development of a more fragmented landscape pattern. Great landowners – estates, collective and state farms, large agricultural enterprises – have enlarged and joined land patches.

3.2.2 Woodlands

Forest is one of the most important natural resources in Estonia. The coastal areas of some regions were most heavily forested, due to sandy soils that were unsuitable for cropland. The conditions for clear cutting were also unfavourable, due to the value of forests in protecting soil, and usually these areas have been forested for centuries (Figure 7). Moreover, several acts protecting coastal forests were adopted already in earlier years. In 1764, a law was passed that declared woods within 50 fathoms of the island and mainland coasts to be protected zones, in order to preserve their silhouettes and to make it easier for seafarers to orient themselves. In 1839 the protected zone was widened to 150 fathoms. With the Forest Act adopted in 1888, a right to designate protected forest in the coastal zone was issued, with the aim of preventing sand removal from dune areas (Örd 2000).

The changes in woodlands are much smaller than the changes in agricultural landscapes. These changes mainly indicate different stages in management of forests: 1. from clearing to young stand; 2. from young stand to mature forest. A big part of coastal forests was designated as protected forest in the second half of 20th century, which meant that clear-cutting was prohibited.

During the last decades, in the vicinity of towns, coastal landscapes have become subject to increasing utilisation pressure, and the coastal forests have been replaced with private residences and summerhouses. This process is continuing nowadays. The most susceptible to recreation is the pine heath forest.

Until now the coastal landscapes of Estonia have been diverse, and the majority of the ecosystems are quite close to the original natural habitat. Most of the coastal area is under protection.

The coastal landscapes belong to the national landscapes of Estonia, and are seen as expressing the unique, but at the same time the most typical features of the country. The reasons for that are geographical location, diversity of climatic conditions, variety of bedrock and soil properties, the relatively big share of natural landscapes, and persistence of traditional extensive methods of land-use in Estonia until the middle of 20th century. In the second half of the 20th century, closed border zone of the Soviet Union was the major factor in the low utilisation of coastal areas.

The database of coastal landscape monitoring areas reflects diversity of coastal landscape and shows the land cover changes in the 20th century.

3.3   Current Tendencies of Landscape Changes

Landscape diversity and its dynamics reflect to a large extent the natural processes, as well as human influences.

As a result of land uplift, the rise of water level is not so important in our coastal areas. Most of the coastal villages are located at a distance from the coastline today. Thus the settlements in coastal areas of Estonia, except for the ports, are not in direct danger in case of a moderate rise in the sea level (Kont et al. 1997). A rising sea level will bring along greater changes in coastal wetlands, where halophilous habitats would be predicted to move to inland areas.

Current socio-economic changes (going over to a market economy, land privatisation, and changing land ownership) have put increased pressure on the coastal landscapes we know as valuable and traditional. The cessation of traditional activities led to ecological and visual degradation of agricultural land, as well as to a reduction of diversity of woodlands. Land-use changes have caused habitat loss and alteration, and landscape changes influence a wide range of countryside interests.

The most serious problem is connected with semi-natural communities. The perishing of traditional agricultural methods in the rural landscape is the reason for the great decrease of the area of meadow habitats. The majority of semi-natural communities that have survived in Estonia are unmanaged today, as a result of which their area may decrease even more in the coming years. The main factor responsible for decreases in biological diversity is changes in agriculture. For example, grazing or haymaking regulated the growth of reeds at the seashore meadows. It has been suggested that a suitable grazing capacity for coastal meadows is 1-1.5 head of cattle, 2.5-3 sheep, or 0.5-1 horse per hectare of meadow area.

For several protected areas on the Estonian coast, a management plan has been prepared, and management for conservation objectives is being carried out. Since 2001, nature management subsidies are being paid for the maintenance of the semi-natural communities in protected areas and in potential Natura 2000 sites beyond the protected areas.

Recreational uses have given coastal areas a steadily increasing value, while their traditional uses (e.g., as pastures or meadows or fishing sites) and the original settlements connected with such traditional sources of livelihood have declined. Processes such as uncontrolled recreation pose a serious threat to landscape diversity.

Therefore, more and more attention must be paid to social aspects and human activities. It is important to know how major human activities and uses can affect the coastal landscape and to know the relationships among different users, such as the people who live in and use the coastal zone, policymakers and managers whose decisions and actions affect the behaviour of coastal people, and members of the scientific community (1990).

Changes in a landscape are acceptable if they are in harmony with its fundamental character. The current landscape pattern and tendencies of landscape change are connected with the aims of nature protection and development of recreation along the coasts.

The future of landscape diversity also will depend on national and international land-use policies. As Estonia belongs to less productive region of agriculture, compared with other European countries, we can expect a decrease in the share of open landscapes after the country joins the European Union.

4. CONCLUSIONS

The coastal landscapes of Estonia are diverse and the majority of their ecosystems are quite close to the original natural habitat. The ancient topography and lithological composition of deposits have played important role in the formation of the pattern of coastal landscape. Due to land uplift, changes in coastal landscape pattern are of short-term duration in geological scale. The great variety in many characteristics reflects the recency of the area, and contacts of two different environments – sea and land. Centuries-old human land-use has played an important role in the formation of the present-day coastal landscape. The monitoring of coastal landscape has gathered more information of the present structure of the Estonian landscapes and the main trends of changes.

ACKNOWLEDGEMENTS

This research was funded by the Estonian Ministry of Education (grant No. 0282121so2) and conducted within the framework of the National Environmental Monitoring Programme. We are grateful to Prof. Ain Haas for improving the English version the manuscript.

REFERENCES

Carter, R.W.G. (1990). *Coastal Environments. An Introduction to the Physical, Ecological and Cultural Systems of Coastlines.* Academic Press.
Davis, J. C. (1986). *Statistics and Data Analysis in Geology.* 2nd ed. New York: John Wiley.

Dramstad, W.E., Fry, G., Fjellstad, W.J., Skar, B., Helliksen, W., Sollund, M.-L.B., Tveit, M.S. & Geelmuyden, E. (2001). Integrating Landscape-Based Values – Norwegian Monitoring of Agricultural Landscapes. *Landscape and Urban Planning,* 57, 257-268.

Forman, R.T.T. (1997). *Land Mosaics: The Ecology of Landscapes and Regions.* Cambridge: Cambridge University Press.

Hellström, K. (2002). Agricultural reforms and policies reflected in the farming landscapes of Hiiumaa from 1850 to 2000. *Acta Universitatis Agriculturae Sueciae. AGRARIA 325.* Doctoral dissertation. Swedish University of Agricultural Sciences.

Kessel, H. & Miidel, A. (1973). On the Late and Post-Glacial Crustal Movements in Estonia. *Proceedings of the Academy of Sciences. Chemistry and Geology,* 22, 257-264.

Kokovkin, T. & Loodla, K. (1998). Changes in Cultural Landscapes on Kassari Island since the Early 18th Century. *Proceedings of Estonian Acadademy of Sciences. Biology and Ecology,* 47, 114-125.

Kont, A., Ratas, U. & Puurmann, E. (1997). Sea-Level Rise Impact on Coastal Areas of Estonia. *Climatic Change,* 36, 175-184.

Kukk, T. & Kull, K. (1998). Diversity of Estonian Plant Communities and Problems of Their Protection. In V. Lilleleht (Ed.), *The Natural Diversity of Estonia and its Protection* (pp. 69-80). Tartu-Tallinn.

Kukk, T. & Ploompuu, T. (1992). *Empetrum hermaphroditum. Eesti Loodus,*4, 250-252.

Lepart, J. & Debussche, M. (1992). Human Impact on Landscape Patterning: Mediterranean Examples. In A.J. Hansen & F. Castri di (Eds.), *Landscape Boundaries: Consequences for Biotic Diversity and Ecological Flows* (pp. 79-106). New York etc: Springer-Verlag.

Märtson, K. (1996). *Dependence of Estonian Seashore Meadows' Species Diversity upon Environmental Parameters.* BSc thesis. Manuscript. Institute of Botany and Ecology of University of Tartu. Tartu.

Orviku, K., Jaagus, J., Kont, A., Ratas, U. & Rivis, R. (2003). Increasing Activity of Coastal Processes Associated with Climate Change in Estonia. *Journal of Coastal Research,* 19, 364-375.

Pielou, E. (1975). *Ecological Diversity.* New York: Wiley.

Puurmann, E., Ratas, U. & Rivis, R. (2002). Monitoring of Coastal Landscapes. In A. Roose (Ed.), *Environmental Monitoring of Estonia 2001* (pp. 114-116 & 186-187). Tartu: Tartu University Press.

Ratas, U. & Puurmann, E. (1995). Human Impact on the Landscape of Small Islands Structure to the Islets of West Estonian Archipelago. *Journal of Coastal Conservation,* 1, 119-126.

Ratas, U., Puurmann, E. & Kokovkin, T. (1996). Long-Term Changes in Insular Landscapes on the Example of Vohilaid, West Estonia. In J.-M. Punning (Ed.), *Estonia. Geographical Studies* (pp. 90-106). Tallinn: Estonian Academy Publishers.

Ratas, U., Puurmann, E. & Rivis, R. (1997). Monitoring of the Coastal Landscape of Estonia. In T. Frey (Ed.), *Problems of Contemporary Ecology* (pp. 191-196). Tartu: OÜ Greif.

Rivis, R., Ratas, U. & Kont, A. (2002). Some Implications of Coastal Processes Associated with Climate Change on Harilaid, Western Estonia. In F.V. Gomes, F.T. Pinto & L. das Neves (Eds.), *The Changing Coast. Littoral 2002* (6th International Symposium) Proceedings, Vol. II (pp. 133-139). Porto: Marca AG.

Vetik, R. (Ed.) (2002). *Estonian National Report on Sustainable Development 2002.* Tallinn: Akadeemiatrükk.

Örd, A. (2000). *Protected Forests and their Management.* Tartu: Tartu University Press.

Chapter 26

MANAGEMENT STRATEGIES IN FOREST LANDSCAPES IN NORWAY

Terje Skjeggedal[1], Tor Arnesen[2], Guri Markhus[3] & Per Gustav Thingstad[4]

[1] Nord-Trøndelag Research Institute, Norway
[2] Eastern Norway Research Institute, Norway
[3] Faculty of Social Sciences and Natural Resources, Nord-Trøndelag University College, Norway
[4] Museum of Natural History and Archaeology, Norwegian University of Science and Technology, Norway

1. INTRODUCTION

1.1 Perspectives

The main purpose of this work is to present some views on the management of landscape change in forest landscapes from the perspectives of ecological integrity and management regimes, using a transdisciplinary approach. This gives us an opportunity to describe the practice of various management regimes and discuss future management strategies. The presentation is based on the project *The Battlefield of Regimes*, financially supported by the Research Council of Norway (Skjeggedal 2001; Skjeggedal et al. 2001). This project is mainly concerned with the management of landscape change in forested areas in an ecological perspective.

Only 0.7 percent of Norway is developed as villages and towns, and about 3 percent is arable land. The rest is mainly forest or mountainous terrain. Nature conservation areas protect about 10 percent of the country. Although the Planning and Building Act applies to the whole country, it is mainly used in developed areas. Separate sectoral legislation like the Agricultural Act, the Forestry Act and the Nature Conservation Act regulates the agricultural and forestry infrastructure, the utilisation of natural

425

H. Palang et al. (eds.), European Rural Landscapes:
Persistence and Change in a Globalising Environment, 425-444.
© 2004 *Kluwer Academic Publishers. Printed in the Netherlands.*

resources, and protected areas. Nearly 40 percent of the land area is forested. The last 50 years have brought major changes to the forest landscape. Clear felling has dominated and numerous forest roads have been constructed. This increased exploitation has resulted in fragmentation and depletion of the old-growth forest.

Ecological systems are complex, dynamic and self-organising, changing continually through time. Concepts like *stability, balance,* and *anticipatory management* only make sense in very limited space frames and time frames and are not suitable for the management of landscapes. They are closely linked to a perspective of nature just as nature, separate from human activity, and explanations in terms of cause and effect models. The premise is that it is possible to predict and anticipate the consequences of decisions. The Nature Conservation Act, which in Norway regulates the protection of "undisturbed" areas, supports and stimulates these static, nature-orientated management strategies.

There are several reasons for changing the attitudes to landscape management. Firstly, there is no "correct" state as such for an ecosystem in a given situation. Several states are possible, and science cannot tell which is ecologically best; it can only say what the various states represent. The decision on which of the states should be preferred is necessarily a decision on values (Kay & Regier 2000). Ecosystem management will thus always be value-laden, reflecting – openly or covertly – given interests, a sort of risk and threat management where, despite incomplete information, we have to assess the possibility of an action having an adverse effect. Secondly, the term "undisturbed" is awkward in landscape management. All areas are more or less affected by human activity, so by "undisturbed" we can at best mean "considered undisturbed". Some disturbances are for cultural reasons considered to "belong" to the landscape, or to be irrelevant from a specific point of view, for instance old mountain huts in Norwegian landscapes.

1.2 The Concept of Ecological Integrity

We are thus seeking a concept for landscape management that takes account of our criticism of the traditional strategies, a concept that may also enable us to go further in a transdisciplinary evaluation of the consequences of the ongoing changes in forest landscapes. The concept of ecological integrity is a promising alternative, but little experience has so far been gained in practising it. The term *ecological integrity* is used in different ways (Woodley et al. 1993; Forman 1995; Westra & Lemons 1995; Crossley 1996; Kay & Regier 2000; Parks Canada Agency 2000; Andreasen et al. 2001). There is often no significant difference from the earlier use of *stability, balance* and *natural conditions,* and the definitions seem to more or less exclude human activity and human values from the concept of ecological integrity. Hence, they do not satisfy our criticism of current landscape management strategies. However, other writers do draw a connection between man and nature. One of these is Regier who says:

> The notion of ecosystem integrity is rooted in certain ecological concepts combined with certain sets of human values. The relevant normative goal of human-environmental relationships is to seek and maintain the integrity of a combined natural/cultural ecosystem which is an expression both of ecological understanding and an ethic that guides the search for proper relationships (Regier 1993:3).

This means that ecological integrity is a characterisation of ecosystems with reference to self-organisation and is thus a consideration of human values associated with maintaining or enhancing these features of ecosystems. Assessments of ecological integrity should not consider natural events alone, but also cultural values and human capability to handle changing ecological conditions (Regier 1993; Reid 1996).

A concept of ecological integrity is not meaningful in a cause and effect model. Dealing with complex cause and effect explanations cannot lead to adequate predictions about the future, in part because the fundamental distinction between cause and effect dissolves. In essence, ecological integrity is about self-organisation of ecological systems, which is characterised by emergence, surprise, inherent uncertainty and limited predictability. Self-organisation is not about maintaining the ecosystem in a specific state or configuration. It is about maintaining the integrity of the process of self-organisation (Kay & Regier 2000).

Ecological integrity is more a concept for observation and interpretation than a blueprint for how to handle integrity in practice. To achieve this, one may develop a set of indicators. The indicators should reveal changes in nature and make it possible to assess whether the changes should be perceived as threats to integrity or not (cf. Munn 1993). In our study we have chosen to concentrate on the parameter of fragmentation of forest landscapes and to use a patch-corridor-matrix model. This model is classical and is described by, for example, Forman (1995) and will not be repeated here. In recent decades, the fragmentation of large habitats or land areas into smaller parcels has become a major environmental issue. The fragmentation of landscapes is affecting both their structure and their function, and leads to habitat loss, reduction of habitat patch size and increasing isolation of habitat patches. There are numerous alternatives for measuring different characteristics of fragmentation (Forman 1995; Farina 1997; Leitão & Ahern 2001). In our project, we have chosen the following measures: total habitat area of old-growth forest, core habitat area (buffered for 100-metre edge zones), number of patches and main patch size.

1.3   The Concept of Management Regimes

When the management of landscape change is being considered, not only formal laws, but also informal and customary laws and cultural norms and standards of behaviour are important parts of the institutional framework that constitutes "the rules of the game". Moreover, since these institutions do not actually perform any action themselves, we have to consider them in combination with the organisations and the persons, the actors. For this combination, we use the term *regime* (Sevatdal 1999). Regimes are defined here as stable coalitions or groups of actors having informal and formal positions through which they continually influence policy by means of the resources to which they have access (Stone 1989). *Regime* also has an element of power, which is very appropriate in the management of landscape change. Regimes are often grouped around acts. In this brief presentation, we will examine just three of the most important regimes in Norwegian nature management related to the Planning and Building Act, the Forestry Act and the Nature Conservation Act.

The purpose of the Planning and Building Act (PBA) is to coordinate activities on different levels of management and provide a basis for decisions concerning the use and protection of resources and development. In the PBA, the areas between developed and protected areas are labelled "agricultural areas, nature areas and areas for open-air recreation" (ANR). The ANR is a category in the Municipal Master Plan. According to the PBA, individual persons and groups who are affected must be given the opportunity to participate actively in planning processes. Normally, the borough council can approve the Municipal Master Plan. The PBA is therefore often labelled the Local Authority Act. The urbanised areas have been the primary focus of land-use planning, with architects and engineers as the dominating professions in the local authorities.

Although the main purpose of the Forestry Act is to regulate the forest industry, it also considers recreation, landscape, species, hunting and fishing. The Forestry Act has one type of plan, the management plan, which concerns a single property and is not made public. It is a matter between just the forest owner and the local forest authority. In principle, felling is a right and responsibility of the forest owner. Nearly all those employed in the forestry regime, at all levels, are educated only in forestry.

The Nature Conservation Act (NCA) mainly preserves "undisturbed" nature from human activity by designating national parks, protected landscapes, nature reserves and protected habitats. In contrast to the PBA, the NCA is administered by the state, represented by the Directorate for Nature Management and the County Governors. The local authorities are just invited to express their opinions during the process of determining a conservation plan. Most employees in the nature conservation regime are biologists and nature management officers.

2. METHODS

2.1 Study Areas

Our topics call for case studies as the research method (Flyvbjerg 1991; Yin 1994). However, since case studies do not give statistically representative results, they do not permit generalisation in a statistical sense. On the other hand, they do offer an opportunity to create projects that are "analytically" representative. Empirical results may be considered with reference to their original theoretical basis and thus give relevant contributions to the discussion on landscape management in rural areas.

We have studied the changes in the outfields over the last 40 years in two boroughs in Norway. Here we will concentrate attention on our two study areas in the rural borough of Lierne, at Berglia (64°15' N, 13°30' E) and Raudberga (64°25'N, 13°55'E) (Figure 1). Lierne, in eastern central Norway, is a mountainous district with scattered settlements. It is a wilderness borough and an "extreme case", using the terminology of Flyvbjerg (1991). Lierne extends over nearly 3000 km², but has only 1,500 inhabitants. Only one village has more than 200 inhabitants. Some small areas are protected under the terms of the Nature Conservation Act as nature reserves, but the Norwegian Government has presented plans that will protect about 30 percent of Lierne in national parks. The other local authority in the project is Molde in western Norway, extending over 362 km². About 80 percent of its 23,000 inhabitants live in the town of Molde. Molde is not a wilderness borough, but is interesting in the context of our project because of the outfields located near the town. These areas generate a lot of attention and discussion regarding landscape management and, hence, valuable data for our project.

The landscape at both Berglia and Raudberga is dominated by spruce forest mixed with birch and some pine, rowan, alder and aspen. Due to felling since the 1950s, significant parts of these wooded areas now consist of open, or partially open, clear-felled areas interspersed with young, productive woodland. The tree line (woodland line) varies between 500 and 650 m above sea level. There are also some bogs, watercourses and low alpine habitats. Some farms and cultivated land also occur, particularly in the

Berglia area. The two areas chiefly border up to alpine habitats. As is typical for many Fennoscandian boreal forest landscapes, our study areas were already quite fragmented in their pristine condition, as only 65 percent of the 156 km^2 below the tree line were "originally" covered by forest in the Berglia area and 72 percent of 94 km^2 at Raudberga. Figure 2 shows the state of one of the study areas, Berglia, in the 1950s and 1970s, and the current state (the 1999 situation).

Figure 26-1. The location of the two study areas, Berglia and Raudberga, in Lierne, Norway. (the shaded area).

2.2   A Transdisciplinary Approach

The topics of the project require knowledge and research methods from different subjects and a combination of natural and social sciences in transdisciplinary teamwork. It should be obvious that planning and management to improve ecological conditions cannot be carried out without taking into account human values and interests (Fry 2001). Our qualifications to handle these problems are both transdisciplinary education and practice, with biology, philosophy, geographical information systems and land-use planning as the main subjects. All of us are using the outfields for recreation and research, and we have no education or practice that ties us to these areas through forestry or other industrial activities. Of course, our background does not dictate our conclusions, but it gives premises for our attitude. Like the management regimes, researchers have to choose observations with regard to their suitability for their purposes. No observation of landscapes can be objective or neutral, but it depends on the values and interest of the observer.

The landscape changes in the two study areas have been recorded using sets of vertical aerial photographs that have been examined with a simple pocket stereoscope to record the situation in the 1950s, 1970s and 1990s. The observations have been digitised and transferred to digitised soil maps. Obviously, this method is superficial. For instance, only open and partially clear-felled areas can be recorded without counting the tree density. Hence, the current amount of old-growth forest is somewhat overestimated, since we have not been able to register selected felling. Our method only reveals the clear-felled areas. Since the purpose is to analyse changes on the landscape level, i.e. areas that are several square kilometres in size, we nevertheless assume that the observations provide an adequate basis for our conclusions. The calculations have mainly been done by means of the GIS program Arc View, version 3.1.

To analyse ecological integrity, it is necessary to recognise aspects of both nature and human influence that are fundamentally interwoven and cannot really be separated. However, we have to sort the data. Therefore, the first step is to record landscape changes merely from a biological perspective and evaluate whether they can be considered as potential threats to ecological integrity. To take into account cultural and social values, the next step is to interpret whether the potential biological threats should be considered as real threats. This is obviously not a purely scientific question, but the biological considerations also depend on the choice of, for instance, the type of species and the scale in space and time.

Some bird species in substantially fragmented landscapes may have much greater requirements than the mean size of the remaining patches. However, their home ranges usually match the patch size scale of the forest fragments (Andrén 1997), and their occurrence correlates well with different landscape structure indices (Dramstad et al. 2001). The bird species populations associated with old-growth boreal forest use large areas. In addition, they are sensitive to changes and function in landscape structure in time and space. Judging by the requirements for "landscape species" given by Sanderson et al. (2002) and Brooker (2002), they might also serve as an indicator for monitoring the response to alterations in the boreal forest landscape for many other vulnerable species of various taxa. Therefore, to demonstrate the possible ecological consequences of the ongoing alterations in the boreal forest landscape, we have chosen the bird guild associated with old-growth forest (e.g., *Accipiter gentilis*, *Tetrao urogalus*, *Picoides tridactylus*, *Regulus regulus* and *Perisoreus infaustus*, cf. Table 1 in Thingstad et al. 2003) as the "landscape species" to determine whether the human activities impinge on the ecological integrity. On the other hand, more generalist species seem to perform well and even prosper under the new conditions in the present-day, highly fragmented forest landscape (Väisänen et al. 1986; Andrén 1994). However, since their preservation is not an urgent topic in the debate on ecological conditions in the boreal forest, we have not considered them as being applicable as "landscape species".

When old-growth forest landscapes become more fragmented as a result of, for instance, road building and felling (cf. McGarigal et al. 2001), the old-growth bird guild will suffer not only due to habitat losses, but also as a consequence of various negative edge effects penetrating the remaining old-growth patches (Angelstam 1992; Edenius & Sjöberg 1997; McCollin 1998; see also Lahti 2001). Hence, the density of the guild will be somewhat reduced in a buffer zone towards these edges. We have therefore buffered the old-growth forest patches by a 100-metre zone (a moderate penetration of the edge effect) towards clear-felled areas, bogs, lakes, roads, cultivated land, farms, houses and cabins. The buffering procedure was only applied where the old-growth patches bordered other types of habitat larger than 1 ha, as smaller patches were not expected to create significant edge effects. Power lines, deciduous woodland and patches of productive forest were not expected to create noticeable edge effects either, and were therefore not buffered against (cf. Figure 3). The isolation effect seems to be of little importance for birds, and is therefore not taken into consideration in our analyses.

The practice of various management regimes with regard to the landscape changes in Lierne and Molde has been investigated by analysing various planning documents and actual individual cases on the municipal level, and by interviewing employees in management regimes at the local and county authority levels, and in national bodies. The documents we have studied were as far as possible a complete collection of the most important documents pertaining to each regime in the period since 1965, with focus on the last few years. The individual cases we have studied are those that have been dealt with comprehensively and thus give extensive documentation of the arguments put forward by the management regimes. On several occasions during the course of the project, we have discussed the topics we are considering with employees in the management regimes, thus getting to know them and their opinions well. The interpretations presented here are based on the last interviews, which employed a form of questionnaire that was used more as a basis for interpretations than as a list of questions.

3. RESULTS AND DISCUSSION

3.1 Landscape Change and Management Strategies

The total area covered by old-growth forest stands, and in particular old-growth core areas more than 100 metres from any edge effects from surrounding "hostile" habitats, have been significantly impoverished in recent decades in both our study areas (cf. Figures 2 and 3, which show one of the study areas, Berglia).

Clear felling was in its infancy in 1958, but by 1999 the landscape had changed character. Forest roads divided the area and clear felling had mainly reduced the old-growth areas. In 1958, there were only 1.5 km of forest roads in Berglia, but by 1999 their total length was 69.1 km. This also means that accessibility to wilderness areas has considerably increased. In Raudberga, there has been a significant increase in the number of cabins. In Berglia, the total extent of old-growth areas has been reduced from 98.1 to 71.6 km^2, and of old-growth core areas from 27.4 to 12.1 km^2, the mean core area being reduced from 14.5 to 6.0 km^2. In Raudberga, the total extent of old-growth areas has been reduced from 62.2 to 33.8 km^2, and of old-growth core areas from 19.6 to 9.1 km^2, the mean core being reduced from 17.8 to 8.2 km^2 (cf. Tables 1-4). Clear felling gained a foothold earlier at Raudberga, where the forest is mainly owned by large companies, than at Berglia, which mostly comprises smaller private properties whose owners showed some initial resistance to this new felling method. At the current forest succession stage, much more productive forest is therefore growing up at Raudberga.

Table 26-1. Landscape changes in Berglia, Lierne, Norway in 1958, 1978 and 1999.

	1958	1978	1999
All old-growth areas (km^2)	98.1	86.0	71.6
Clear-felled areas (km^2)	0.2	8.0	18.7
Partially clear-felled areas (km^2)	1.8	5.8	1.6
Productive forest	0	0	7.1
Cabins (number)	24	22	27
Forest roads (km)	1.5	25.7	69.1

From a biological point of view, there is no doubt that the landscape changes can be considered a potential threat to the ecological conditions. The negative edge effect applied in the analyses has caused a reduction rate in the old-growth bird guild that is significantly greater than a one-to-one relationship with the habitat loss, which would be expected if the loss of habitat area was the only effect. These aspects are discussed in detail by Thingstad et al. (2003) and will not be repeated here. According to our definition of *ecological integrity*, potential threats do not necessarily mean real threats, either from a biological or a sociological point of view.

However, we lack adequate data to discuss this at present. So far we have only studied the fragmentation of old-growth forest caused by felling and forest road construction, which is just one part of the complex history of human exploitation of the forest resources. Moreover, we have only illustrated the situation using one category of species, the old-growth bird guild. Nevertheless, in our approach it is sufficient to identify potential threats, because the next step is to examine whether the management regimes have observed and registered potential threats and if so have considered whether they should be looked upon as real threats requiring special management efforts.

Table 26-2. Forest statistics for Berglia, Lierne, Norway, showing the development for habitat types used by the bird guild associated with old growth in 1958, 1978 and 1999.

	1958	*1978*	*1999*
All old-growth areas (km^2)	98.1	86.0	71.6
Old-growth buffer areas (km^2)	70.7	66.3	59.5
Old-growth core areas (km^2)	27.4	19.7	12.1
Old growth: number of core areas	189	209	202
Mean core area ± S.E. (ha)	14.5 ± 5.1	9.4 ± 2.5	6.0 ± 1.3

Table 26-3. Landscape changes in Raudberga, Lierne, Norway in 1958, 1971 and 1999.

	1958	*1971*	*1999*
All old-growth areas (km^2)	62.2	45.1	33.8
Clear-felled areas (km^2)	4.3	20.5	9.8
Partially clear-felled areas (km^2)	1.6	2.6	0.9
Productive forest	0	0	23.4
Cabins (number)	14	25	66
Forest roads (km)	18.9	24.9	37.4

Despite the great landscape changes over the last 30-40 years, the management regimes seem to show little awareness of them. We cannot see that their ecological observations pay any regard to the potential threats to ecological integrity. The Planning and Building Act regime is orientated towards buildings and, for Lierne and Molde where neither scattered residential building nor building of cabins takes place to any great extent, it has little relevance in connection with "agricultural areas, nature areas and areas for open-air recreation" (ANR).

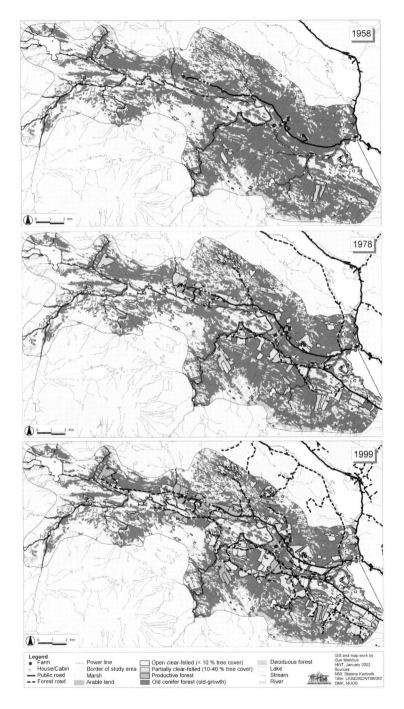

Figure 26-2. The spatial habitat configurations at Berglia in Lierne, Norway, at three different times, 1958, 1978 and 1999.

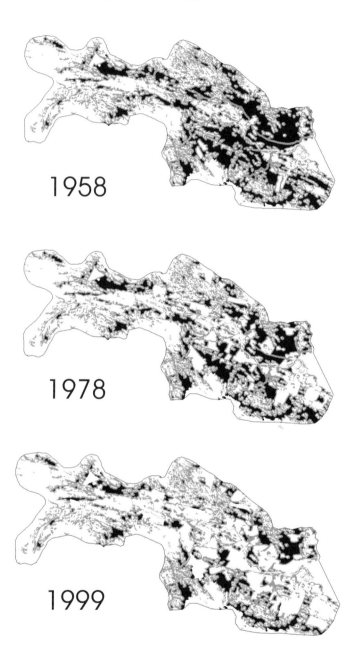

Figure 26-3. The spatial configuration of the old-growth forest patches with a 100-metre buffer zone at Berglia in Lierne, Norway, at three different times, 1958, 1978 and 1999. Black indicates the core areas and grey the buffer zones (the edge effect).

Table 26-4. Forest statistics for Raudberga, Lierne, Norway, showing the development for habitat types used by the bird guild associated with old growth in 1958, 1971 and 1999.

	1958	1971	1999
All old-growth areas (km^2)	62.2	45.1	33.8
Old-growth buffer areas (km^2)	42.6	33.7	24.7
Old-growth core areas (km^2)	19.6	11.5	9.1
Old growth: number of core areas	110	114	112
Mean core area ± S.E. (ha)	17.8 ± 5.8	10.1 ± 2.1	8.2 ± 2.3

While the nature conservation regime obviously sees threats over large areas, the Nature Conservation Act (NCA) deals primarily with "undisturbed" areas. This regime has very few strategies or measures for managing areas that are undergoing anthropogenic change. Like the ANR, the NCA permits little action in such areas.

The forestry regime is also influenced by the distinction made by the nature conservation regime between "disturbed" and "undisturbed" nature. The environmental trend in forestry during the 1990s mainly concerned species and habitats that should be protected, leaving the remainder of the forest for "proper" forestry. In Norway, only 1 percent of the productive forest is protected as nature reserves. The *Living Forest* project (*Levende Skog* 1998) defined standards for the certification of forestry, and a plan has been drawn up for registering key habitats throughout the country. The intentions of the *Living Forest* project are to bring landscape ecological considerations into forestry, but so far we have not seen much of this new practice in our studies. We may therefore conclude that forestry mainly takes place without significant consideration for the environment, as understood in the concept of ecological integrity. The reason is not necessarily lack of interest, the subject is simply not on the agenda because it is mainly the protected areas that should be looked after.

Another typical aspect of the management of the ANR is that ecological and psychological data are mixed together even though there are fundamental differences between landscape ecology and "landscape psychology", meaning how people perceive their surroundings. Whereas biological knowledge can to a certain extent be specified quite precisely, it is not possible to state universal demands for human experience of nature, which will vary depending on values and interests and can only be generalised to a small extent.

Each of the management regimes seems to have its own, and the "right", understanding of the landscape. They show little awareness of the need for their views on nature not necessarily needing to be the views of society at large. They act almost as if there exists one common and objective way of judging nature, which should be shared by everyone. Since the understanding of landscape has to be subjective, it is obvious that management of nature also requires knowledge in social sciences.

An interesting observation is that forest seems to survive almost everything and still remains forest. After felling, including huge clear-felled areas, some type of forest will almost always grow up again. Planting, natural rejuvenation and management give a foundation for new forest. On the other hand, our data also show that there have been activities that obviously can be considered as threats to species and diversity in the woodlands. The reflections made by the forestry regime do not seem to include all the ecological issues that should be considered. From an ecological point of view, the question is what happens to the total biological diversity in the area, not only the existence of the forest. The main challenges are how to ensure that the habitat qualities for old-growth associated species are maintained within the regrowth forest stand. These factors argue strongly for the case that others than foresters should also take part in the process of drawing the limits for forestry.

3.2   Who should be Responsible for Managing the ANR?

An important question is whether more power to the Planning and Building Act (PBA) would give more environment-friendly practices in the ANR. Our documentation of the situation gives no basis for answering this. However, the results give an opportunity to interpret and discuss the suitability of the various management regimes that relate to the ANR. For this discussion we have chosen six criteria. Three of the criteria – *tradition, expert knowledge* and *process competence* – consider different kinds of competence, and the other three – *participation, local government* and *national means* – represent general management principles.

The PBA regime has very little tradition in land-use planning that does not concern development. The classical issue is to consider development interests against specified protection interests. The land between developed and protected areas is left out and gains little attention. Except for the planning of cabins, only the forestry regime has practised land-use planning in the ANR. However, apart from for instance wildlife and recreational interests, the forestry regime has a history showing that it mainly serves industrial interests. The nature conservation regime, with its focus on undisturbed areas, has little tradition for land-use planning and hence not for planning in the ANR either.

The criterion of *expert knowledge* follows that of tradition. Even though the PBA has a wide purpose paragraph, the PBA regime is still dominated by architects and engineers with little competence in natural and social sciences. In this respect, the forestry regime is somewhat more capable as regards ecological competence, but its competence in social sciences is weak. The nature conservation regime obviously has competence in natural sciences.

However, its one-sided focus on undisturbed nature leaves little space for the ANR, and its knowledge is scarcely used there. Like both the PBA and the forestry regimes, the nature conservation regime has little competence in social sciences.

Only the PBA regime has both legislation and long, extensive practice in leading *processes* with wide participation from individual persons and groups who are affected, as well as other management actors. The forestry regime mainly has contact with forest owners. Even though the nature conservation regime now to a certain extent has the same legislation for the planning process as the PBA, it lacks experience and knowledge of how to practice it.

Participation is closely linked to process competence. Only the PBA and the NCA have legal demands on participation. At present, the PBA is the only regime that is capable of accomplishing planning processes with wide participation in a kind of collaborative planning exercise. The Forestry Act is orientated towards the forest owners and forestry management plans are not even open for public inspection.

With respect to the *local government* criterion, the PBA and the Forestry Act differ from the NCA since their authority is placed at the municipal level. The PBA is particularly closely linked to local authorities. On the other hand, the NCA is the law of the state, with the Department of Environmental Affairs at the County Governors' Offices as an important actor.

The criterion of *national means* is often in opposition to local government, such as in the ongoing conflicts regarding nature conservation and carnivore protection in Norway. The NCA should, by definition, take care of national means. The forestry regime is also closely linked to national means, even though its power is to a large extent placed at the municipal level. The PBA is also concerned with national means, but above all it gives much power to local authorities.

Table 5 summarises the suitability of the management regimes for managing the ANR. The distinction is only between "well suited" and "less suited" and the criteria are not weighted. It shows that none of the regimes satisfy all the criteria that need to be fulfilled to be well suited for managing the ANR. Neither the PBA nor the nature conservation regimes have tradition and expert knowledge appropriate for the ANR, but the PBA regime is strong in process competence. The forestry regime has some biological knowledge, but like the other regimes it is rather weak in social sciences. With regard to environmental issues, all the regimes are orientated towards either development, like the PBA regime, or protected areas. The forestry regime does, of course, take account of the ANR, but in our opinion it concentrates too much on just the very small areas that should be left "undisturbed".

Table 26-5. Different management regimes characterized by suitability for managing the agricultural areas, nature areas and areas for open-air recreation (ANR).

Management regime	Criteria					
	Tradition	Expert knowledge	Process competence	Partici-pation	Local government	National means
Planning and Building Act	-	-	+	+	+	+/-
Forestry Act	+	+/-	-	-	+/-	+/-
Nature Conservation Act	-	+/-	-	+/-	-	+

In the ANR, actors with different values and interests are mixed together, in both the public-sector management and the local communities. One way to handle this problem may be to involve the affected parties in the management process, in a kind of collaborative planning exercise (Healy 1997). Process competence is therefore essential to co-ordinate and clarify the land use. In this perspective, the PBA regime is obviously the one that is best qualified. This is an important argument for giving more responsibility for major land-use planning to the PBA, not only in urban areas, but also in the ANR. This does not mean that a single regime, for instance the PBA regime, should have all the power. One regime would hardly be able to handle all the different management problems in the ANR. However, we believe the PBA regime should be given more power than it has today, in order to co-ordinate the planning processes and formulate the plans. If the PBA regime is to be given more responsibility for the ANR, it would need to be equipped with biological and agricultural knowledge, in addition to its traditional architectural and engineering competence. In this perspective, it is not a good solution to have a single act controlling both planning and buildings, and thus tying land-use planning more to development than to protection. Another important premise for giving more power to the PBA is that it would need better possibilities for providing long-term, predictive regulations. Municipal Master Plans are now going to be evaluated for possible alterations every fourth year, and this may create uncertainties.

4. CONCLUSIONS

In this paper, we have shown that the concept of ecological integrity may offer a suitable perspective for describing and analysing landscape changes in rural areas. In contrast to the traditional perspective where nature is considered static, "undisturbed", pure, a biological phenomenon without relation to society, the concept of ecological integrity gives an opportunity to consider nature as dynamic, "disturbed", influenced by society, and also to view it from a transdisciplinary perspective. Hence, management strategies must focus more on maintaining the capacity to adapt to change than predicting and anticipating the consequences of decisions.

The landscape changes in the ANR over the last 30-40 years are considerable and very complex, and forestry has prime responsibility. People are in part directly responsible for the changes (e.g., felling, road building and planting). Other changes, like rejuvenation after felling, are a response to human activity, and yet others result from continual processes in nature. This gives very complicated and varied cause and effect relationships, which cannot be explained in terms of simple cause and effect models.

However, the landscape changes represent potential threats to the ecological integrity. This does not necessarily mean that the threats are real and the landscape will change in an undesirable direction. The point is that the changes are so extensive and the implications so uncertain that it should be considered whether they may give undesirable results.

The landscape changes have been offered very little attention by the management regimes. We cannot see that consideration has been given to potential threats to ecological integrity. Attention has been more directed towards protecting species, habitats and more or less "undisturbed" areas than to ecological evaluations of landscapes and the use of "disturbed" areas. The management regimes have little interest, competence or means to perform ecological evaluations of landscapes and take actions in forested areas.

The management regimes do not treat potential threats as potential threats. Evaluation of potential negative effects of human activity is mainly concentrated around species and habitats in areas that are protected against encroachments, little attention being given to the ANR. The reasoning in favour of protection is complex, natural science arguments being mixed together with how nature is experienced and used. Generally, the reasoning based on natural science is understood as neutral, objective and absolute, without pausing to reflect that there will always be different values and interests depending on the analyses and views on nature.

Process competence is a necessary premise for managing the ANR. In the ANR, many different values and interests are mixed together, within the management system, within the local communities and, not least, within the meeting between the management system and the local communities (e.g., *vis-à-vis* national means). Hence, it is important to have competence to coordinate and define land use in areas where serious conflicts often arise, like the ANR. Processes do not always lead to consensus, but are most important for achieving legitimacy for decisions. The different participants in the planning process have to perceive both the process and the results as being fair. With reference to processes and transdisciplinary teamwork, the PBA regime is clearly best as regards formal legislation, tradition and competence. Nevertheless, the differences between urban and rural areas are considerable. Hence, the PBA regime is not without reservation qualified to "take over" the ANR, since it has serious defects, regarding both tradition and competence.

ACKNOWLEDGEMENTS

We are indebted to Ann Norderhaug at the Norwegian Crop Research Institute and Jørund Aasetre at the Nord-Trøndelag Research Institute for helpful comments, to two anonymous referees, one for constructive remarks and the other for making us aware of the possibility that our objective with this transdisciplinary approach might be miscomprehended, and to Richard Binns who has improved the English.

REFERENCES

Andreasen, J.K, O'Neil, R.V., Noss, R. & Slosser, N.C. (2001). Considerations for the Development of a Terrestrial Index of Ecological Integrity. *Ecological Indicators*, 1, 21-35.

Angelstam, P. (1992). Conservation of Communities – The Importance of Edges, Surroundings and Landscape Mosaic Structure. In L. Hansson (Ed.), *Ecological Principles of Nature Conservation*. London: Elsevier Applied Science.

Andrén, H. (1994). Effects on Habitat Fragmentation on Birds and Mammals in Landscapes with Different Proportions of Suitable Habitat: a Review. *Oikos*, 71, 355-366.

Andrén, H. (1997). Habitat Fragmentation and Changes in Biodiversity. *Ecological Bulletin*, 46, 171-181.

Brooker, L. (2002). The Application of Focal Species Knowledge to Landscape Design in Agricultural Lands Using the Ecological Neighbourhood as a Template. *Landscape and Urban Planning*, 60, 185-210.

Crossley, J.W. (1996). Managing Ecosystems for Integrity: Theoretical Considerations for Resource and Environmental Managers. *Society and Natural Resources*, 9, 465-481.

Dramstad, W.E., Fjellheim, W.J., Skar, B., Helliksen, W., Sollund, M.L.B., Tveit, M.S., Geelmuyden, A.K. & Framstad, E. (2001). Integrating Landscape-Based Values – Norwegian Monitoring of Agricultural Landscapes. *Landscape and Urban Planning*, 57, 257-268.

Edenius, L. & Sjöberg, K. (1997). Distribution of Birds in Natural Landscape Mosaics of Old-Growth Forest in Northern Sweden: Relations to Habitat Area and Landscape Context. *Ecography*, 20, 425-431.

Farina, A. (1997). *Principles and Methods in Landscape Ecology*. London: Chapman & Hall Ltd.

Flyvbjerg, B. (1991). *Rationalitet og makt. Bind 1. Det konkretes vitenskap.* København: Akademisk forlag.

Forman, R.T.T. (1995). *Land Mosaics. The Ecology of Landscapes and Regions.* Cambridge: Cambridge University Press.

Fry, G.L.A. (2001). Multifunctional Landscapes – Towards Transdisciplinary Research. *Landscape and Urban Planning*, 57, (3-4), 159-168.

Healy, P. (1997). *Collaborative Planning. Shaping Places in Fragmented Societies.* Basingstoke: Macmillan Press Ltd.

Kay, J.J. & Regier, H. (2000). Uncertainty, Complexity, and Ecological Integrity: Insights from an Ecosystem Approach. In P. Crabbe, A. Holland, L. Ryszowski & L. Westra (Eds.), *Implementing Ecological Integrity: Restoring Regional and Global Environmental and Human Health.* Kluver, NATO Science Series, Environmental Security, 121-156.

Leitão, A.B. & Ahern, J. (2001). Applying Landscape Ecological Concepts and Metrics in Sustainable Landscape Planning. *Landscape and Urban Planning*, 59, 65-93.

Levende Skog (1998). Standardutredninger fra Levende Skog. Levende Skog. Oslo.

Lahti, D.C. (2001). The "Edge Effect on Nest Predation" Hypothesis after Twenty Years. *Biological Conservation*, 99, 365-374.

McCollin, D. (1998). Forest Edges and Habitat Selection in Birds: a Functional Approach. *Ecography*, 21, 247-260.

McGarigal, K., Romme, W.H., Christ, M. & Roworth, E. (2001). Cumulative Effects of Roads and Logging on Landscape Structure in the San Juan Mountains, Colorado (USA). *Landscape Ecology* 16, 327-349.

Munn, R.E. (1993). Monitoring for Ecosystem Integrity. In S. Woodley, J. Kay & G. Francis (Eds.), *Ecological Integrity and the Management of Ecosystems* (pp. 105-115). Cambridge: Cambridge University Press.

Parks Canada Agency (2000). "Unimpaired for Future Generations?" Protecting Ecological Integrity with Canada's National Parks. *Report of the Panel on the Ecological Integrity of Canada's National Parks.* Ottawa.

Regier, H.A. (1993). The Notion of Natural and Cultural Integrity. In S. Woodley, J. Kay & G. Francis (Eds.), *Ecological Integrity and the Management of Ecosystems* (pp. 3-18). Cambridge: Cambridge University Press.

Reid, W.V. (1996). Beyond Protected Areas: Changing Perceptions of Ecological Management Objectives. In R.C. Scaro & D.W. Johnsen (Eds.), *Biodiversity in Managed Landscapes. Theory and Practice* (pp. 442-453). Oxford: Oxford University Press.

Sanderson, E.W., Redford, K.H., Vedder, A., Coppolillo, P.B. & Ward, S.E. (2002). A Conceptual Model for Conservation Planning Based on Landscape Species Requirements. *Landscape and Urban Planning*, 58, 41-56.

Sevatdal, H. (1999). Real Estate Planning; An Applied Academic Aubject. *Kart og Plan*, 59, 258-266.

Skjeggedal, T. (2001). Landscape Management Between Developed and Protected Areas. In Ü. Mander, A. Printsmann & H. Palang (Eds.), *IALE European Conference 2001. Development of European Landscapes. Conference Proceedings* (pp. 494-498). Volume II. Publications Instituti Geographici Universitatis Tartuensis, 92.

Skjeggedal, T., Arnesen, T., Markhus, G., Saglie, I.-L. & Thingstad, P.G. (2001). *Regimenes slagmark. Om arealutnytting og forvaltningsregimer i LNF-områder.* NTF-rapport 2001: 3. Nord-Trøndelagsforskning. Steinkjer.

Stone, C.N. (1989). *Regime Politics.* Kansas: University Press of Kansas.

Thingstad, P.G., Skjeggedal, T. & Markhus, G. (2003). Human-Induced Alteration of Two Boreal Forest Landscape in Central Norway, and Some Possible Consequences for Avian Fauna. *Journal of Nature Conservation*, 11, 145-156.

Väisänen, R.A., Järvinen, O. & Rauhala, P. (1986). How are Extensive, Human-Caused Habitat Alternations Expressed on the Scale of Local Bird Populations in Boreal Forests? *Ornis Scandinavia*, 17, 282-292.

Westra, L. & J. Lemons (Eds.) (1995). *Perspectives on Ecological Integrity*. London: Kluwer Academic Publishers.

Woodley, S., Kay, J. & Francis, G. (Eds.) (1993). *Ecological Integrity and the Management of Ecosystems*. Cambridge: Cambridge University Press.

Yin, R.K. (1994). *Case Study Research. Design and Methods.* 2nd Ed. Thousand Oaks: Sage Publications, Inc.

Chapter 27

PAST LANDSCAPE USE AS AN ECOLOGICAL INFLUENCE ON THE ACTUAL ENVIRONMENT

Hansjörg Küster
Institute for Geobotany, Hannover University, Germany

1. INTRODUCTION

In beginning a paper on outdoor ecology, normally the stand or study area is described, concerning where the research has been carried out. The stand is characterised by climatic, hydrological, lithological and pedological factors. They are regarded as natural in the sense of being non-anthropogenic. There is also the widespread – but not explicitly written – opinion that natural factors are or were stable up to the point of time when intensive land-use on the present-day scale began.

In this paper, it will be stressed that stands are not changed by actual land-use only. Stands are not characterised by stable ecological conditions, and they have already been markedly changed by previous land-use. Detailed knowledge concerning past land-use is therefore necessary in order to describe the actual quality of a stand.

Examples will be presented to show how stands were influenced or changed by previous land-use, and how their ecological qualities shifted markedly. Attention will focus on:

1. The change of nutrients concerning agriculture in general;
2. The different influences of primary and secondary succession on landscape development;
3. The different influences of a settlement's economy in regards to landscape development.

H. Palang et al. (eds.), European Rural Landscapes:
Persistence and Change in a Globalising Environment, 445-454.

2.   CHANGING CONTENTS OF NUTRIENTS IN SOILS

It is commonly thought that plants should grow, die and decay on the same soil, so that nutrients remain on the same stand during the development of an ecosystem. This is the foundation for a concept based on the cyclical turnover of nutrients; but this is only in theory. In reality, the litter of nutrients originating from decayed plants is not kept stable at an actual stand, but instead is transported by wind, water and animals. Therefore, there are places where the contents of nutrients are reduced and others where they are accumulated. A symbiosis between plants and micro-organisms (e.g., an actinorhiza between *Alnus* and *Frankia*, see Schwintzer & Tiepkema 1990) leads eventually to an accumulation of ammonium and/or nitrate in the soils underneath the plants.

Greater amounts of nutrients are transported in the course of land-use. Each harvest of crops leads to a removal of nutrients from a field, so that it becomes poorer in nutrients. Minerals such as magnesium, potassium, nitrate, phosphate, sulphate, etc, are parts of harvested plants, and are normally not reintroduced into fields. On the other hand, nutrients are accumulated where men collected litter, especially in and near settlements.

Grazing has a similar result. It reduces the amounts of minerals in a pasture, and minerals are accumulated on dung heaps. In meadows, the amounts of nutrients are reduced by harvesting. Through woodland management, timber and wood (and their minerals) are removed from woodland stands, and over time minerals accumulate in places where fires are managed and ash is produced, and in places where timber is decayed.

Thus, within each settlement an accumulation of nutrients takes place, whereas in used areas the mineralic content of soils is reduced. In many stands, a reduction of nutrients through the removal of plant materials does not pose an obvious problem, because further nutrients are made available to plants through pedological processes; such as the destruction of clay minerals. If alder woodland is felled, its soil is naturally manured by nitrates from the actinorhiza, so that it remains possible to manage grassland for many decades on former alder carr stands without manuring. The grasslands are perfectly manured by nitrates already occurring in the soil.

Stands are manured to avoid shortages of mineralic nutrients in the soils. But this does not mean that each stand in the vicinity of a settlement receives the same portion of manure, i.e. that the reduction of mineralic nutrients is compensated for as precisely as possible. Also, fields are normally manured more intensively than grazing areas and woodlands. Meadows were manured by watering systems (Hoppe 2002) in the past, and later by litter and mineralic manure as well.

By tracing high amounts of phosphates in the soil, it is possible to show where prehistoric settlements or cemeteries were located (Gebhardt 1976, 1982). Thus, it is clear that previous land-use and previous accumulation of minerals within settlements changes the contents of nutrients in soils. The actual quality of a stand is clearly influenced by its former use. Stands used earlier for cropping, haymaking and woodland management became poorer in nutrients. High amounts of nutrients in soils are presently available where settlements or even dung heaps were situated in former times.

There are numerous other important influences of land-use on stands. When woodlands are cleared and cropping is started, evapotranspiration is reduced, which is also the case when areas are grazed. Therefore, land-use can lead to desertification in arid areas (Mensching & Seuffert 2002). Also, other climatic parameters are changed when woodlands are cleared (Frenzel 1977, 1979) – on open land, the local climate is markedly more continental than inside woodlands (Geiger 1961).

3. PRIMARY AND SECONDARY SUCCESSIONS

Ecosystems change when natural parameters change, and they also change when management is intensified, reduced or stopped. After the last glaciation, woodlands developed for the first time. This followed a cold phase, when natural conditions did not allow the occurrence of any woodland in most parts of Europe. During such a primary succession, a soil and its vegetation developed in a parallel way. In the beginning of the succession, the soils were atypical for woodlands, and developed woodlands did not occur nearby. At all places of greater area, similar successions took place. Many woodlands were cleared later to allow for agriculture and pasture in an artificially open landscape.

The successions after the glaciation were different from those which took place after land-use was practiced and later given up. The soils were already typical forest soils, but shifts and modifications in their mineralic contents had taken place during the land-use phase. Well-developed woodlands with their biodiversity occurred nearby. When land-use is given up, a secondary succession (Fischer 1987) takes place. It is not characterised by a parallel development of soil and vegetation, but more by the rapid development of vegetation alone, according to a status quo state of the soil – which is not totally in a natural state, but rather influenced by both former natural developments and former land-use. Plant species present in the seedbank of the soil and plant species occurring on the stands nearby spread and colonise the abandoned farmland. Abandoned farmland is also attractive to wild animals. They graze on the fallow and introduce fruits and seeds of zoochoric plants; such as rowan, blackberry, oak and beech. As a result, secondary successions do not always lead to the formation of the same vegetation as primary successions; and the species composition is changed by each phase of land-use followed by a phase of fallow and secondary succession (Richter 1997).

4.   SELF-SUBSISTING SETTLEMENTS INCORPORATED INTO TRADE SYSTEMS – DIFFERENT INFLUENCES ON THE STANDS

Archaeologists (Hvass 1982; Kossack 1982, 1997; Lüning 1989) have shown that self-subsisting settlements, which are more or less economically independent, were not as stable as settlements tied to economic networks. After some decades, the settlements were given up and rebuilt at another site. This process, called *shifting cultivation*, could be triggered by a shortage of nutrients in fields causing low yields, or by a shortage in wood and/or timber. Shortage of timber in the vicinity of a settlement may have been most decisive in many parts of Europe. After having managed an area for some decades, all tall trees had been removed and new trees had not yet grown. If a house burnt down or was dilapidated, timber was not available locally, but only farther away at another site (Küster 1998). It may have been easier for settlers to move to another site where timber still was available, and to clear a new patch of woodland. Then at the abandoned site, a secondary woodland succession took place. Therefore, in periods and areas of self-subsisting settlements, land-use did not only cause deforestation, but secondary successions took place as well (Küster 2001); leading to different vegetation compositions and stand qualities. The more secondary successions occurred on a stand, the more its species composition and ecological quality changed.

If settlements were part of trade systems, their sites had to be stable. Trade can only be developed in stable settlement structures, i.e. the inhabitants of the settlement are not forced to leave the site. It is necessary to provide the settlements with products that are lacking. If a crop yield is too low, people must hope it is possible to buy additional crop or flour on the market. If timber is in demand, it should be delivered via economic exchange. On the other hand, an integration of economic networks means that a surplus of agrarian products, wood and timber should be produced whenever possible to sell these products on the market. Only people who normally sell something can hope to receive special products that they lack.

The influence of stable settlements on their environments was markedly different from that of shifting cultivation. In the vicinity of stable settlements, secondary woodland successions did not occur. The segments of the environment (fields, grazing areas and woodlands) were used in the same manner for a very long time. Only in the vicinity of stable settlements could large-scale investments in water mills, flooded meadows, orchards, and vineyards could be expended – they did not exist in a system of shifting cultivation. Woodland degradation was much more intensive in the vicinity of a stable settlement. The withdrawal of mineralic nutrients from the soils under woodland and grazing areas lasted for centuries, uninterrupted by phases of secondary succession. It was therefore much more intensive, and the main reason for the formation of heathlands since the Middle Ages (Hüppe 1993; Behre 2000). On the other hand, nutrients were accumulated in and nearby settlements since centuries ago. They were also brought to some fields and hay meadows in order to manure them. In a landscape with stable structures, a permanent accumulation of nutrients in a "nucleus" region of a settlement is obvious, whereas soils became poorer in the commons, the area outside the "nucleus" region.

In the industrial era, additional manure became available through potassium mining and large-scale phosphate, nitrate and ammonium production. Manure became abundantly available, so that yields of fields and hay meadows increased enormously. Large parts of former heathlands could also be cultivated intensively. Other parts of infertile areas were afforested, as they were no longer demanded for agriculture. They were not manured. As a result, many forest stands are very poor in nutrients today.

5. CONSEQUENCES FOR LANDSCAPE RESEARCH

The few remarks above show that there are new tasks for historical geography and palynology to undertake in order to provide a better basis for understanding stands and their dynamic characters. This is a precondition for describing and planning present and future land-use in an adequate way.

Nowadays, the aim of sustainable development is often proclaimed. It is certainly illusionary to expect complete sustainability in land-use, as it is not possible to practise agriculture and keep the amounts of nutrients on equal levels in all agrarian stands – nor can forestry be practised without withdrawing nutrients from the stands by removing woods or timber. But what must be stressed is that it is very important to use all resources as efficiently as possible. All fossil resources are limited. And it is not desirable to collect remnants of used resources as litter at places where they change ecosystems markedly; e.g., in some parts of the pedosphere, in the atmosphere, and in the hydrosphere, especially in lakes and oceans.

Soils are analysed using pedochemical methods in order to receive information on their fertility. But this is insufficient. For most-efficient land-use, the potentials of the soils must also be known. One should know how many nutrients were available in soils in earlier times, before intensive land-use began, and how the contents of nutrients in soils developed through time. Also, the development of acidity levels must be known, as the availability of nutrients is influenced by the pH level of the soils (Larcher 2001).

6.   A NEW TASK FOR HISTORICAL GEOGRAPHY

So one should find as many clues as possible about former land-use on each stand. Previous agriculture on a present-day farming area could result in a reduction of mineralic nutrients if the stands were not manured regularly; but the opposite can also be the case. This depends on the agrarian system that was practised. Some nutrients may still be released by clay mineral decay in fertile soils, so that less mineralic manure might be needed than is actually applied. But this may be necessary on fields installed on former heathlands, where natural nutrients cannot be released by clay mineral decay, and are therefore no longer available.

Also, woodland stands must be regarded differently. Whereas mineralic nutrients may be available and well-balanced in ancient woodlands, there can be marked shortages of nutrients in some afforested areas. If historical geographers can determine that agriculture was practised on an actually-wooded stand, it is likely that farming ended because yields became too low. That means the availability of nutrients for trees is also very limited. It may be that some nutrients are totally lacking. It is not sustainable to only establish a forest on former farmland, when shortages of mineralic nutrients are not compensated for. A shortage in clay minerals on such stands also results in a low stability of the soil structure – spruce woodlands with their flat root systems are not kept stable on such stands, and are destroyed during storms.

It is very important to work with archived historical maps in order to receive detailed information on previous land-use, and the possible withdrawal or accumulation of nutrients that can be induced by this. Remnants of former land-use practises still visible in fields but not recorded in maps give important additional hints as to the history of land-use and the development of mineralic nutrients in soils. Ancient ditches may tell where grasslands have been drained and meadows have been watered in order to accumulate nutrients. If localities of ancient settlements and other areas where nutrients were accumulated (such as dung heaps) are known, this may also tell where nutrients are presently available in higher amounts.

Archaeologists must be integrated into this work. The significance of a prehistoric settlement at a concise site means that the contents of nutrients were changed in the past, too, which may have an influence on the development of a stand up to the present.

Archaeologists can also trace the influence of past soil erosion. If only the bottoms of prehistoric postholes are preserved in the soils, it can be concluded that major parts of the soils have been eroded since the time that the site was inhabited. The opposite is the case if postholes, ditches or wells are still completely preserved in the soils, or even if the past soil surface is visible underneath the actual soil. This means that soil materials were accumulated at a site after it had been inhabited.

7. A NEW TASK FOR PALYNOLOGY

It will never be possible to trace all former land-use phases at specific sites, including former settlements themselves, because many of them may be eroded or so well hidden in the soil that they are not traceable. Also, maps and written sources do not tell every detail about former land-use, as the sources may be incomplete. In other cases, written sources are not preserved or not collected or made available in archives. This certainly does not mean that it is negligible or unnecessary to collect as many data on former land-use as possible. But one must be aware that it will not be possible to collect the complete story about each stand and what might have influenced its quality.

It is possible to test to which extent the history of local land-use is traced using pollen analysis. Pollen is permanently deposited in accumulating sediments of lakes, fens and bogs. Deposited cereal pollen grains may have come from fields that belonged to known settlements, but also from fields cultivated by unknown settlers. It will never be possible to locate settlements and fields by pollen analysis. But it is feasible to collect the total cereal pollen influx to sediments, which means that a complete history of land-use is only obtainable by applying this method, but not via history, archaeology and historical geography.

With pollen analysis, it is also conceivable to determine when land-use began, and when its major strategies changed. It is most important to date the period when shifting cultivation was replaced by permanent cultivation. This change resulted in the end of any allowance of a secondary woodland succession, and in the intensification of land-use. This date can also not be determined by applying exclusively historical geographical, historical and archaeological methods. Traces of settlements or field systems found in maps, written sources or in environments must by characterised by pollen analytical results – these can tell whether the settlements or field systems were already integrated into a stable land-use system or not (Küster 2004).

It is important to look at sedimented cereal pollen grains to see when the change from shifting to permanent cultivation occurred. Clear gaps in their accumulation may be clues as to shifting cultivation, i.e. to phases between periods of agriculture when crops were not cultivated. But as the number of distributed cereal pollen grains recorded in pollen profiles is often small, other hints as to whether shifting or permanent cultivation was practised must be found through the careful interpretation of pollen diagrams.

To decide whether shifting or permanent cultivation was practised, it is not possible to look only at the presence of pioneer tree species (such as birch) in the pollen diagrams. Birch does not occur only on abandoned fields and settlement sites, but also on patches of temporarily less-intensively used land and ruins inside a principally permanently-used area.

What is different between shifting and permanently-used systems is the occurrence of tree species that expand after the pioneer phases of woodlands. They indicate that the area has been out of use for a longer time span. Oak and beech will not expand on extensively-used grazing areas and not on ruins, but in the course of a secondary woodland succession, when men do not return to a place, pioneer birch wood develops.

This is clearly visible in the pollen diagrams. Not only could the identical types of woodlands develop again – which equalled those that had been cut before the foundation of the settlement – but also other tree species could expand. They possibly were favoured by the fact that new woodlands developed without any concurrence to other tree species that occurred in the areas, when agriculture was practised. During the phase of prevailing shifting cultivation in Europe, beech expanded in many parts of central Europe (Küster 1997), as did spruce in the Western Alps (Markgraf 1970) and in northern Europe (Moe 1970).

In contrast to this, permanent settlement and land-use caused beech and fir to become rarer. These tree species were not able to regenerate in woodlands that were intensively used. On the other hand, quickly-regenerating wood species such as hornbeam, lime and hazel became more frequent (Pott 1981), as well as pioneer species like birch and pine, and also heathland species such as Calluna.

Therefore, different phases of land-use are clearly discernible in pollen diagrams:

1. A phase without land-use;
2. A phase with shifting cultivation characterised by a weak presence of cereal pollen grains and a furtherance of beech or spruce, indicating secondary woodland succession, and;

3. A phase of permanent cultivation characterised by a stronger presence of cereal pollen grains and a furtherance of quickly-regenerating wood and heathland species. Beech and sometimes also spruce were becoming rarer during this phase, a consequence of more intensive management.

8. CONCLUSIONS

Ecology fails if it only describes actual natural components of stands. Most stands were or are influenced by human impact, which made or makes them culturally influenced forever. Through human impact, nutrients have been reduced or accumulated. And through human impact, vegetation development is directed in a different way than it was before. Nature can develop on all stands if man does not interact with them, but the natural development was or is influenced by cultural interaction in any case. Most stands are characterised by both natural developments and cultural interactions, so that it is rather schematic to only distinguish between natural and cultural landscapes. Most landscapes are influenced by both natural and cultural factors; i.e. natural development or change and the aim of man to keep patches of "his" environment stable to guarantee that crops, wood and timber can be received from the same fields every year.

Ecologists must be aware of this fact. They must include the knowledge of historical geography, history, archaeology and palynology to judge stands in a more detailed and correct way. There is not only a historical or academic interest in the analysis of past land-use. It must be clear that the actual stand is a result of natural parameters and of past land-use. If one looks only at natural factors, a stand cannot be described appropriately. Actual and past managements changed the quality of a stand enormously. This quality influences the identity of a stand and of a landscape. It developed through time, and was influenced by both natural change and human interference.

In this context, planning must be understood as an actual human reaction to natural developments on stands already influenced by numerous human interactions in the past – and not as an answer to nature alone.

REFERENCES

Behre, K.-E. (2000). Der Mensch öffnet die Wälder – zur Entstehung der Heiden und anderer Offenlandschaften. *Rundgespräche der Kommission für Ökologie 18: Entwicklung der Umwelt seit der letzten Eiszeit*, 103-116.

Fischer, A. (1987). *Untersuchungen zur Populationsdynamik am Beginn von Sekundärsukzessionen.* Dissertationes Botanicae 110. Berlin, Stuttgart.

Frenzel, B. (1977). Postglaziale Klimaschwankungen im südwestlichen Mitteleuropa. *Erdwissenschaftliche Forschung*, 13, 297-322.

Frenzel, B. (1979). L'homme comme facteur géologique en Europe. *Bulletin de l'Association française pour l'Etude du Quaternaire*, 4, 191-199.

Gebhardt, H. (1976). Bodenkundliche Untersuchung der eisenzeitlichen Ackerfluren von Flögeln-Haselhörn, Kr. Wesermünde. *Probleme der Küstenforschung im südlichen Nordseegebiet*, 11, 91-100.

Gebhardt, H. (1982). Phosphatkartierung und bodenkundliche Geländeuntersuchungen zur Eingrenzung historischer Siedlungs- und Wirtschaftsflächen der Geestinsel Flögeln, Kr. Cuxhaven. *Probleme der Küstenforschung im südlichen Nordseegebiet*, 14, 1-9.

Geiger, R. (1961). *Das Klima der bodennahen Luftschicht*. Braunschweig.

Hoppe, A. (2002). Die Bewässerungswiesen Nordwestdeutschlands – Geschichte, Wandel und heutige Situation. *Abhandlungen aus dem Westfälischen Museum für Naturkunde* 64(1). Münster.

Hüppe, J. (1993). Entwicklung der Tieflands-Heidegesellschaften Mitteleuropas in geobotanisch-vegetationsgeschichtlicher Sicht. *Berichte der Reinhold-Tüxen-Gesellschaft*, 5, 49-75.

Hvass, S. (1982). Ländliche Siedlungen der Kaiser- und Völkerwanderungszeit in Dänemark. *Offa*, 39, 189-195.

Kossack, G. (1982). Ländliches Siedlungswesen in vor- und frühgeschichtlicher Zeit. *Offa*, 39, 271-279.

Kossack, G. (1997). *Dörfer im nördlichen Germanien, vornehmlich aus der römischen Kaiserzeit. Lage, Ortsplan, Betriebsgefüge und Gemeinschaftsform*. München.

Küster, H. (1997). The Role of Farming in the Postglacial Expansion of Beech and Hornbeam in the Oak Woodlands of Central Europe. *The Holocene*, 7(2), 239-242.

Küster, H. (1998). *Geschichte des Waldes*. München.

Küster, H. (2001). Natur, Umwelt, Landschaft. Definitionsversuche aus der Sicht der Vegetationsgeschichte. *Albersdorfer Forschungen zur Archäologie und Umweltgeschichte*, 2, 11-19.

Küster, H. (2004). Cultural landscapes. An Introduction. In M. Dieterich & J. van der Straaten (Eds.), *Cultural Landscapes and Land Use: the Conservation Society Interface*. London: Kluwer, in press.

Larcher, W. (2001). *Ökophysiologie der Pflanzen*. 5th Ed. Stuttgart: Ulmer.

Lüning, J. (1989). Siedlung und Kulturlandschaft der Steinzeit. *Siedlungen der Steinzeit. Spektrum der Wissenschaft*, 7-11.

Markgraf, V. (1970). Palaeohistory of the Spruce in Switzerland. *Nature*, 228, 249-251.

Mensching, H.G. & Seuffert, O. (2002). Zukünftige Wüsten – Folgen der Desertifikation. In U. Joger & U. Moldrzyk (Eds.), *Wüste* (pp. 138-145). Darmstadt.

Moe, D. (1970). The Post-Glacial Immigration of *Picea Abies* into Fennoscandia. *Botaniska Notiser*, 123, 61-66.

Pott, R. (1981). Der Einfluß der Niederwaldwirtschaft auf die Physiognomie und die floristisch-soziologische Struktur von Kalkbuchenwäldern. *Tuexenia*, 1, 233-242.

Richter, M. (1997). *Allgemeine Pflanzengeographie*. Stuttgart.

Schwintzer, C.R. & Tiepkema, D. (Eds.) (1990). *The Biology of Frankia and Actinorhizal Plants*. San Diego.

Chapter 28

CAN LANDSCAPES BE READ?

Mats Widgren
Department of Human Geography, Stockholm University, Sweden

1.   INTRODUCTION

This paper takes as its starting point a question which can be formulated like this: Through reflection and deconstruction, is it at all possible at this time to maintain the idea that landscapes can be read and analysed in a scientific manner?[1] It is appropriate to ask this question in the context of the Permanent European Conference for the Study of the Rural Landscape (PECSRL). Throughout the history of this conference, the idea that landscapes can be explained in a way that stands over and above local, national and ethnic understandings has formed an important line of thought. What was sometimes in the 1960s and 1970s referred to as the "modern" school of cultural landscape research was thus based on the idea of cultural landscape studies as an international, comparative *science*. Here, I deliberately use the word *science*, not simply the Swedish *vetenskap* or the German *Wissenschaft* – but *science* as in natural science (cf. Schaefer 1953: 236).

This modern, post-war cultural landscape research was based on the following three components:

1. The objects of study were the "forms" in rural landscapes. The comparative perspective played an important role, so consequently it was considered important to establish an international terminology (Uhlig 1967, 1972).

---

[1] My use of landscape "reading" refers to the everyday practice of landscape reading, which is also the first step in a scientific analysis of landscapes. This paper is thus not intended to contribute to the discussion of whether landscapes can be read as signifying systems (cf. Duncan 1990).

H. Palang et al. (eds.), European Rural Landscapes:
Persistence and Change in a Globalising Environment, 455-465.

2. The method was morphogenetic and aimed at uncovering the *origin* and *development* of forms in the agrarian landscape.

3. The explanatory framework was heavily influenced by *evolutionary thinking*. Agrarian landscapes were seen as progressing from one stage to another (see for example Krenzlin 1958).

2. THE FIRST CHALLENGE – FORM VERSUS PROCESS

During the last three decades, this "modern" cultural landscape research has experienced two general criticisms, which have questioned its fundaments in different ways. These challenges question in their respective fashions whether landscapes can be read at all, cross-culturally and comparatively, as part of an international research agenda.

In a volume on field systems published in 1973, Alan Baker and Robin Butlin provided a voice for skepticism against the morphogenetic methods (Baker & Butlin 1973). And later, in a review of the publication of the eleventh meeting of PECSRL, David Austin expressed his uncertainty in his *Doubts About Morphogenesis* (Austin 1985).

In his overview of the history of the Permanent European Conference for the Study of the Rural Landscape, Alan Baker divided the intellectual history of landscape studies into three phases: traditional, modern and postmodern. Studies based on the morphogenetic approach were viewed as a subgroup within traditional studies, and characterised as "a few but arguably significant attempts to confront general issues and to address general problems relating to the origins and transformations of European rural landscapes" (Baker 1988: 9). It seems that Baker distinguished modern from traditional based on the use of quantitative methods (cf. also his and Butlin's arguments on the merits of quantitative methods instead of qualitative, Baker & Butlin 1973).

However, seen in retrospect the morphogenetic, evolutionistic approach to studying agrarian landscapes is much easier to understand as a part of the modernist paradigm, regardless of the use of quantitative or qualitative methods. The critique against the morphogenetic school can thus be perceived rather as the first signals of a more profound critique against the spatial, geometrical, and morphological approach that dominated quantitative geography *as well as* the modern school of cultural landscape research in the 1960s and early 1970s (cf. Helmfrid's paper in this volume).

In the early 1970s, Gunnar Olsson voiced his critique against the spatial analytical school in much the same way that Baker, Butlin and Austin were criticising the morphogenetic approach, casting doubt on the whole programme of geography as a spatial and morphological discipline. The form-process dichotomy (which, indeed, lies at the heart of the problem with morphogenetic approaches) was scrutinised by Olsson on epistemological grounds, and he considered that the spatial analytical school carried "the seed of its own destruction" (Olsson 1974).

Looking at some of the morphogenetic studies of rural landscapes, it is easy to understand how such a critique could be voiced – they sometimes showed little interest in social processes and social theory. Classes and power, as well as social processes and social relations more generally, often played a minor role in explanations of landscape change under the morphogenetic paradigm. As a reaction to this, the critique against the morphological approach sometimes led historical geographers to abandon a focus on landscape and instead concentrate more on general economic and social history.

3. THE SECOND CHALLENGE – LANDSCAPES AS WAYS OF SEEING

The second challenge experienced by the modern landscape research agenda was formulated by – among others – Denis Cosgrove and Stephen Daniels. In their preface to *The Iconography of Landscape,* they stated that: "A landscape is cultural image, a pictorial way of representing, structuring and symbolizing surroundings" (Cosgrove and Daniels 1988: 1). From that period on, the view that the concept of landscape is much more related to ways of seeing – rather than to settlement forms, field patterns and other physical structures – has gained ground in many humanistic disciplines. The term *landscape* is now often used to mean environmental perception.[2]

It is in line with such a view of landscapes that Relph argues that:

Trying to investigate places and landscapes by imposing standardized methods is like … judging wines by measuring their alcohol content – the information may be accurate but it seriously misrepresents the subject matter (Relph 1989: 49).

The analogy comparing the investigations of landscapes with the judging of wines is to me a very appropriate and interesting one. Wine tasting is usually described as very structured and standardised method of investigating wines. It includes a formal procedure progressing from looking and smelling to tasting. Some of the sensory observations can actually be

---

[2] See for example Wall (2002: 99): "By 'landscape' I mean the way in which people realise their world and how they connect to it."

done at a laboratory, while others are dependent on values and judgments; some of which are shared by smaller or larger groups of experts (inter-subjective), while others are more personal, intuitive and subjective.[3] Alcohol content is actually *one* of the properties of wine which is interesting even for wine experts, and which should preferably be analysed using standardised procedure.

The practice of wine tasting can thus, quite contrary to Relph's view, serve as a basis for the development of standardised, structured and (dare I write it?) *scientific* approaches to landscape research. My grounds for such an approach are based on three different but related arguments. The first takes its starting point in the landscape *concept*. The second looks at the *practice* of landscape reading. I finally argue that the need for a structured analysis of landscapes can only be derived from an understanding of the present social context of landscape studies, and that we therefore also have to discuss the *relevance* of reading landscapes.

4.   LANDSCAPE AS A CONCEPT

In much recent Anglophone literature on landscapes, the understanding of landscapes as scenery and therefore as ways of seeing has become a taken-for-granted starting point. In his article *Recovering the Substantive Nature of Landscape*, Kenneth Olwig, however, uncovered the etymology of the landscape concept, and was able to show that the English connotation of landscape as scenery was developed through Dutch landscape paintings based on an older German concept; which refers to the territory, the conditions of that territory, and the customs and rules with which the land was governed (Olwig 1996). Olwig has further developed this theme and has discussed the historical and political role of the two landscape concepts (Olwig 2002).

The understanding of landscapes as lived-in territories is in line with one of the Swedish usages of the word *landskap*, referring to the old, pre-medieval lands that preceded the state in the Nordic countries. Similar words exist in most languages, often alluding to land, soil, earth, and people and nation. On the other hand, the specific Anglophone landscape concept is missing in – or has been only recently added to – many other languages of the world (for Estonian, see Peil 1999: 3). It is also interesting to note that Luig and Oppen (1997) use the German landscape concept as an important background for understanding and reading African landscapes.

---

[3] *Vinprovning* (article in *Nationalencyklopedin*).

It is thus possible to argue that today we are dealing with three interrelated concepts of land and landscapes that landscape studies must relate to (see Table 1).

Table 28-1. Landscape concepts.

Landscape as scenery	Representation
	Idea (mental construction)
	"A way of seeing"
Landscape as institution	Customary law
	Social order, land rights
	"A way of communicating, a way of acting"
Land as resource	Land use
	Production
	Capital

The first concept deals with the understanding of landscapes as scenery, and derived from that as *ways of seeing*. Through the English usage, this understanding of landscape has become a part not only of international language, but through the discourse's dominant role many authors have also shown that this landscape concept has become a force for ideological control and exclusion (cf. e.g., Neumann 1998; Mitchell 2002).

Secondly, we have the Germanic landscape concept, which focuses not only on the physical or cognitive "appearance" of the land, but also places as much emphasis on the people of that land and the social institutions that govern it.

Finally, landscape is often used to indicate land, and the ways in which it has been transformed by labour and serves as a basis for both biological production and the accumulation of wealth.

Of these three concepts of landscapes, the first one is by far the most ethnocentric, in that is tied to a specific cultural and social context; the second and the third lend themselves more easily to comparison and understanding across linguistic and cultural boundaries. It is somewhat a paradox that the critical tradition in geography has been so closely associated with the first landscape concept, where the representation is in focus (cf. Cosgrove 2003) – while the landscape concept of the old German morphological research (for example) was much closer to an understanding of landscapes as land, and the historical materialist definition of geography a study of those conditions (both naturally occurring and humanly created) that provide the material basis for the reproduction of social life (Harvey 1984).

I would thus argue that a cross-cultural comparative understanding of landscapes must be based on a combination of the three concepts, rather than simply following the present emphasis on the Anglophone concept only.

5.   THE PRACTICE OF LANDSCAPE READING

A second source of support concerning a structured way of reading landscapes takes its starting point in the everyday reading of landscapes. People do read landscapes and landscape representations daily. Landscape images form an important part of the media flow. Advertisements, propaganda, rock videos, etc., all make efficient use of landscapes in conveying ideas and feelings, and thus make use of our everyday understanding and subconscious reading of landscape sceneries. This becomes obvious when testing a number of landscape slides on a group of first-year Swedish university students. The landscapes in Figures 1 and 2 – on the west coast of Sweden and in England respectively – are immediately located correctly by many of the students. I then used to ask a follow-up question: "How do you know that?" One correct answer to that question could be, in the case of Figure 1: "I have been there". And in the case of Figure 2: "I have seen *Emmerdale Farm* on TV" (Students do, however, usually think that I expect a more serious, academic argument). The important argument here is that the practice of landscape reading is based on recognition and on reading the whole image in one glance, in the same way that you read faces of people.

Figure 28-1. Herrön, Bohuslän, Sweden (1980). Photo: M. Widgren.

It has been argued that experts read landscapes in much the same way (Nesheim 1998). This expert knowledge is, however, often acquired on the basis of years of more-formal analytical procedure – much in the same way that an experienced medical doctor can often make a diagnosis quickly due to his or her previously accumulated knowledge, while a doctor-in-training might need to put more emphasis on a more formal procedure.

Figure 28-2. Gt Asby, Cumberland. Photo: M. Widgren.

Some parts related to such a formal procedure of reading landscapes become clear when you try to go beyond the immediate recognition and attempt to look into the details of the images. You then have to look at *forms* that you see and then discuss the possible *functions* of these forms – procedures well-established in morphological studies of landscapes, but also near at hand in an everyday reading of landscapes. When reading landscape images and real-world landscapes, we also make use of the fact that landscape is *process* – the result of past processes, as well as the reflection of ongoing processes (labour, seasonal change, expansion and contraction).

But of course, this way of reading landscapes also has its limitations. Foreign students in Sweden do not recognise the landscape in Figure 1 because they usually have not been there – they do not share the cultural background of the Swedish students. And similarly, Swedish students have difficulties in understanding Figure 3, which is an excerpt from a Zimbabwean bill. Because of the different cultural backgrounds, they do not immediately understand the symbolic meaning of the barbed wire fence. Swedes do not always share the notion that "a barbed wire fence implies ownership", which an English textbook on landscapes tells us in a chapter on metaphors and meanings (Atkins et al. 1998: 220). In the Swedish landscape, a barbed wire fence is often (and legally) passed in order to find a nice picnic place or to pick berries. Landscapes and landscape elements may remain unintelligible to many of us because the social and cultural *context* is foreign to us, or because the context of the representation is unknown. This is the argument that followers of the postmodern and cultural turn of landscapes studies are rightly emphasising – landscapes as a way of seeing.

Figure 28-3. Zimbabwean banknote (excerpt).

Does this awareness of the context of landscape reading mean that the concepts of *form, function* and *process* have lost their role in a structured, comparative reading of landscapes? I would argue that the opposite is the case. The concepts *form* and *function* remain essential parts of the reading of landscapes, provided that we have an understanding of the different contexts that define their function. The *form* is a barbed wire fence, but its function is not – as in many Swedish landscapes – merely to separate different land uses and to prohibit cattle from destroying crops. In the Zimbabwean context, it definitely implies not only ownership, but also a prohibition on trespassing. One must be aware of the different contexts in order to understand the function, but it is the basic questions of form and function – so basic to landscape morphology approaches – that form the starting points from which the enquiry is put into the context.

I would therefore suggest that the four concepts of *form, function, process* and *context* may constitute a starting point, a checklist for a critical, formalised and structured reading of landscapes.

6. THE RELEVANCE OF LANDSCAPE STUDIES

In the 1950s, when the first steps were taken towards the present European networks of landscape studies, the concept of cultural landscapes was little known outside academia. Since then, the concept has gained in importance and has become a central part of the political agendas in many countries. Through the European Union, subsidies to preserve cultural values in the landscape are distributed – often on the basis of standpoints and results from academic research on landscapes, and often with a heavy emphasis on the scenery aspect of landscapes. Another sign of the political interest in landscapes is the European Landscape Convention, part of the ongoing work being done by the Council of Europe. Perceptions of what is a valuable historical landscape do therefore direct flows of Euros in different directions and affect the livelihoods of farmers all over Europe. Whether we like it or not, the research on landscape history feeds into arguments on heritage, valuation, uniqueness, etc. Landscape history matters.

If we look at environmental issues on a global scale, it is easy to see that conceptions of environmental problems are often based on – and popularised through – landscape representations. The picture of a dead cow on clay desiccated into polygons symbolised the Sahel drought in the 1970s and 1980s. The image represented what was mostly a generally accepted view of desertification at the time, including its extent and its causes. Landscape history later showed that the explanations of desertification accepted in the 1970s are no longer acceptable today. The picture and its background is scrutinised by McCann in his book *Green Land, Brown Land, Black Land: An Environmental History of Africa, 1800-1990.*

Fairhead & Leach's work *Misreading the African Landscape* (1996) is another good example that demonstrates the relevance and importance of landscape reading, and the problems of previous superficial misreadings. Based on a close empirical reading of West African savanna landscapes, they argue that what previous observers saw as small, remaining forest islands in a previously-forested savanna were in fact the result of people's efforts in planting trees around their settlements in what was formerly a less-forested area. Fairhead & Leach turned the established landscape history of the area upside down and placed a big question mark on years of donor projects and international programmes of reforestation in the area. They are anthropologists, but the work they have done can equally be considered as path-breaking historical geography. Their critique of established environmental myths was not based primarily on discourse analysis, but on painstaking and detailed empirical work in the field and in archives. Faced with the questions occupying many culturally-oriented landscape researchers today – those of cultural construction and representation – they wrote: "On the one hand we are dealing with landscape and its history as representation, but on the other hand we are attempting to reveal its empirical 'reality' – facts or events" (1996: 15-16). The key to uncovering the myths, the representations and the discourses on land and landscape thus rests in the land itself, not only in the representations.

7.   CONCLUSIONS

It has been argued that there is room for a dialectic synthesis between the modern and the postmodern approaches to landscapes. The recent emphasis on context, representation and on landscapes as different ways of seeing does not necessarily stand alone in contrast to a morphological approach to landscapes. Considering on the one hand the politically important role of arguments based on representations of landscape, and on the other the realities of land as a source for power and conflicts, one can see that a critical and empirically-based materialist landscape history certainly has a role to play.

REFERENCES

Atkins, P., Simmons, I. & Roberts, B. (1998). *People, Land and Time.* London: Arnold.

Austin, D. (1985). Doubts about Morphogenesis. *Journal of Historical Geography*, 11, 201-209.

Baker, A.R.H. & Butlin, R.A. (Eds.) (1973). *Studies of Field Systems in the British Isles.* Cambridge: Cambridge University Press.

Baker, A. (1988). Historical Geography and the Study of the European Rural Landscape. *Geografiska annaler Ser B*. Vol 70B No 1, 5-16.

Cosgrove, D. & Daniels, S. (1988). *The Iconography of Landscape.* Cambridge: Cambridge University Press

Cosgrove, D. (2003). Landscape: Ecology and Semiosis. In H. Palang & G. Fry (Eds.), *Landscape Interfaces: Cultural Heritage in Changing Landscapes* (pp.15-21). Dordrecht: Kluwer Academic.

Duncan, J.S. (1990). *The City as Text: The Politics of Landscape Interpretation in the Kandyan Kingdom.* Cambridge: Cambridge University Press.

Fairhead, J. & Leach, M. (1996). *Misreading the African Landscape. Society and Ecology in a Forest-Savanna Mosaic.* Cambridge: Cambridge University Press.

Harvey, D. (1984). The History and Present Condition of Geography: An Historical Materialist Manifesto. *Professional Geographer*, 3, 1-11.

Krenzlin, A. (1958). Blockflur, Langstreifenflur und Gewannflur als Funktion agrarischer Nutzungsssysteme in Deutschland. *Berichte zur dt. Landeskunde*, 20, 2, 250-266.

Luig, U. & v. Oppen, A. (1997). Landscape in Africa: Process and Vision. *Paideuma*, 43, 7-45.

McCann, J. (1999). *Green Land, Brown Land, Black Land: An Environmental History of Africa 1800-1990.* London: Eurospan.

Mitchell, D. (2002). Dead Labor and the Political Economy of Landscape – California Living, California Dying. In K. Anderson, M. Domosh, S. Pile & N. Thrift (Eds.), *Handbook of Cultural Geography* (pp. 233-249). London: Sage.

NE – Nationalencyklopedin, 19. (1996). Stockholm: Bra Böcker.

Nesheim, O. (1998). Hva kan landskapseksperten? Om det inre bildet av landskapet. In A. Norderhaug (Ed.), *Nordisk landskapsseminar, Sogndal 1996* (pp. 39-46). Rapport nr. 7.

Neumann, R.P. (1998). *Imposing Wilderness. Strugles of Livelihood and Nature Preservation in Africa.* Berkeley: University of California Press.

Olsson, G. (1974). The Dialectics of Spatial Analysis. *Antipode*, 50-61.

Olwig, K. (1996). Recovering the Substantive Nature of Landscape. *Annals of the Association of American Geographers*, 86, 630-653.

Olwig, K. (2002). *Landscape, Nature and the Body Politic: From Britain's Renaissance to America's New World.* Madison: University of Wisconsin Press.

Peil, T. (1999). *Islescapes. Estonian Small Islands and Islanders Through Three Centuries.* Acta Univeritatis Stockholmiensis 8. Doctoral dissertation. Stockholm: Almqvist & Wiksell.

Relph, E. (1989). Responsive Methods, Geographical Imagination and the Study of Landscapes. In A. Kobayashi & S. Mackenzie (Eds.), *Remaking Human Geography* (pp. 149-163). Boston, Mass.: Unwin Hyman.

Schaefer, F. (1953). Exceptionalism in Geography: A Methodological Examination. *Annals of the Association of American Geographers*, 43, 226-49.

Uhlig, H. (1967). *Flur und Flurformen, Types of field patterns, Le finage agricole et sa structure parcellaire.* Gießen: Lenz Verlag.

Uhlig, H. (1972). *Die Siedlungen des ländlichen Raumes, Rural settlements, L'habitat rural.* Gießen: Lenz Verlag.

Wall, Åsa (2002). Borderline Viewpoints. The Early Iron Age Landscapes of Henged Mountains in East Central Sweden. *Current Swedish Archaeology.* 10.

Chapter 29

THE PERMANENT CONFERENCE AND THE STUDY OF THE RURAL LANDSCAPE
A Retrospect

Staffan Helmfrid
Department of Human Geography, Stockholm University, Sweden

1. INTRODUCTION

It is the accomplishment of Xavier de Planhol, at the age of 30 years old, to have organised the first European multidisciplinary scholarly meeting in the field of *géographie et histoire agraires*. It happened in 1957, perhaps not incidentally the time of the formation of the European Communities (Treaty of Rome). He also later actively promoted the development towards a "Permanent Conference". As an informal founding father – from 1971, he was elected secretary-general of the Conference – he participated in 10 consecutive meetings, at eight of them with papers.

It has been over 45 years since this first meeting, and it can be considered as having taken place "long ago", historically speaking. This is especially so because the very object of our interest, the rural landscape, has undergone in this period the most rapid and fundamental transformation ever in the long history of rural Europe. This transformation process was brought to the foreground in the agenda at Liège (1969) and Perugia (1973).

As a survivor from the Nancy conference, I can confirm that we gathered in 1957 in another world, in a Europe whose great variety of rural landscapes were still visibly characterised by centuries-old farms, social structures and modes of production. The transformation of the last 50 years has left behind many old structures in fields and settlements, having a new or no function in a highly urbanised world. They may cheat the eye as an optical illusion of "merry old Europe". In reality, remaining traditional landscape elements –

H. Palang et al. (eds.), European Rural Landscapes:
Persistence and Change in a Globalising Environment, 467-482.

be they buildings, villages, field patterns or vegetational relics – today survive only as cultural heritage under institutional care.

In short, the scholars who came together at Nancy in 1957 had grown up within living traditional landscapes, the origin and history of which was the focus of their scientific interest. In that world, the landscapes of 18th and 19th century land reforms – such as enclosures in England and *udskifte* and *laga skifte* in Scandinavia – were still the modern landscapes of Europe.

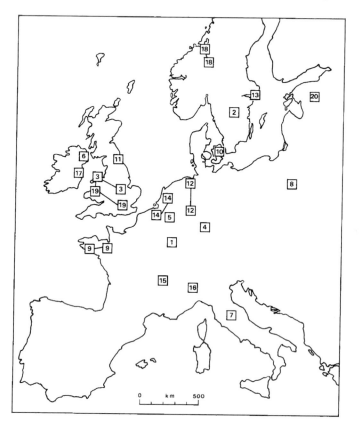

Figure 29-1. Locations of the former sessions of the PECSRL. For numbers, see the list of conference proceedings at the end of this chapter.

In my retrospect, let me first recall the formal history of the symposia. We can start by looking at where this migratory conference has met so far. The concentration of meetings held in Northwestern Europe no doubt reflects where this field of study has been most intensively cultivated throughout the 20th century. The first invitations to go southwards (to Perugia, Italy) and eastwards (to Warszawa, Poland) occurred in the 1970s, but were not followed up until 1994 (Torino, Italy) and 2002 (Tartu, Estonia).

Geography departments in the United Kingdom have organised no less than four meetings, France three, and Germany, Sweden and Italy two each. All in all France, Germany and the UK have provided the greatest number of active participants, as well as papers printed. However, there have been some major shifts in national representation through recent decades. German participation at the meetings, with a few exceptions, sharply declined from the middle of the 1970s. At that time, so-called "modern geography" absorbed the interest of German geographers, as it had earlier in Sweden and the UK. In addition, a new interdisciplinary forum – *the Arbeitskreis für genetische Siedlungsforschung in Mitteleuropa*, centered in Bonn – was created in 1974. French participation last peaked at Lyon (1992), only to cease after that meeting. British participation has been continuous but shifting in volume. Besides the UK, only Sweden has been represented at all 20 symposia, Tartu included. In the middle of the 1990s, another significant shift in national representation occurred. From 1996, Norway dominated after 40 years of being almost completely absent, as the flourishing Trondheim school of Michael Jones entered the scene. At the same time, the Baltic countries and other parts of the former Soviet Union broadened the national spectrum.

The identity of a Permanent Conference, through a series of independent symposia, depends on continuity in participation and in topics for discussion. All in all, about 525 persons have participated in one or more of the past 19 meetings. The backbone of continuity is formed by those who came back after their first meeting. 88 persons have participated at least three times, 22 of these at least seven times, two of them ten times (Xavier de Planhol and Jean Peltre), two of them 11 times (Anneliese Krenzlin and Hans-Jürgen Nitz), one 13 times (Robin Butlin) and one 14 times (Staffan Helmfrid).

But what is more important is the continuity in active participation as manifested in printed papers. Jean Peltre – who died before his contribution to the Torino meeting in 1994 had appeared in print – holds the record of nine papers, followed by Robin Butlin, Xavier de Planhol, Maria Kielczewska-Zaleska and Johannes Renes with eight; Glanville Jones, Viggo Hansen, Hans-Jürgen Nitz, Karl Erik Frandsen and Tim Unwin with seven; and Pierre Flatrès, Sölve Göransson, Brian Roberts, Halina Szulc, J. Pinard, Michel Cabouret and John Chapman with six papers each.

A group of approximately 20 scholars gave character to the first 25 years of meetings, and another 10 of a newer generation set the standards for the last 20 years. Only five veterans participated during both periods.

A great number of "one-time participants" at meetings coincides with an extensive local participation of research students.

Besides the benefits gained from the exchange of ideas and the observations during field excursions, the printed proceedings represent the legacy of the conference. So far 17 volumes have appeared, from the meeting in Nancy to the one in Røros/Trondheim. Bibliographically, alas, there is no uniform series of proceedings. It is not even easy to collect a complete set. In Sweden, no public library holds all proceedings published, and one of them (from the meeting in Lyon) is presently not available in any public library at all. Each meeting organiser has published the papers in locally available periodicals, yearbooks or other publication series. This makes it easier to deal with financial problems, but at the same time reduces the visibility of a conference.

The name "Permanent European Conference ..." was introduced in 1971, appearing on the title page of the Perugia (1973) proceedings. The Torino (1994) symposium was the first to be officially numbered on the title page (i.e. the "16th session of PECSRL"), and the Dublin (1996) report is the first to be named (i.e. "Proceedings of the 17th session ..."). With the exception of the Belfast (1971) proceedings, until 1977 all were organised alphabetically by authors' names. The Belfast papers and all proceedings since Rennes (1977) have been organised more ambitiously in a thematic manner, according to the programme of the symposium.

The publications mirror in a way the process of a step-by-step-institutionalisation of the conference. The creation of a home page, an e-mail network, and an institutionalised secretariat mark a further step towards formalised structure at the beginning of the new millennium. Along these lines, congratulations are due to the efficient organisers of the Tartu-Otepää meeting for their excellent work. However, further steps of institutionalisation should be considered carefully. Science needs a balance between organised structures and "chaos". Institutions may hide intellectual stagnation in their field of research and survive merely as social events.

1.1 First Conclusions

From what has been observed regarding the formal history of the PECSRL, it can be divided into four periods, in a vague manner characteristic of all historical periodisation. Major changes occurred around 1980, as a new generation entered the scene and the founding fathers/mothers left. In the period before 1980, the first four symposia (1957-1966) stand out as independent, occasional initiatives, but with reference to issues raised at Nancy (1957). The meetings from Liège (1969) onwards represent a planned series of symposia. In 1971, the name PECRSL was adopted. In the 1990s, a marked shift in thematic focus as well as national representation occurred.

2. A CRITICAL REVIEW 1987

For the Stockholm meeting in 1987, Alan Baker was commissioned to review the previous 12 symposia. He did it in a thoroughly-investigated, thoughtful and critical evaluation under the title *Historical Geography and the Study of the European Rural Landscape*. His aim was to inspire a new direction of the Permanent Conference in accordance with recent developments within historical geography, and even to convert the Conference into an *Association of European Historical Geographers*. His proposal failed to meet broad understanding among the participants. My own reason not to support the idea was the firm conviction that the study of rural landscape is much more than just historical geography, even if historical geography must remain the basic discipline. At the same time, historical geography as conceived within geography is much more than just the study of rural landscape.

A background to Alan Baker's critical evaluation was the paradigmatic shift towards a more analytical and theoretical approach in the social sciences, a shift which since the middle of the 1960s also had reached geography in Britain, later than in the United States and Scandinavia, but earlier than in France and Germany; this was then followed by post-modernism. The first critical remarks had been heard at the Durham/Cambridge symposium in 1981, where Anngret Simms talked about "... the intellectual challenge. The future health of the PECSRL would depend largely on whether it adapted its aims in the light of progress being made in other branches of historical geography ..."

The main target for Alan Baker's criticism was the dominance of papers of particular cases of what he calls empirical *formal-morphogenetic studies*, and a lack of process-oriented analysis and theory. He was right. This reflects the fact that there is no selection process involved in making the final programmes of these symposia. This may reduce the attraction to advanced scholars; however, it gives less-advanced students an opportunity to meet internationally.

Alan Baker's point of view is illustrated by the way he groups the papers according to three different styles, seen as stages in the modern history of social science: *traditional* (*humanistic*, in which he, in my mind, falsely includes *morphogenetic studies*); *modern* (*scientific* including positivist approach and quantitative analysis); and *postmodern* (*meaning and intentionality, search for symbolic significance, the social meaning of landscapes, place-synthesis*). Alan Baker found very few papers of the *modern* type, based on quantitative analysis and deductive theory; as well, the *postmodern* type contributions, putting full emphasis on human beings as subjects and agents in landscape forming, were negligible.

Alan Baker's paper has obviously had an impact, especially his demand for a wider range of perspectives, such as ecological and locational points of view. But in my mind, he did not do justice to the contributions presented in the early period symposia. His references are restricted to English and French language papers, and to measure the impact of the proceedings he looks for citations in British and French books. In any case, Alan Baker neglected important contributions to earlier symposia – I think, for linguistic reasons, especially in regards to papers published in German. Here let me just mention Helmut Jäger, Anneliese Krenzlin, Wolfgang Hartke or Hans-Jürgen Nitz.

While Alan Baker says he wants to see the Permanent Conference crafting a cooperative venture, he fails to observe that this has already been done, the result being a three-volume publication in multilingual scientific terminology on the study of agrarian landscapes. It was initiated at the first symposia, and published in the 1970s after the substantial efforts of Harald Uhlig and Kay Lienau, and financed by German sources. One should keep in mind that a joint venture in an international network of scholars is no easy thing to manage and finance.

I think the classification of our rural landscape studies according to the periodically shifting styles, or *-isms*, in social sciences is not the most fruitful way of describing the intellectual history of the symposia. Paradigmatic shifts show an epidemiological diffusion pattern. What is fashionable in one discipline/country for a period of time may already be out of date in another, and never heard of in a third. Looking back, one often realises there were more new words than new knowledge produced.

Let us examine instead how the *thematic focus* of symposia has changed over time, as an answer to questions arising from the ongoing transformations of landscape itself over the past 50 years.

3.   A SCIENTIFIC REVOLUTION IN RURAL LANDSCAPE RESEARCH

Around 1900, German scholars had formulated some general models of explanation for the distribution of the different types of rural settlements and field patterns observed in Europe. The most inspiring and longest lasting impact was made by the ethno-deterministic theory of August Meitzen, who also introduced the basic German language terminology into the field. Due to the domination of German scholars at the time, it was widely adopted as scientific terminology, translated or not into other languages. Common to all early models and theories was the basic idea that the rural settlement types and field patterns still visible in the early 20th century – and documented in cadastral plans of the 19th and earlier centuries – were primary forms, maintained since the first permanent agrarian colonisation of the different territories. The chronology of this historical process of colonisation was thus an important field of investigation in the first half of the 20th century, with names like Otto Schlüter and Robert Gradmann deserving mention. Like other grand generalisations of the late 19th century, the Meitzen theory was dismantled by numerous empirical local studies in Germany in the early 20th century, discouraging further attempts at generalisations. All of this was published in German only, and remained little-known in other language areas.

A uniquely rich variety of traditional rural landscapes and settlement types in a territory divided by dominant historical frontiers like the Roman "limes" and the Slavo-Teutonic "limes sorabicus", in a nation whose universities were at the forefront of science and scholarship worldwide; this setting explains the dominance of classical German *Siedlungsgeographie* in the early 20th century. This could be called "traditional" in the 1950s. The direct background to the Nancy meeting, however, was the post-WWII revival of German historical geography, with results questioning the traditional *Siedlungsgeographie*. Agrarian historical geography was now a field of research to which a substantial number of the leading German geographers devoted themselves more and more, in cooperation with leading archaeologists and historians. Göttingen, Münster, Marburg, Giessen, Frankfurt, and Würzburg formed a network of neighbouring universities engaged in vivid interaction within this field.

Thus, around the middle of the 20th century, in the decade before the meeting in Nancy, nothing less than a scientific revolution occurred in the study of rural landscapes. A new insight, which already had been won in Scandinavian research since the 1930s (but was unknown abroad as it had been published only in national languages, see Helmfrid 1962), brought about a boom in multidisciplinary landscape research in Germany. The signal had been given in a paper by Wilhelm Müller-Wille, published in 1944, but it was fully observed only after the war. Müller-Wille introduced in continental research what he called the *topographic-genetic analysis* of *the field pattern of villages*. The uncovering of evolutionary stages in the history of each village raised new questions, and demanded a concentrated and time-consuming investigation of a great number of cases to allow new generalisations. This new type of work-intensive micro-analysis was supported by a number of techniques developed since the early decades of the century, such as pollen analysis, phosphate analysis, air photo interpretation, ^{14}C-dating, and metrological analysis, not to mention new techniques in excavation. Following pioneering research in Denmark (Hatt) and England, remains of prehistoric and medieval deserted fields and settlements were widely explored (as well as in Germany), to be compared with patterns in the oldest cadastral plans. The field remains – witnesses of both advances and retreats in cultivation and colonisation at different periods of time – were extensively investigated, mapped, and at key points, excavated. The new German school of thought, or schools, aroused great interest in neighbouring countries. In Scandinavia, the new wave met open doors and confirmed what was already going on. In the UK, with traditionally little interest in rural landscape studies among geographers, German geographers looked for comparable evidence (Jäger, Uhlig i.a.)

Inspired by Wolfgang Hartke, two French geographers, Étienne Juillard and André Meynier, published in the *Münchener Geographische Hefte* (1956) under the title *Die Agrarlandschaft in Frankreich. Forschungsergebnisse der letzten 20 Jahre,* an overview of French research done since Marc Bloch's seminal work *Les caractères originaux de l'histoire rurale française* (1931). The French landscape was thus described in Hartke's translation – a very demanding endeavor – preceding the French invitation to Nancy.

3.1 Nancy

In preparing for the Nancy *Colloque international de géographie et histoire agraires,* Étienne Juillard, André Meynier, Xavier de Planhol and G. Sautter published an updated and extended version of this book in French, under the title *Structures agraires et paysages ruraux.* The book is divided into an introduction and three chapters, representing the three major French landscape regions, or *civilisations agraires*, as identified by Marc Bloch: the northeastern *open-field*, the western *bocage*, and the southern Mediterranean. The introduction to the book reflects the considerations behind the invitation to Nancy:

> La recherche scientifique française s'est consacrée tardivement à la génèse des paysages ruraux...Fait symptomatique: le mot de 'Siedlungskunde' n'a pas encore en francais d'équivalent qui satisfasse tout le monde. Alors qu'en Allemagne Meitzen et ses continuateurs avaient déjà réalisés un fécond rapprochement entre des disciplines diverses, il a fallu en France attendre Marc Bloch, qui, en 1931, dans une brillante synthèse sut grouper les apports des archéologues, des linguistes, des historiens et des géographes, pour présenter une série d'hypothèses générales et orienter des recherches nouvelles, aussi bien géographiques qu'historiques. (The scientific study in France has only tardily approached the study of origin and development of rural landscapes ... Typically the word *Siedlungskunde* has as yet no satisfactory counterpart in French ... When in Germany already Meitzen

and his followers had realized a fruitful rapprochement between different disciplines, it has not succeeded in France to follow Marc Bloch, who in 1931 in a brilliant synthesis gathered results from archaeology, linguists, historians and geographers to formulate a series of general hypotheses and inspire new research, geographical as well as historical) (Juillard et al. 1957).

A statement of Marc Bloch's that the fundamental characters of French rural landscapes had been formed in early prehistoric time by anonymous populations provoked much dispute from geographers, not the least Roger Dion, who was studying in more detail the variations and local history of settlements. Marc Bloch concluded later (1941) that the *"énigme des régimes agraires"* ("the riddle of agrarian regimes") could not be solved in Britain, Germany or France, but only in Europe. By 1957, the glorious generalisations of Bloch – like those of Meitzen – had been abandoned, and new theories were looked for. Bloch had himself already realised that more detailed studies were necessary (1934):

> Plus il poursuivait sa réflexion, plus il constatait la nécessité d'analyses plus précises, notamment dans la domaine des sols, des techniques agricoles, des structures sociales. (The more he thought about it, the more he realized the necessity of more precise analysis, especially regarding the soils, the agrarian techniques, the social structures).

3.2   From Nancy to Würzburg (1957-1966)

Within the focus of the Nancy colloquium stood the confrontation between the new German schools, and the ideas of Marc Bloch and the French schools of history and geography. Contributions from other countries witnessed about the role of classical German *Siedlungsgeographie*, but also of more independent, if not isolated, scholarly traditions, especially in the UK. There, Hoskins' *The Making of the English Landscape* (1955) had only recently raised interest in the topic, and the rich sources of documentary, map and field evidence were about to be explored by geographers.

The first step towards the integration of European rural landscape studies was hampered by the lack of a coherent and well-defined terminology that was transferable between languages. Thus, discussions at Nancy were dominated by questions such as "What do you mean by xxx? What is yyy?."

It was evident at Nancy that much work had to be done in order to knit all incompatible pieces of knowledge – from local studies and from time periods all around Europe – together to form something resembling a coherent field of research. Participants were stimulated to react within their own research on the perspectives opened by the theories and generalisations developed in Germany, which were based upon a rapidly growing number of research reports. In my own doctoral thesis (1962), I made an effort to integrate Swedish studies of landscape history into the European context, and vice-versa.

As the Secretary-General of the XIX International Geographical Congress of the Nordic countries (1960), I took the chance to organise as part of the congress programme a symposium on the *Morphogenesis of the Agrarian Cultural Landscape*, in Vadstena, a medieval town in the middle of my own field research area. There was a strong German representation, including a number of leading geographers. If France dominated at Nancy, Germany did so at Vadstena.

Two memorable results came out of the Vadstena symposium. Firstly the contribution made by Anneliese Krenzlin – later unquestionably for more than 30 years the "grand old lady" of the meetings – on the origin and evolution of the complex *"Haufendorf mit Gewannflur"* characterising southern Germany and adjacent territories, which provoked a heated debate that had to be postponed until a special conference could be held in Germany. No other topic had been more severely discussed in classical *Siedlungsgeographie* than this one, and now Anneliese Krenzlin seemed to provide a final solution. In contrast to longstanding theories about the very old age of these villages, they instead were proven to have reached their mature form relatively late, as a result of socio-economic processes and laws of inheritance. The continued German debate later inspired German colleagues to invite the colleagues to the Würzburg symposium in 1966.

The second Vadstena result of lasting value was the general support for the idea proposed by Harald Uhlig, to form a working group with the task of creating a well-defined and generally applicable terminology in the main languages. Uhlig was bold enough to take on for himself the time-consuming role of leading the group. It met for working sessions at the following symposia until a three-volume multilingual terminology lexicon was completed and published in the 1970s. It would be recommendable also to a new generation of students to consult these volumes. They contain a systematic, scientific terminology based on definitions and logic. Much confusion could be avoided with their usage.

After Vadstena in 1960, the *XX International Geographical Congress* in London (1964) provided the next opportunity to meet in a new environment. Harry Thorpe and his students, especially Brian Roberts, organised as part of the congress a symposium with direct reference to Nancy and Vadstena. It gathered in Bangor and ended in Leicester, introducing both Welsh and English landscape studies. An effort was made at the IGU General Assembly to establish a commission or working group of European Historical Geography with Helmut Jäger as chairman, but without success in this global organisation. So the next invitation, to Würzburg for 1966, was again the result of local initiative.

At Würzburg the determination to continue was strong. One of the "newcomers" there, Frans Dussart, was invited to the next meeting at Liège.

3.3   From Liège to Roskilde (1969-1979)

The Liège symposium was the beginning of a planned series of meetings. As a further step towards creating a formal structure for this series, the Belfast (1971) meeting adopted the name *Permanent European Conference for the Study of the Rural Landscape*. Xavier de Planhol, now at Sorbonne and formally elected secretary-general, was a guarantor of continuity who worked in the position for another ten years.

At Liège, the process and results of the ongoing agrarian revolution and urbanisation was put on the agenda and was treated in one-third of all papers.

The Perugia meeting (1973) was the first in the Mediterranean region. Henri Desplanques, the organiser, presented an excellent survey of the rich variety in traditional landscapes around the Mediterranean, already undergoing dramatic changes. The shift in focus from history to present time was strengthened and reflected in a majority of papers.

At Rennes (1977), there was a marked shift in national representation. The founding of an *Arbeitskreis für genetische Siedlungsforschung in Mitteleuropa*, in 1974, from that time on weakened the PECSRL as a European forum for landscape history.

3.4   From Durham/Cambridge to Baarn/Ghent (1981-1990)

The Durham/Cambridge symposium 1981 in many respects marks a turning point in the history of our conference. De Planhol had resigned, and the number of participants from the Nancy-Vadstena generation had declined as prominent members of a new generation made their débuts. The agenda added rural industry to the programme. The symposium also tried a new way of working, reducing paper sessions in favour of field seminars.

In an analytical survey of the contributions to this symposium, the first outspoken critical questioning of what was referred to in the United Kingdom as the *retrogressive method* was heard in J.G. Hurst's lecture. As Alan Baker writes: "Hurst's arguments illustrate the problems faced by all scholars using morphogenetic analysis; simplistic retrogressive projections are all too easy – but more evidence and more experience will add depth to the interpretations ... and arguments linking a given morphology with a documented phase of settlement often appear very tenuous when subjected to rigorous testing." Of course, a retrogressive reconstruction of landscape patterns cannot be reversibly interpreted as a deterministic series of evolutionary stages. The decision-making process in landscape change has to be considered. Unfortunately, written evidence of this process is seldom available for earlier times. The fact remains – we must also consider the "unborn" possible landscapes.

At the 1981 symposium, Hans-Jürgen Nitz reported his findings on a 9th-century Carolingian state colonisation. Nitz was to come back with more and more evidence, bringing to light traces of systematic state colonisation, as well as from later epochs and other territories. The Rastede/Hagen symposium in 1985, organised by Nitz, consequently put aspects of power in the foreground, such as state power impacts of the 16th-17th centuries and commercial power impacts in the Middle Ages.

A new theme was also added in Stockholm (1987), namely studies of the agroecosystem in different environments, focusing on the nutrient use efficiency under traditional systems of cultivation (Gunilla Olsson).

In his introduction to the meeting at Baarn/Ghent (1990), Jelier Vervloet stated that "… also historical geographers have to *discuss the future of our cultural landscape* ... Much of the traditional European cultural landscapes are rapidly disappearing. Yet only a very limited number of historical geographers are trying to think about *possibilities to preserve* the most valuable phenomena …" The tone was set for the following symposia. The group found a proper spokesman for this new mission, and Tim Unwin accepted taking over the coordinating role as the new secretary-general. The interregnum after de Planhol was over.

3.5   From Lyon to Tartu/Otepää (1992-2002)

Tim Unwin began as co-organiser of the Lyon symposium in 1992. He made a broad, synthesising *"avant-propos"*, a plaidoyer for the cultural dimension in landscapes from a holistic and perceptual point of view, stressing the esthetic values in landscape. Thus, he stressed the English understanding of the word *landscape* firstly as *scenery*, in contrast to the unmistakably geographic term used in German texts – *Agrarlandschaft*. The undefined understanding of the word *landscape* causes no little confusion when using English language in conference discussions. Landscape history in the scientific geographical sense was a new concept in English historical geography in the 1950s. Unwin talks about interpreting the meaning of landscapes, *meaning landscape* understood as a category of fine art in the great tradition of English landscape architecture.

At the same symposium Michael Jones stressed the question of how to use the cultural heritage in landscape for the future. The issue of the preservation of cultural heritage in a rapidly changing world raised questions of criteria for selecting valuable landscapes to preserve. Historical geographers, historians and archaeologists can identify historical values to protect. When it comes to esthetic values, however, psychologists and art historians no doubt have better tools and theories than do geographers.

The 1994 meeting at Torino was organised by Paola Sereno, an active participant since 1975. In her excellent introduction to the proceedings, she observed the shift in focus over time towards more process-oriented studies. For this second symposium in Italy, and following the ideas from Rastede, she put focus on the decision-making levels relevant to ongoing change in landscape. All around the meeting place we saw a landscape in rapid and dynamic change, from small-scale variability to large-scale monotony and old farm buildings in ruins.

The Dublin symposium in 1996 confirmed the fundamental shift in themes, as well as in participation, since Lyon. Norway provided the biggest national representation. From now on only papers in English language were invited, no doubt reducing the number of scholars interested in contributing. Participation from Germany and France had ceased totally. The themes announced for the meeting included *ethnicity in the landscape, change in perception and use of rural resources,* as well as *theoretical and methodological aspects.* A few titles from the paper sessions may illustrate the very long distance travelled since Nancy: *Visible landscape and mental landscape, Landscape ideals and value judgments, Landscape-cultural landscape: philosophical and ethical perspectives.*

Landscape values and perceptions were one of the main topics also at Röros/Trondheim in 1998, organised by Michael Jones. More strongly than before, the botanical aspects of cultural landscapes (biodiversity) were emphasised in the field excursions, as well as in papers. Contributions were also invited on the topic of the *role of landscape in the constitution of national and regional identities,* a delicate research topic indeed in the context of rising nationalism (*"Blut und Boden"*).

To the renewal of participation contributed a group of Estonians, a happy sign of the reunion of this nation with the European scholarly community. For the first time also Lithuania, Russia and Belarus were represented.

Also, in the Egham/Aberystwyth symposium 2000, organised by Tim Unwin, Norwegian-Estonian dominance was obvious. Among the themes offered at this meeting, we find *symbolic wildscapes.* What had been signaled at Lyon was completed in this meeting. *Identity* and *meaning* were the key concepts, and texts were full of terms like *esthetics, ethics, values and choices,* even *mytho-poetic,* as well as *moral account of landscape.* The focus had shifted from the study of the landscape in its physical appearance and contents, to the study of images and emotions in the human minds, not unlike the transition from Enlightenment to Romanticism.

Summarising my observations on the thematic development of the 19 symposia under scrutiny, I draw the conclusion that the most relevant way of describing the history of PECSRL is found in the shifts of thematic focus in response to fundamental changes in rural landscapes themselves, and the problems they raise in modern society. As turning points, I have identified Belfast (1971), Durham/Cambridge (1981), and Lyon (1992). The main themes have logically moved from the basic questions of origin and evolution to the decision-making processes behind changes, and further towards analysing the recent and ongoing dramatic landscape transformations on the one hand, to issues of landscape management and the application of historical geography in the selection and care of landscapes on the other hand.

4. FUTURE OF THE PECSRL

Coming to an end, I dare to draw some conclusions for the future. The vitality of the PECSRL depends on the scholarly benefits potential participants anticipate from the agenda of each symposium, not the least who they expect to meet. A thematic concentration raises the attraction for active researchers. A loosely knit network of the PECSRL-type has proven to be astonishingly successful in surviving paradigmatic shifts in relevant disciplines, as well as the challenges from a rapid transformation of society and landscape. A major value of symposia in our field of interest is the well-prepared and research-based field excursion. The history, recent changes and future of rural landscapes (*Agrarlandschaften*) should remain the main concern of the conference, to make it an expert network regarding questions on the preservation of historic values in Europe's rural landscapes, a rapidly disappearing cultural heritage.

REFERENCES

Bloch, M. (1931). *Les caractères originaux de l'histoire rurale francaise*. Paris.
Burggraaff, P. (1988). Bemerkungen zur genetischen Siedlungsforschung in Luxemburg. *Genetische Siedlungsforschung in Mitteleuropa und seinen Nachbarräumen*, 1-2, 469-481.
Gradmann, R. (1913). *Das Ländliche Siedlungswesen des Königreichs Württemberg*. Stuttgart.
Hatt, G. (1949). Oldtidsagre. Det Kongelige. Danske Videnskbs Selskab. Arkeol.-Kunsthist. Skr. Bd II:1. København.
Helmfrid, S. (1962). Östergötland 'Västanstång'. Studien ueber die Altere Agrarlandschaft und ihre Genese. *Geografiska Annaler*, Vol. XLIV, 1-2.

Hoskins, W.G. (1955). *The Making of the English Landscape*. London: Penguin History S.

Jäger, H. (1953). Methoden und Ergebnisse Siedlungskundlicher Forschung. *Zeitschrift für Agrargeschlite und Agrarsoziologie, 1*.

Juillard, E., Meynier, A., de Planhol, X. & Sautter, G. (1957). Structures agraires et paysages ruraux. *Un quart de siècle de recherches francaises. Annales de l'Est, Nancy*.

Krenzlin, A. (1958). Blockflur, Langstreifenflur und Gewannflur als Ausdruck agrarischer Wirtschaftsformen in Deutschland. *Ber. Z. Dtsch. Landesk.*, 1958, 2.

Krenzlin, A., (1962). Zur Genese der Gewannflur in Deutschland. *Geografiska Annaler*,1-2.

Meitzen, A. (1895). *Siedelung und Agrarwesen der Westgermanen und Ostgermanen, der Kelten, Römer, Finnen und Slawen*. I-III+Atlas. Berlin.

Mueller-Wille, W. (1944). Langstreifenflur und Drubbel. Ein Beitrag zur Siedlungsgeographie Westgermaniens. *DArchLandesVolksforschung 8*, 1944. 9-44.

Schlueter, O. (1908). *Vorlesungen ueber Allgemeine Siedlungs- und Verkehrsgeographie*. Berlin.

Uhlig, H. (1956). Die Kulturlandschaft. Methoden der Forschung und das Beispiel Nordostengland. *Kölner Geographische. Arbeiten.*, H. 9/10. Köln. Geographischen Institut der Universität zu Köln

Uhlig, H. & Lienau, K. (1972-1978). *Materialien zur Terminologie der Agrarlandschaft I-III*: I. Types of field patterns, Giessen 1978, II. Rural Settlements, Gießen 1972, III. Rural Population, Giessen 1974.

Conference Proceedings

1. Nancy 1957: *Géographie et histoire agraires*. Actes du colloque international organisé par la Faculté des lettres de l'Université de Nancy, 2-7 septembre 1957. X. de Planhol (Ed.), *Annales de l'Est*, Mémoire no 21, Nancy 1959, 452 p.

2. Vadstena 1960: *Morphogenesis of the Agrarian Cultural Landscape,* Papers of the Vadstena symposium at the XIX International Geographical Congress, August 14-20, 1960. S. Helmfrid (Ed.), *Geografiska Annaler*, 43, Stockholm 1961, 328 p.

3. Bangor-Leicester 1964: Jones, G.R.J. & Thorpe, H., Symposium S 4a: *The Rural Landscape and its Evolution*, July 11th to 20th, 1964. J.W. Watson (Ed.), *20th International Geographical Congress, Congress Proceedings*, London 1967, p 221-229.

4. Würzburg 1966: *Beiträge zur Genese der Siedlungs- und Agrarlandschaft in Europa*. Rundgespräch vom 4 bis 6 Juli 1966 in Wuerzburg, veranstaltet von der Deutschen Forschungsgemeinschaft unter Leitung von H. Jäger, A. Krenzlin und H. Uhlig. *Erdkundliches Wissen*, H. 18, *Geographische Zeitschrift*, Wiesbaden 1968, 211 p.

5. Liège 1969: *L'Habitat et les paysages ruraux d'Europe*. Comptes rendus du Symposium tenu à Liège du 29 juin au 5 juillet 1969. Volume publiée par les soins de F. Dussart dans *Les Congrès et Colloques de l'Université de Liège*, 58, Université de Liège, 1971, 472 p.

6. Belfast 1971: R.H. Buchanan, R.A. Butlin and D. McCourt (Eds.) (1976). *Fields, Farms and Settlement in Europe*. Papers presented at a Symposium, Belfast, July 12-15, 1971, Ulster Folk and Transport Museum, Cultra Manor, Holywood, Co. Down. 161 p.

7. Perugia 1973: *I paesaggi rurali Europei*. Atti del convegno indetto a Perugia dal 7 al 12 Maggio 1973 dalla Conférence Permanente Pour l'Étude du Paysage Rural. H. Desplanques (Ed.) (1975). Perugia *Deputazione di Storia Patria per l'Umbria. Appendici al Bollettino*, No 12. 613 p.

8. Warszawa 1975: M. Kielczewska-Zaleska de M.J. Grzeszezak & Mme T. Lijewska (Eds.) (1978). *L'Évolution de l'habitat et des paysages ruraux d'Europe. Actes de la Conférence tenue à Varsovie en septembre 1975.Geographia Polonica*, 38, Polish Academy of Sciences, Institute of Geography and Spatial Organization, Warszawa, 1978. 304 p.

9. Rennes-Quimper 1977: (1979). *Paysages ruraux européens.* Travaux de la Conférence Européenne Permanente pour l'Étude du Paysage Rural, Rennes-Quimper, 26-30 septembre 1977, publiés par les soins de P. Flatrès, Université de Haute Bretagne, Rennes. 629 pp.

10. Roskilde 1979: K.-E. Frandsen, S. Gissel & V. Hansen (Eds.) (1981). *Collected Papers Presented at the Permanent European Conference for the Study of the Rural Landscape,* held at Roskilde, Denmark, 3-9 June 1979. Copenhagen. 226 p.

11. Durham-Cambridge 1981: B.K. Roberts & R.E. Glasscock (Eds.) (1983). *Villages, Fields and Frontiers.* Studies in European Rural Settlement in the Medieval and Early Modern Periods. Papers presented at the meeting of the PECSRL at Durham and Cambridge, 10-17 September 1981, *British Archaeological Reports, International Series,* 185, Oxford. 433 p.

12. Rastede-Hagen 1985: H.-J. Nitz (Ed.) (1987). *The Medieval and Early-Modern Rural Landscape of Europe under the Impact of the Commercial Economy.* Papers presented at the meeting held at Rastede and Hagen, FRG, 2-9 September 1985, Department of Geography, University of Göttingen. 328 p.

13. Stockholm 1987: U. Sporrong (Ed.) 1990, *The Transformation of Rural Society, Economy and Landscape.* Papers from the 1987 Meeting of the PECSRL at Stockholm, *Meddelanden serie B,* 71, Department of Human Geography, Stockholm University. 326 p. Selected papers also published in *Geografiska Annaler,* 70B, 1988, 237 p.

14. Baarn-Gent 1990: Verhoeve, A, & J.A.J. Vervloet (Eds.) (1992). *The Transformation of the European Rural Landscape: Methodological Issues and Agrarian Change 1770-1914.* Wageningen: Winand Staring Centre. *Tijdschrift van de Belgische Vereniging voor Aardrijskundige Studien,* LXI, nr 1, Brussels 1992.

15. Lyon 1992: Bethemons, J. (Ed.) (1192). *L'avenir des paysages ruraix européens: entre gestion des héritages et dynamique du changement.* Colloque de Lyon, 9-13 juin 1992. Université Lumière Lyon 2/Université Jean Moulin Lyon 3 (Laboratoire de Géographie Rhodanienne), Lyon (Collection Les Chemins de la Recherce, No hors série). 321 p.

16. Torino 1994: P. Sereno & M.L. Sturani (Eds.) (1994). *Rural Landscape between State and Local Communities in Europe, Past and Present.* Proceedings of the 16th Session of the Standing European Conference for the Study of the Rural Landscape. Torino 12-16 September 1994. Edizioni dell'Orso. Torino. 321 p.

17. Dublin 1996: F.H.A. Aalen & M. Hennessy (Eds.) (2001). *Proceedings of the PECSRL.* Papers from the 17th Session, Trinity College, Dublin 1996, Department of Geography, Trinity College, University of Dublin. 119 p.

18. Røros-Trondheim 1998: G. Setten, T. Semb & R. Torvik (Eds.) (1998). *Shaping the Land.* Proceedings of the Permanent European Conference for the Study of the Rural Landscape, 18th session in Røros and Trondheim, Norway, September 7th-11th. *Papers from the Department of Geography, University of Trondheim, New Series A,* Three Volumes. (Selection of papers in *Norwegian Journal of Geography,* Vol. 53, Nr 2-3).

19. Egham-Aberystwyth 2000. T. Unwin & T. Spek. (Eds.) (2003). *European Landscapes: from mountain to sea.* Proceedings of the 19th session of the Permanent European Conference for the Study of the Rural Landscape (PECSRL) at London and Aberystwyth (UK) 10-17 September 2000. Tallinn: Huma Publishers. 238 p.

20. Tartu-Otepää 2002: H. Palang, H. Soovāli, M. Antrop & G. Setten (Eds.) (2004). *European Rural Landscapes: Persistence and Change in a Globalising Environment.* Dodrect, Boston, London: Kluwer Academic Publishers (A selection of papers in a special issue of *Landscape and Urban Planning,* to be published in 2004).

University of Plymouth Library

Subject to status this item may be renewed
via your Voyager account

http://voyager.plymouth.ac.uk

Exeter tel: (01392) 475049
Exmouth tel: (01395) 255331
Plymouth tel: (01752) 232323